$1 \, kN/m^2 = 1 \, kPa$

$1 \, PSI = 6.895 \, kPa$

$1 \, LBF = 4.448 \, N$

$1 \, MN =$

$N = kg \cdot m/s^2$

$\gamma_w = 1000 \, kg/m^3$

PRINCIPLES OF
SOIL DYNAMICS

THE PWS-KENT SERIES IN ENGINEERING

Askeland, *The Science and Engineering of Materials*, Second Edition
Borse, *FORTRAN 77 and Numerical Methods for Engineers*, Second Edition
Bronzino, *Biomedical Engineering and Instrumentation*
Clements, *Microprocessor Systems Design*, Second Edition
Das, *Principles of Foundation Engineering*, Second Edition
Das, *Principles of Geotechnical Engineering*, Second Edition
Das, *Principles of Soil Dynamics*
Duff, *Concepts of Technical Graphics*
Fisher, *Power Electronics*
Fleischer, *Engineering Economy: Capital Allocation Theory*
Gere and Timoshenko, *Mechanics of Materials*, Third Edition
Glover and Sarma, *Power Systems Analysis and Design*
Janna, *Introduction to Fluid Mechanics*, Second Edition
Johnson and Karim, *Digital Design*
Karim, *Electro-Optical Devices and Systems*
Keedy, *Introduction to CAD Using CADKEY*, Second Edition
Knight, *The Finite Element Method in Machine Design*
Logan, *A First Course in the Finite Element Method*, Second Edition
McGill and King, *Engineering Mechanics: Statics*, Second Edition
McGill and King, *Engineering Mechanics: An Introduction to Dynamics*, Second Edition
McGill and King, *Engineering Mechanics: Statics and An Introduction to Dynamics*, Second Edition
Poularikas and Seely, *Signals and Systems*, Second Edition
Poularikas and Seely, *Elements of Signals and Systems*
Reed-Hill and Abbaschian, *Physical Metallurgy Principles*, Third Edition
Reynolds, *Unit Operations and Processes in Environmental Engineering*
Sack, *Matrix Structural Analysis*
Schmidt and Wong, *Fundamentals of Surveying*, Third Edition
Segui, *Fundamentals of Structural Steel Design*
Shen and Kong, *Applied Electromagnetism*, Second Edition
Sule, *Manufacturing Facilities*
Weinman, *VAX FORTRAN*, Second Edition
Weinman, *FORTRAN for Scientists and Engineers*

PRINCIPLES OF
SOIL DYNAMICS

Braja M. Das

Southern Illinois University at Carbondale

PWS-KENT Publishing Company

BOSTON

PWS–KENT
Publishing Company

20 Park Plaza
Boston, Massachusetts 02116

PWS-KENT Publishing Company is a division of Wadsworth, Inc.

Library of Congress Cataloging-in-Publication Data
Das, Braja M.,
    Principles of soil dynamics / Braja M. Das.
      p. cm.
    Includes bibliographical references and index.
    ISBN 0-534-93129-4
    1. Soil dynamics. I. Title.
  TA710.5.D27 1992
624.1′5136—dc20   92-2856
                         CIP

This book is printed on recycled, acid-free paper.

Printed in the United States of America
    95 96 97—10 9 8 7 6 5 4 3 2

Sponsoring Editor: Jonathan Plant
Assistant Editor: Mary Thomas
Production and Cover Design: Helen Walden
Manufacturing: Ellen Glisker
Cover Printer: John Pow Company
Text Printer/Binder: Arcata Graphics/Halliday

To Janice and Valerie

Contents

Preface

During the past three decades, considerable progress has been made in the area of soil dynamics. Soil dynamics courses have been added or expanded for graduate-level study in many universities. The knowledge gained from the intensive research conducted all over the world has gradually filtered into the actual planning, design, and construction process of various types of earth-supported and earth-retaining structures.

Principles of Soil Dynamics is based on a book originally published by Elsevier Science Publishing Company (New York) in 1983. The new book provides updated coverage throughout, including machine foundations on piles (Chapter 11) and seismic stability of earth embankments (Chapter 12). Chapter 4 now includes comprehensive coverage of the properties of dynamically loaded soils, and the major test procedures used.

In developing this text, simplicity of presentation for clear understanding was my main consideration. For that reason, a major portion of the text is assigned to the treatment of fundamental concepts of the subjects being discussed. Also, more recently developed materials were drawn from published literature and incorporated into the text.

Both systems of units (English and SI) are used throughout the book. A number of worked-out examples are included, which I believe are essential for the student. Practice problems are given at the ends of most chapters, and a list of references is included at the end of each chapter. I also believe the text will be of interest to researchers and practitioners.

Acknowledgments

Thanks are due to my colleagues, Vijay K. Puri and Bruce A. DeVantier, for their valuable suggestions and assistance during the preparation of this text. I am indebted to my wife, Janice, for typing the manuscript as well as preparing the final figures. I would like to thank the following reviewers: Dr. D. V. Morris, Texas A & M University; Vijay K. Puri, Southern Illinois University; and Shamim Rahman, North Carolina State University. Thanks are also due to the staff of PWS-KENT Publishing Company for their help during the publication process.

Braja M. Das
Carbondale, Illinois

Chapter 1

INTRODUCTION

1.1   General Information

Soil mechanics is the branch of civil engineering that deals with the engineering properties and behavior of soil under stress. Since the publication of the book *Erdbaumechanik auf Bodenphysikalischer Grundlage* by Karl Terzaghi (1925), theoretical and experimental studies in the area of soil mechanics have progressed at a very rapid pace. Most of these studies have been devoted to the determination of soil behavior under static load conditions, in a broader sense, although the term *load* includes both static and dynamic loads. Dynamic loads are imposed on soils and earth structures by several sources, such as earthquakes, bomb blasts, operation of machinery, construction operations, mining, traffic, wind, and wave actions. It is well known that the stress-strain properties of a soil and its behavior depend upon several factors and can be different in many ways under dynamic loading conditions as compared to the case of static loading. Soil dynamics is the branch of soil mechanics that deals with the behavior of soil under dynamic load, including the analysis of the stability of earth-supported and earth-retaining structures.

During the last 30 years, several factors, such as damage due to liquefaction of soil during earthquakes, stringent safety requirements for nuclear power plants, industrial advancements (for example, design of foundations for power generation equipment and other machinery), design and construction of offshore structures, and defense requirements, have resulted in a rapid growth in the area of soil dynamics.

1.2   Nature and Type of Dynamic Loading on Soils

The type of dynamic loading in soil or the foundation of a structure depends on the nature of the source producing it. The operation of a reciprocating or a rotary machine typically produces a dynamic load pattern, as shown in Figure 1.1a. This dynamic load is more or less sinusoidal in nature and may be idealized, as shown in Figure 1.1b. The impact of a hammer on a foundation produces a transient loading condition in soil, as shown in Figure 1.2a. The load typically increases with time up to a maximum value at time $t = t_1$ and drops to zero after that. The case shown in Figure 1.2a is a single-pulse load. A typical loading

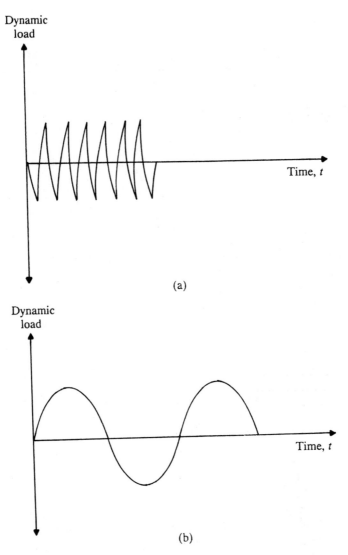

Figure 1.1   (a) Typical load versus time record for a low-speed rotary machine;
(b) sinusoidal idealization for (a)

pattern (vertical acceleration) due to a pile-driving operation is shown in Figure
1.2b. Dynamic loading associated with an earthquake is random in nature.
Figure 1.3 shows the accelerogram of the El Centro, California, earthquake of
May 18, 1940 (north-south component).

For consideration of land-based structures, earthquakes are the important
source of dynamic loading on soils. This is due to the damage-causing potential
of strong motion earthquakes and the fact that they represent an unpredictable
and uncontrolled phenomenon in nature. The ground motion due to an earthquake may lead to permanent settlement and tilting of footings and, thus, the

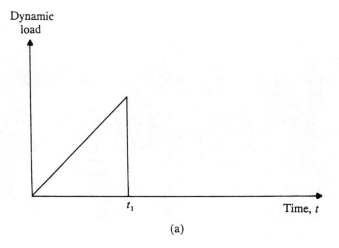

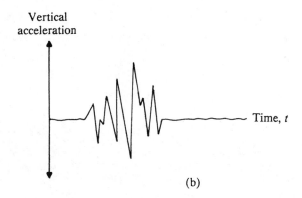

Figure 1.2   Typical loading diagrams: (a) transient loading due to single impact of a hammer; (b) vertical component of ground acceleration due to pile driving

structures supported by them. Soils may liquify, leading to buildings sinking and lighter structures such as septic tanks floating up (Prakash, 1981). The damage caused by an earthquake depends on the energy released at its source, as discussed in Chapter 7.

For offshore structures, the dynamic load due to storm waves generally represents the significant load. However, in some situations the most severe loading conditions may occur due to the combined action of storm waves and earthquake loading. In some cases the offshore structure must be analyzed for the waves and earthquake load acting independently of each other (Puri and Das, 1989; Puri, 1990).

The loadings represented in Figures 1.1, 1.2 and 1.3 are rather simplified presentations of the actual loading conditions. For example, it is well known that earthquakes cause random motion in every direction. Also, pure dynamic loads do not occur in nature and are always a combination of static and dynamic

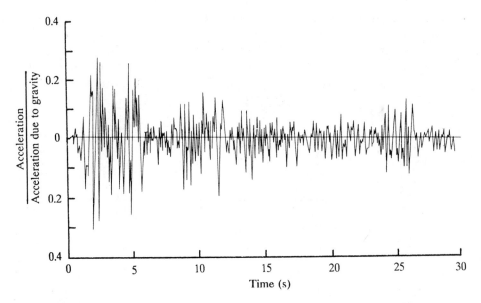

Figure 1.3   Accelerogram of El Centro, California, earthquake on May 18, 1940—
N-S component

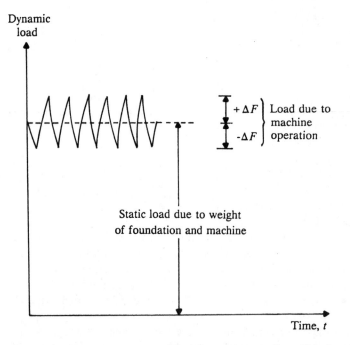

Figure 1.4   Schematic diagram showing loading on the soil below the foundation
during machine operation

loads. For example, in the case of a well-designed foundation supporting a machine, the dynamic load due to machine operation is a small fraction of the static weight of the foundation (Barkan, 1962). The loading conditions may be represented schematically by Figure 1.4. Thus in a real situation the loading conditions are complex. Most experimental studies have been conducted using simplified loading conditions.

1.3   Importance of Soil Dynamics

The problems related to the dynamic loading of soils and earth structures frequently encountered by a geotechnical engineer include, but are not limited to, the following:

1. Earthquake, ground vibration, and wave propagation through soils
2. Dynamic stress, deformation, and strength properties of soils
3. Dynamic earth pressure problem
4. Dynamic bearing capacity problems and design of shallow foundations
5. Problems related to soil liquefaction
6. Design of foundations for machinery and vibrating equipment
7. Design of embedded foundations and piles under dynamic loads
8. Stability of embankments under earthquake loading.

In order to arrive at rational analyses and design procedures for these problems, one must have an insight into the behavior of soil under both static and dynamic loading conditions. For example, in designing a foundation to resist dynamic loading imposed by the operation of machinery or an external source, the engineer has to arrive at a special solution dictated by the local soil conditions and environmental factors. The foundation must be designed to satisfy the criteria for static loading and, in addition, must be safe for resisting the dynamic load. When designing for dynamic loading conditions, the geotechnical engineer requires answers to questions such as the following:

1. How should failure be defined and what should be the failure criteria?
2. What is the relationship between applied loads and the significant parameters used in defining the failure criteria?
3. How can the significant parameters be identified and evaluated?
4. What will be an acceptable factor of safety, and will the factor of safety as used for static design condition be enough to ensure satisfactory performance or will some additional conditions need to be satisfied?

The problems relating to the vibration of soil and earth-supported and earth-retaining structures have received increased attention of geotechnical engineers in recent years, and significant advances have been made in this direction. New theoretical procedures have been developed for computing the response of foundations, analysis of liquefaction potential of soils, and design of retaining walls and embankments. Improved field and laboratory methods for determining dynamic behavior of soils and field measurements to evaluate the performance of prototypes deserve a special mention. In this text an attempt has been made to present the state-of-the-art information available on some of the important problems in the field of soil dynamics. Gaps in the existing literature, if any, have also been pointed out. The importance of soil dynamics lies in providing safe, acceptable, and time-tested solutions to the problem of dynamic loading in soil, in spite of the fact that the information in some areas may be lacking and the actual loading condition may not be predictable, as in the case of the earthquake phenomenon.

REFERENCES

Barkan, D. D. (1962). *Dynamics of Bases and Foundations.* McGraw-Hill Book Company, New York.

Prakash, S. (1981). *Soil Dynamics.* McGraw-Hill Book Company, New York.

Puri, V. K. (1990). "Dynamic Loading of Marine Soils," *Proceedings*, 9th International Conference on Offshore Mechanics in Arctic Engineering, ASME, Vol. 1, pp. 421–426.

Puri, V. K., and Das, B. M. (1989). "Some Considerations in the Design of Offshore Structures: Role of Soil Dynamics," *Proceedings*, Oceans '89, Vol. 5, Seattle, Washington, pp. 1544–1551.

Terzaghi, K. (1925). *Erdbaumechanik auf Bodenphysikalischer Grundlage*, Deuticke, Vienna.

Chapter 2

FUNDAMENTALS OF VIBRATION

2.1   Introduction

Satisfactory design of foundations for vibrating equipments is based on displacement considerations. Displacement due to vibratory loading can be classified under two major divisions:

1. Cyclic displacement due to the elastic response of the soil-foundation system to the vibratory loading

2. Permanent displacement due to compaction of soil below the foundation

In order to consider the first condition described, it is essential to know the nature of the unbalanced forces in a foundation such as shown in Figure 2.1.

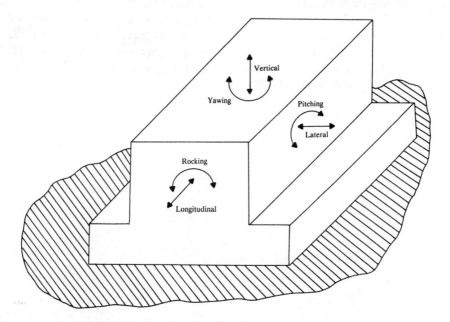

Figure 2.1   Six modes of vibration for a foundation

Note that a foundation can vibrate in any or all six possible modes. For ease of analysis, each mode is considered separately and design is carried out by considering the displacement due to each mode separately. The mathematical considerations of the displacement of foundations can be made by treating soil as a viscoelastic material. This can be explained with the aid of Figure 2.2a, which shows a foundation subjected to a vibratory loading in the vertical direction. The parameters for the vibration of the foundation can be evaluated by treating the soil as equivalent to a spring and a dashpot which supports the foundation as shown in Figure 2.2b. This is usually referred to as a *lumped parameter* vibrating system.

In order to solve the vibration problems of lumped parameter systems, one needs to know the fundamentals of vibration engineering. Therefore, a brief review of the mathematical solutions of simple vibration problems is presented. More detailed discussion regarding other approaches to solving foundation vibration problems and evaluation of basic parameters such as the spring constant and damping coefficient are presented in Chapter 5.

2.2   Fundamental Definitions

Following are some fundamental definitions that are essential in the development of the theories of vibration.

Free Vibration: Vibration in system under the action of forces inherent in the system itself and in the absence of external impressed forces.

Forced Vibration: Vibration of a system caused by an external force.

Degree of Freedom: The number of independent coordinates required to describe the solution of a vibrating system. For example, the position of the mass m in Figure 2.3a can be described by a single coordinate z, so it is a *single degree of freedom system*. In Figure 2.3b, two coordinates (z_1 and z_2) are necessary to describe the motion of the system; hence this system has *two* degrees of freedom. Similarly, in Figure 2.3c, two coordinates (z and θ) are necessary, and the number of degrees of freedom is two.

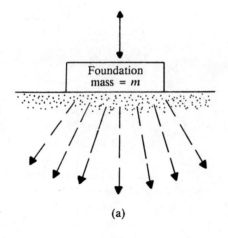

(a)

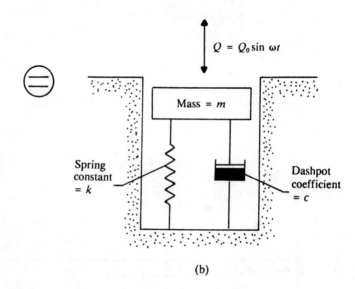

(b)

Figure 2.2   A lumped parameter vibrating system

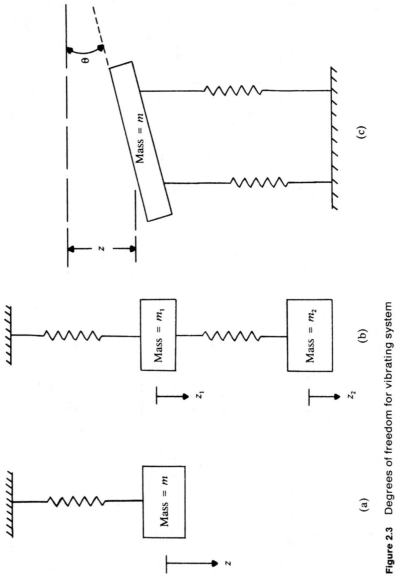

Figure 2.3   Degrees of freedom for vibrating system

SYSTEM WITH A SINGLE DEGREE OF FREEDOM

2.3   Free Vibration of a Spring-Mass System

Figure 2.4 shows a foundation resting on a spring. Let the spring represent the elastic properties of the soil. The load W represents the weight of the foundation

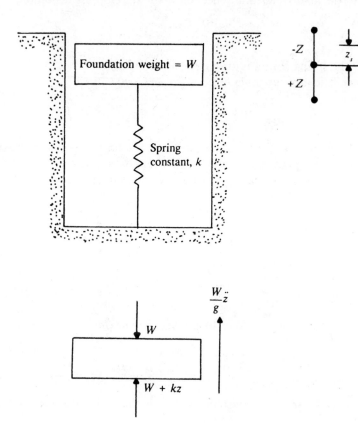

Figure 2.4 Free vibration of a mass-spring system

plus that which comes from the machinery supported by the foundation. If the area of the foundation is equal to A, the intensity of load transmitted to the subgrade can be given by

$$q = \frac{W}{A} \tag{2.1}$$

Due to the load W, a static deflection z_s will develop. By definition,

$$k = \frac{W}{z_s} \tag{2.2}$$

where k is the spring constant for the elastic support.

The coefficient of subgrade reaction k_s can be given by

$$k_s = \frac{q}{z_s} \tag{2.3}$$

If the foundation is disturbed from its static equilibrium position, the system will vibrate. The equation of motion of the foundation when it has been disturbed through a distance z can be written from Newton's second law of motion as

$$\left(\frac{W}{g}\right)\ddot{z} + kz = 0$$

or

$$\ddot{z} + \left(\frac{k}{m}\right)z = 0 \tag{2.4}$$

where g is the acceleration due to gravity, $\ddot{z} = d^2z/dt^2$, t is time, and m is mass $= W/g$.

In order to solve Eq. (2.4), let

$$z = A_1 \cos \omega_n t + A_2 \sin \omega_n t \tag{2.5}$$

where A_1 and A_2 are both constants and ω_n is the undamped natural circular frequency.

Substitution of Eq. (2.5) into Eq. (2.4) yields

$$-\omega_n^2(A_1 \cos \omega_n t + A_2 \sin \omega_n t) + \left(\frac{k}{m}\right)(A_1 \cos \omega_n t + A_2 \sin \omega_n t) = 0$$

or

$$\omega_n = \sqrt{\frac{k}{m}} \tag{2.6}$$

The unit of ω_n is in radians per second (rad/s). Hence,

$$z = A_1 \cos\left(\sqrt{\frac{k}{m}}t\right) + A_2 \sin\left(\sqrt{\frac{k}{m}}t\right) \tag{2.7}$$

In order to determine the values of A_1 and A_2, one must substitute the proper boundary conditions. At time $t = 0$, let

Displacement $z = z_0$

and

$$\text{Velocity} = \frac{dz}{dt} = \dot{z} = v_0$$

Substituting the first boundary condition in Eq. (2.7),

$$z_0 = A_1 \tag{2.8}$$

Again, from Eq. (2.7)

$$\dot{z} = -A_1 \sqrt{\frac{k}{m}} \sin\left(\sqrt{\frac{k}{m}}\, t\right) + A_2 \sqrt{\frac{k}{m}} \cos\left(\sqrt{\frac{k}{m}}\, t\right) \tag{2.9}$$

Substituting the second boundary condition in Eq. (2.9)

$$\dot{z} = v_0 = A_2 \sqrt{\frac{k}{m}}$$

or

$$A_2 = \frac{v_0}{\sqrt{k/m}} \tag{2.10}$$

Combination of Eqs. (2.7), (2.8), and (2.10) gives

$$z = z_0 \cos\left(\sqrt{\frac{k}{m}}\, t\right) + \frac{v_0}{\sqrt{k/m}} \sin\left(\sqrt{\frac{k}{m}}\, t\right) \tag{2.11}$$

Now let

$$z_0 = Z \cos \alpha \tag{2.12}$$

and

$$\frac{v_0}{\sqrt{k/m}} = Z \sin \alpha \tag{2.13}$$

Substitution of Eqs. (2.12) and (2.13) into Eq. (2.11) yields

$$\boxed{z = Z \cos(\omega_n t - \alpha)} \tag{2.14}$$

where

$$\boxed{\alpha = \tan^{-1}\left(\frac{v_0}{z_0 \sqrt{k/m}}\right)} \tag{2.15}$$

$$\boxed{Z = \sqrt{z_0^2 + \left(\frac{v_0}{\sqrt{k/m}}\right)^2} = \sqrt{z_0^2 + \left(\frac{m}{k}\right) v_0^2}} \tag{2.16}$$

The relation for the displacement of the foundation given by Eq. (2.14) can be represented graphically as shown in Figure 2.5. At time

$t = 0,$ $\qquad z = Z\cos(-\alpha)$ $\qquad = Z\cos\alpha$

$t = \dfrac{\alpha}{\omega_n},$ $\qquad z = Z\cos\left(\omega_n \dfrac{\alpha}{\omega_n} - \alpha\right)$ $\qquad = Z$

$t = \dfrac{\frac{1}{2}\pi + \alpha}{\omega_n},$ $\qquad z = Z\cos\left(\omega_n \dfrac{\frac{1}{2}\pi + \alpha}{\omega_n} - \alpha\right) = 0$

$t = \dfrac{\pi + \alpha}{\omega_n},$ $\qquad z = Z\cos\left(\omega_n \dfrac{\pi + \alpha}{\omega_n} - \alpha\right) = -Z$

$t = \dfrac{\frac{3}{2}\pi + \alpha}{\omega_n},$ $\qquad z = Z\cos\left(\omega_n \dfrac{\frac{3}{2}\pi + \alpha}{\omega_n} - \alpha\right) = 0$

$t = \dfrac{2\pi + \alpha}{\omega_n},$ $\qquad z = Z\cos\left(\omega_n \dfrac{2\pi + \alpha}{\omega_n} - \alpha\right) = Z$

$\vdots$

From Figure 2.5, it can be seen that the nature of displacement of the foundation is sinusoidal. The magnitude of maximum displacement is equal to Z. This is usually referred to as the *single amplitude*. The peak-to-peak displacement amplitude is equal to 2Z, which is sometimes referred to as the *double amplitude*. The time required for the motion to repeat itself is called the *period of the vibration*. Note that in Figure 2.5 the motion is repeating itself at points A, B, and C. The period T of this motion can therefore be given by

$$T = \frac{2\pi}{\omega_n} \tag{2.17}$$

The *frequency of oscillation f* is defined as the number of cycles in unit time, or

$$f = \frac{1}{T} = \frac{\omega_n}{2\pi} \tag{2.18}$$

It has been shown in Eq. (2.6) that, for this system, $\omega_n = \pm\sqrt{k/m}$. Thus,

$$f = f_n = \left(\frac{1}{2\pi}\right)\sqrt{\frac{k}{m}} \tag{2.19}$$

The term f_n is generally referred to as the *undamped natural frequency*. Since $k = W/z_s$ and $m = W/g$, Eq. (2.19) can also be expressed as

$$f_n = \left(\frac{1}{2\pi}\right)\sqrt{\frac{g}{z_s}} \tag{2.20}$$

Table 2.1 gives values of f_n for various values of z_s.

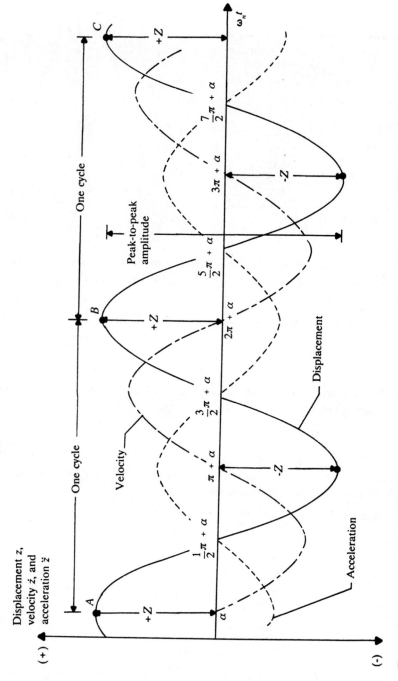

Figure 2.5 Plot of displacement, velocity, and acceleration for the free vibration of a mass-spring system (*Note:* Velocity leads displacement by $\frac{1}{2}\pi$ rad; acceleration leads velocity by $\frac{1}{2}\pi$ rad.)

Table 2.1   Undamped Natural
Frequency[a]

z_s (mm)	Undamped natural frequency f_n (cps)
0.02	111.47
0.05	70.5
0.10	49.85
0.20	35.25
0.50	22.29
1.0	15.76
2	11.15
5	7.05
10	4.98

[a] *Note: g = 9.81 m/s².*

The variation of the velocity and acceleration of the motion with time can also be represented graphically. From Eq. (2.14), the expressions for the velocity and the acceleration can be obtained as

$$\dot{z} = -(Z\omega_n)\sin(\omega_n t - \alpha) = Z\omega_n \cos(\omega_n t - \alpha + \tfrac{1}{2}\pi) \tag{2.21}$$

and

$$\ddot{z} = -Z\omega_n^2 \cos(\omega_n t - \alpha) = Z\omega_n^2 \cos(\omega_n t - \alpha + \pi) \tag{2.22}$$

The nature of variation of the velocity and acceleration of the foundations is also shown in Figure 2.5.

Example 2.1

A mass is supported by a spring. The static deflection of the spring due to the mass is 0.015 in. Find the natural frequency of vibration.

Solution

From Eq. (2.20),

$$f_n = \left(\frac{1}{2\pi}\right)\sqrt{\frac{g}{z_s}}$$

$$g = 32.2 \text{ ft/s}^2, \qquad z_s = 0.015 \text{ in.}$$

So,

$$f_n = \left(\frac{1}{2\pi}\right)\sqrt{\frac{(32.2)(12)}{0.015}} = \underline{25.54 \text{ cps (cycles/s)}} \qquad \blacksquare$$

Example 2.2

For a machine foundation, given weight of the foundation = 45 kN (kilonewtons) and spring constant = 10^4 kN/m, determine

a. the natural frequency of vibration, and
b. the period of oscillation.

Solution

a. $f_n = \dfrac{1}{2\pi}\sqrt{\dfrac{k}{m}} = \dfrac{1}{2\pi}\sqrt{\dfrac{10^4}{(45/9.81)}} = 7.43$ cps

b. From Eq. (2.18),

$$T = \frac{1}{f_n} = \frac{1}{7.43} = \underline{0.135 \text{ s.}}$$ ∎

2.4   Forced Vibration of a Spring-Mass System

Figure 2.6 shows a foundation that has been idealized to a simple spring-mass system. Weight W is equal to the weight of the foundation itself and that supported by it; the spring constant is k. This foundation is being subjected to an alternating force $Q_0 \sin(\omega t + \beta)$. This type of problem is generally encountered with foundations supporting reciprocating engines, and so on. The equation of motion for this problem can be given by

$$m\ddot{z} + kz = Q_0 \sin(\omega t + \beta) \tag{2.23}$$

Let $z = A_1 \sin(\omega t + \beta)$ be a particular solution to Eq. (2.23) $(A_1 = \text{const})$. Substitution of this into Eq. (2.23) gives

$$-\omega^2 m A_1 \sin(\omega t + \beta) + k A_1 \sin(\omega t + \beta) = Q_0 \sin(\omega t + \beta)$$

$$A_1 = \frac{Q_0/m}{(k/m) - \omega^2} \tag{2.24}$$

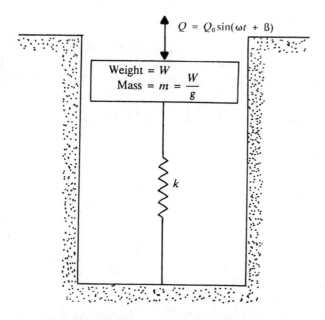

Figure 2.6   Forced vibration of a mass-spring system

Hence, the particular solution to Eq. (2.23) is of the form

$$z = A_1 \sin(\omega t + \beta) = \frac{Q_0/m}{(k/m - \omega^2)} \sin(\omega t + \beta) \qquad (2.25)$$

The complementary solution of Eq. (2.23) must satisfy

$$m\ddot{z} + kz = 0$$

As shown in the preceding section, the solution to this equation may be given as

$$z = A_2 \cos \omega_n t + A_3 \sin \omega_n t \qquad (2.26)$$

where

$$\omega_n = \sqrt{\frac{k}{m}}, \qquad A_2, A_3 = \text{const}$$

Hence, the general solution to Eq. (2.23) is obtained by adding Eqs. (2.25) and (2.26), or

$$z = A_1 \sin(\omega t + \beta) + A_2 \cos \omega_n t + A_3 \sin \omega_n t \qquad (2.27)$$

Now, let the boundary conditions be as follows: At time $t = 0$,

$$z = z_0 = 0 \qquad (2.28)$$

$$\frac{dz}{dt} = \text{velocity} = v_0 = 0 \qquad (2.29)$$

From Eqs. (2.27) and (2.28),

$$A_1 \sin \beta + A_2 = 0$$

or

$$A_2 = -A_1 \sin \beta \qquad (2.30)$$

Again, from Eq. (2.27),

$$\frac{dz}{dt} = \dot{z} = A_1 \omega \cos(\omega t + \beta) - A_2 \omega_n \sin \omega_n t + A_3 \omega_n \cos \omega_n t$$

Substituting the boundary condition given by Eq. (2.29) in the preceding equation gives

$$A_1 \omega \cos \beta + A_3 \omega_n = 0$$

or

$$A_3 = -\left(\frac{A_1 \omega}{\omega_n}\right) \cos \beta \qquad (2.31)$$

Combining Eqs. (2.27), (2.30), and (2.31),

$$z = A_1 \left[\sin(\omega t + \beta) - \cos(\omega t) \cdot \sin \beta - \left(\frac{\omega}{\omega_n}\right) \sin(\omega_n t) \cdot \cos \beta \right] \qquad (2.32)$$

For a real system, the last two terms inside the brackets in Eq. (2.32) will vanish due to damping, leaving the only term for steady-state solution.

If the force function is in phase with the vibratory system (i.e., $\beta = 0$), then

$$z = A_1 \left(\sin \omega t - \frac{\omega}{\omega_n} \sin \omega_n t \right)$$

$$= \frac{Q_0/m}{k/m - \omega^2} \left(\sin \omega t - \frac{\omega}{\omega_n} \sin \omega_n t \right)$$

or

$$z = \frac{Q_0/k}{1 - \omega^2/\omega_n^2} \left(\sin \omega t - \frac{\omega}{\omega_n} \sin \omega_n t \right) \tag{2.33}$$

However, $Q_0/k = z_s =$ static deflection. If one lets $1/(1 - \omega^2/\omega_n^2)$ be equal to M [equal to the magnification factor or $A_1/(Q_0/k)$], Eq. (2.33) reads as

$$z = z_s M \left[\sin \omega t - \left(\frac{\omega}{\omega_n} \right) \sin \omega_n t \right] \tag{2.34}$$

The nature of variation of the magnification factor M with ω/ω_n is shown in Figure 2.7a. Note that the magnification factor goes to infinity when $\omega/\omega_n = 1$. This is called the *resonance condition*. For resonance condition, the right-hand side of Eq. (2.34) yields 0/0. Thus, applying *L'Hôpital's rule*,

$$\lim_{\omega \to \omega_n}(z) = z_s \frac{(d/d\omega)[\sin \omega t - (\omega/\omega_n)\sin \omega_n t]}{(d/d\omega)(1 - \omega^2/\omega_n^2)}$$

or

$$z = \tfrac{1}{2}z_s(\sin \omega_n t - \omega_n t \cos \omega_n t) \tag{2.35}$$

The velocity at resonance condition can be obtained from Eq. (2.35) as

$$\dot{z} = \tfrac{1}{2}z_s(\omega_n \cos \omega_n t - \omega_n \cos \omega_n t + \omega_n^2 t \sin \omega_n t)$$

$$= \tfrac{1}{2}(z_s \omega_n^2 t) \sin \omega_n t \tag{2.36}$$

Since the velocity is equal to zero at the point where the displacement is at maximum, for *maximum displacement*

$$\dot{z} = 0 = \tfrac{1}{2}(z_s \omega_n^2 t) \sin \omega_n t$$

or

$$\sin \omega_n t = 0, \quad \text{i.e.,} \quad \omega_n t = n\pi \tag{2.37}$$

where n is an integer.

For the condition given by Eq. (2.37), the displacement equation (2.35) yields

$$|z_{max}|_{res} = \tfrac{1}{2}n\pi z_s \tag{2.38}$$

where $z_{max} =$ maximum displacement.

It may be noted that when n tends to ∞, $|z_{max}|$ is also infinite, which points

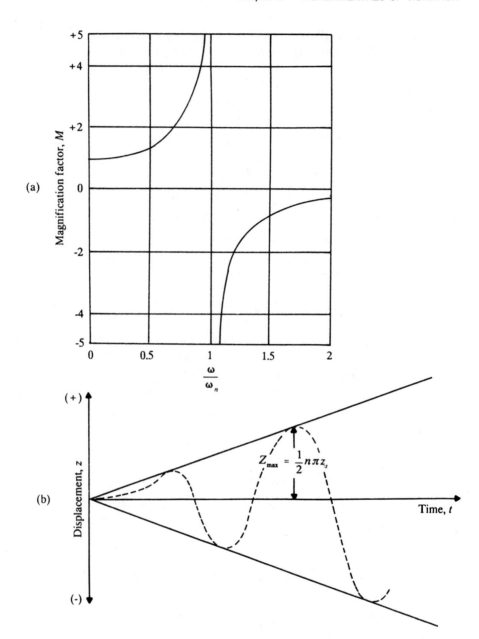

Figure 2.7 Force vibration of a mass-spring system: (a) variation of magnification factor with ω/ω_n; (b) variation of displacement with time at resonance ($\omega = \omega_n$)

out the danger to the foundation. The nature of variation of z/z_s versus time for the *resonance condition* is shown in Figure 2.7b.

Maximum Force on Foundation Subgrade

The maximum and minimum force on the foundation subgrade will occur at the time when the amplitude is maximum, i.e., when velocity is equal to zero. This can be derived from displacement equation (2.33):

$$z = \frac{Q_0}{k} \frac{1}{(1 - \omega^2/\omega_n^2)} \left(\sin \omega t - \frac{\omega}{\omega_n} \sin \omega_n t \right)$$

Thus, the velocity at any time is

$$\dot{z} = \frac{Q}{k} \frac{1}{(1 - \omega^2/\omega_n^2)} (\omega \cos \omega t - \omega \cos \omega_n t)$$

For maximum deflection, $\dot{z} = 0$, or

$$\omega \cos \omega t - \omega \cos \omega_n t = 0$$

Since ω is not equal to zero,

$$\cos \omega t - \cos \omega_n t = 2 \sin \tfrac{1}{2}(\omega_n - \omega)t \sin \tfrac{1}{2}(\omega_n + \omega)t = 0$$

Thus,

$$\tfrac{1}{2}(\omega_n - \omega)t = n\pi; \qquad t = \frac{2n\pi}{\omega_n - \omega} \tag{2.39}$$

or

$$\tfrac{1}{2}(\omega_n + \omega)t = m\pi; \qquad t = \frac{2m\pi}{\omega_n + \omega} \tag{2.40}$$

where m and $n = 1, 2, 3, \ldots$.

Equation (2.39) is not relevant (beating phenomenon). Substituting Eq. (2.40) into Eq. (2.33) and simplifying it further,

$$z = z_{\text{max}} = \frac{Q_0}{k} \cdot \frac{1}{(1 - \omega/\omega_n)} \cdot \sin \left(\frac{2\pi m \omega}{\omega_n + \omega} \right) \tag{2.41}$$

In order to determine the maximum dynamic force, the maximum value of z_{max} given in Eq. (2.41) is required:

$$z_{\text{max(max)}} = \frac{(Q_0/k)}{1 - \omega/\omega_n} \tag{2.42}$$

So,

$$\boxed{F_{\text{dynam(max)}} = k[z_{\text{max(max)}}] = \frac{k(Q_0/k)}{1 - \omega/\omega_n} = \frac{Q_0}{1 - \omega/\omega_n}} \tag{2.43}$$

Hence, the total force on the subgrade will vary between the limits

$$W - \frac{Q_0}{1 - \omega/\omega_n} \quad \text{and} \quad W + \frac{Q_0}{1 - \omega/\omega_n}$$

Example 2.3

A machine foundation can be idealized as a mass-spring system. This foundation can be subjected to a force that can be given as $Q(\text{lb}) = 8000 \sin \omega t$. Given

$$f = 800 \text{ cycles/min}$$

$$\text{Weight of the machine + foundation} = 40{,}000 \text{ lb}$$

$$\text{Spring constant} = 400{,}000 \text{ lb/in.}$$

determine the maximum and minimum force transmitted to the subgrade.

Solution

$$\text{Natural angular frequency} = \omega_n = \sqrt{\frac{k}{m}} = \sqrt{\frac{400{,}000}{40{,}000/(32.2 \times 12)}}$$

$$= 62.16 \text{ rad/s}$$

$$F_{\text{dynam}} = \frac{Q_0}{1 - \omega/\omega_n}$$

But

$$\omega = 2\pi f = 2\pi \left(\frac{800}{60}\right) = 83.78 \text{ rad/s}$$

Thus

$$|F_{\text{dynam}}| = \frac{8000}{1 - 83.78/62.16} = 23{,}000 \text{ lb}$$

Maximum force on the subgrade $= 40{,}000 + 23{,}000 = \underline{63{,}000 \text{ lb}}$

Minimum force on the subgrade $= 40{,}000 - 23{,}000 = \underline{17{,}000 \text{ lb}}$ ∎

2.5   Free Vibration with Viscous Damping

In the case of undamped free vibration as explained in Section 2.3, vibration would continue once the system had been set in motion. However, in practical cases, all vibrations undergo a gradual decrease of amplitude with time. This characteristic of vibration is referred to as *damping*. Figure 2.2b shows a foundation supported by a spring and a dashpot. The dashpot represents the *damping characteristic* of the soil. The dashpot coefficient is equal to c. For free vibration of the foundation (i.e., the force $Q = Q_0 \sin \omega t$ on the foundation is zero), the differential equation of motion can be given by

$$m\ddot{z} + c\dot{z} + kz = 0 \tag{2.44}$$

Let $z = Ae^{rt}$ be a solution to Eq. (2.44), where A is a constant. Substitution of this into Eq. (2.44) yields

$$mAr^2 e^{rt} + cAre^{rt} + kAe^{rt} = 0$$

or

$$r^2 + \left(\frac{c}{m}\right)r + \frac{k}{m} = 0 \tag{2.45}$$

The solutions to Eq. (2.45) can be given as

$$r = -\frac{c}{2m} \pm \sqrt{\frac{c^2}{4m^2} - \frac{k}{m}} \tag{2.46}$$

There are three general conditions that may be developed from Eq. (2.46):

1. If $c/2m > \sqrt{k/m}$, both roots of Eq. (2.45) are real and negative. This is referred to as an *overdamped* case.

2. If $c/2m = \sqrt{k/m}$, $r = -c/2m$. This is called the *critical damping* case. Thus, for this case,

$$c = c_c = 2\sqrt{km} \tag{2.47a}$$

3. If $c/2m < \sqrt{k/m}$, the roots of Eq. (2.45) are complex:

$$r = -\frac{c}{2m} \pm i\sqrt{\frac{k}{m} - \frac{c^2}{4m^2}}$$

This is referred to as a case of *underdamping*.

It is possible now to define a *damping ratio D*, which can be expressed as

$$D = \frac{c}{c_c} = \frac{c}{2\sqrt{km}} \tag{2.47b}$$

Using the damping ratio, Eq. (2.46) can be rewritten as

$$r = -\frac{c}{2m} \pm \sqrt{\frac{c^2}{4m^2} - \frac{k}{m}} = \omega_n(-D \pm \sqrt{D^2 - 1}) \tag{2.48}$$

where $\omega_n = \sqrt{k/m}$.
- For the *overdamped condition* ($D > 1$),

$$r = \omega_n(-D \pm \sqrt{D^2 - 1})$$

For this condition, the equation for displacement (i.e., $z = Ae^{rt}$) may be written as

$$z = A_1 \exp[\omega_n t(-D + \sqrt{D^2 - 1})] + A_2 \exp[\omega_n t(-D - \sqrt{D^2 - 1})] \tag{2.49}$$

where A_1 and A_2 are two constants. Now, let

$$A_1 = \tfrac{1}{2}(A_3 + A_4) \tag{2.50}$$

and

$$A_2 = \tfrac{1}{2}(A_3 - A_4) \tag{2.51}$$

Substitution of Eqs. (2.50) and (2.51) into Eq. (2.49) and rearrangement gives

$$z = e^{-D\omega_n t}\{\tfrac{1}{2}A_3[\exp(\omega_n\sqrt{D^2 - 1}\,t) + \exp(-\omega_n\sqrt{D^2 - 1}\,t)]$$
$$+ \tfrac{1}{2}A_4[\exp(\omega_n\sqrt{D^2 - 1}\,t) - \exp(-\omega_n\sqrt{D^2 - 1}\,t)]\}$$

or

$$z = e^{-D\omega_n t}[A_3\cosh(\omega_n\sqrt{D^2 - 1}\,t) + A_4\sinh(\omega_n\sqrt{D^2 - 1}\,t] \tag{2.52}$$

Equation (2.52) shows that the system which is overdamped will *not oscillate at all*. The variation of z with time will take the form shown in Figure 2.8a.

The constants A_3 and A_4 in Eq. (2.52) can be evaluated by knowing the initial conditions. Let, at time $t = 0$, displacement $= z = z_0$ and velocity $= dz/dt = v_0$. From Eq. (2.52) and the first boundary condition,

$$z = z_0 = A_3 \tag{2.53}$$

Again, from Eq. (2.52) and the second boundary condition,

$$\frac{dz}{dt} = v_0 = (\omega_n\sqrt{D^2 - 1}\,A_4) - D\omega_n A_3$$

or

$$A_4 = \frac{v_0 + D\omega_n A_3}{\omega_n\sqrt{D^2 - 1}} = \frac{v_0 + D\omega_n z_0}{\omega_n\sqrt{D^2 - 1}} \tag{2.54}$$

Substituting Eqs. (2.53) and (2.54) into Eq. (2.52),

$$z = e^{-D\omega_n t}\left[z_0\cosh(\omega_n\sqrt{D^2 - 1}\,t) + \frac{v_0 D\omega_n z_0}{\omega_n\sqrt{D^2 - 1}}\sinh(\omega_n\sqrt{D^2 - 1}\,t)\right] \tag{2.55}$$

- For a *critically damped condition* ($D = 1$), from Eq. (2.48),

$$r = -\omega_n \tag{2.56}$$

Given this condition, the equation for displacement ($z = Ae^{rt}$) may be written as

$$z = (A_5 + A_6 t)e^{-\omega_n t} \tag{2.57}$$

where A_5 and A_6 are two constants. This is similar to the case of the overdamped system except for the fact that the sign of z changes only once. This is shown in Figure 2.8b.

The values of A_5 and A_6 in Eq. (2.57) can be determined by using the initial conditions of vibration. Let, at time $t = 0$,

$$z = z_0, \quad \frac{dz}{dt} = v_0$$

From the first of the preceding two conditions and Eq. (2.57),

$$z = z_0 = A_5 \tag{2.58}$$

Similarly, from the second condition and Eq. (2.57),

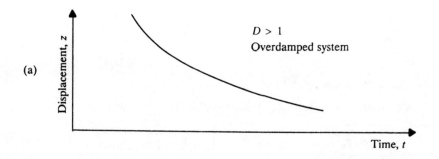

(a)

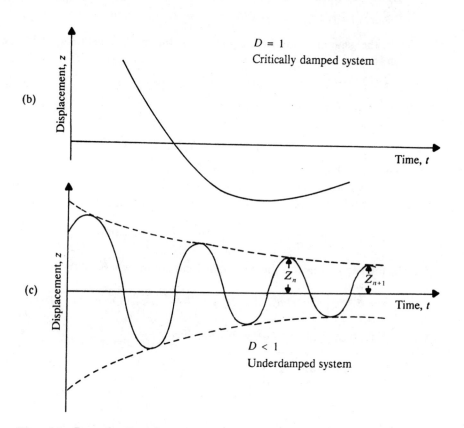

(b)

(c)

Figure 2.8   Free vibration of a mass-spring-dashpot system: (a) overdamped case; (b) critically damped case; (c) underdamped case

$$\frac{dz}{dt} = v_0 = -\omega_n A_5 + A_6 = -\omega_n z_0 + A_6$$

or

$$A_6 = v_0 + \omega_n z_0 \qquad (2.59)$$

A combination of Eqs. (2.57)–(2.59) yields

$$\boxed{z = [z_0 + (v_0 + \omega_n z_0)t]e^{-\omega_n t}} \qquad (2.60)$$

- Lastly, for the *underdamped condition* $(D < 1)$,

$$r = \omega_n(-D \pm i\sqrt{1 - D^2})$$

Thus, the general form of the equation for the displacement $(z = Ae^{rt})$ can be expressed as

$$z = e^{-D\omega_n t}[A_7 \exp(i\omega_n\sqrt{1 - D^2}t) + A_8 \exp(-i\omega_n\sqrt{1 - D^2}t)] \qquad (2.61)$$

where A_7 and A_8 are two constants.

Equation (2.61) can be simplified to the form

$$z = e^{-D\omega_n t}[A_9 \cos(\omega_n\sqrt{1 - D^2}t) + A_{10} \sin(\omega_n\sqrt{1 - D^2}t)] \qquad (2.62)$$

where A_9 and A_{10} are two constants.

The values of the constants A_9 and A_{10} in Eq. (2.62) can be determined by using the following initial conditions of vibration. Let, at time $t = 0$,

$$z = z_0 \quad \text{and} \quad \frac{dz}{dt} = v_0$$

The final equation with these boundary conditions will be of the form

$$z = e^{-D\omega_n t}\left[z_0 \cos(\omega_n\sqrt{1 - D^2}t) + \frac{v_0 + D\omega_n z_0}{\omega_n\sqrt{1 - D^2}} \cdot \sin(\omega_n\sqrt{1 - D^2}t) \right] \qquad (2.63)$$

Equation (2.63) can further be simplified as

$$\boxed{z = Z \cos(\omega_d t - \alpha)} \qquad (2.64)$$

where

$$Z = e^{-D\omega_n t}\sqrt{z_0^2 + \left(\frac{v_0 + D\omega_n z_0}{\omega_n\sqrt{1 - D^2}}\right)^2} \qquad (2.65)$$

$$\alpha = \tan^{-1}\left(\frac{v_0 + D\omega_n z_0}{\omega_n z_0\sqrt{1 - D^2}}\right) \qquad (2.66)$$

$$\boxed{\omega_d = \text{damped natural circular frequency} = \omega_n\sqrt{1 - D^2}} \qquad (2.67)$$

The effect of damping is to decrease gradually the amplitude of vibration with time. In order to evaluate the magnitude of decrease of the amplitude of vibration with time, let Z_n and Z_{n+1} be the two *successive* positive or negative maximum values of displacement at times t_n and t_{n+1} from the start of the vibration as shown in Figure 2.8c. From Eq. (2.65),

$$\frac{Z_{n+1}}{Z_n} = \frac{\exp(-D\omega_n t_{n+1})}{\exp(-D\omega_n t_n)} = \exp[-D\omega_n(t_{n+1} - t_n)] \tag{2.68}$$

However, $t_{n+1} - t_n$ is the period of vibration T,

$$T = \frac{2\pi}{\omega_d} = \frac{2\pi}{\omega_n\sqrt{1 - D^2}} \tag{2.69}$$

Thus, combining Eqs. (2.68) and (2.69),

$$\delta = \ln\left(\frac{Z_n}{Z_{n+1}}\right) = \frac{2\pi D}{\sqrt{1 - D^2}}. \tag{2.70}$$

The term δ is called the *logarithmic decrement*.

If the damping ratio D is small, Eq. (2.70) can be approximated as

$$\delta = \ln\left(\frac{Z_n}{Z_{n+1}}\right) = 2\pi D \tag{2.71}$$

Example 2.4

For a machine foundation, given weight = 60 kN, spring constant = 11,000 kN/m, and c = 200 kN-s/m, determine
a. whether the system is overdamped, underdamped, or critically damped,
b. the logarithmic decrement, and
c. the ratio of two successive amplitudes.

Solution

a. From Eq. (2.47),

$$c_c = 2\sqrt{km} = 2\sqrt{11,000\left(\frac{60}{9.81}\right)} = 518.76 \text{ kN-s/m}$$

$$\frac{c}{c_c} = D = \frac{200}{518.76} = 0.386 < 1$$

Hence, the system is underdamped.
b. From Eq. (2.70),

$$\delta = \frac{2\pi D}{\sqrt{1 - D^2}} = \frac{2\pi(0.386)}{\sqrt{1 - (0.386)^2}} = 2.63$$

c. Again, from Eq. (2.70),

$$\frac{Z_n}{Z_{n+1}} = e^\delta = e^{2.63} = 13.87 \qquad \blacksquare$$

Example 2.5

For Example 2.4, determine the damped natural frequency.

Solution

From Eq. (2.67),

$$f_d = \sqrt{1 - D^2} f_n$$

where f_d = damped natural frequency.

$$f_n = \frac{1}{2\pi} \sqrt{\frac{k}{m}} = \frac{1}{2\pi} \sqrt{\frac{11{,}000 \times 9.81}{60}} = 6.75 \text{ cps}$$

Thus,

$$f_d = (\sqrt{1 - (0.386)^2})(6.75) = 6.23 \text{ cps} \qquad \blacksquare$$

2.6   Steady-State Forced Vibration with Viscous Damping

Figure 2.2b shows the case of a foundation resting on a soil that can be approximated to an equivalent spring and dashpot. This foundation is being subjected to a sinusoidally varying force $Q = Q_0 \sin \omega t$. The differential equation of motion for this system can be given by

$$m\ddot{z} + kz + c\dot{z} = Q_0 \sin \omega t \qquad (2.72)$$

The transient part of the vibration is damped out quickly; so, considering the particular solution for Eq. (2.72) for the steady-state motion, let

$$z = A_1 \sin \omega t + A_2 \cos \omega t \qquad (2.73)$$

where A_1 and A_2 are two constants.
    Substituting Eq. (2.73) into Eq. (2.72),

$$m(-A_1 \omega^2 \sin \omega t - A_2 \omega^2 \cos \omega t) + k(A_1 \sin \omega t + A_2 \cos \omega t)$$
$$+ c(A_1 \omega \cos \omega t - A_2 \omega \sin \omega t) = Q_0 \sin \omega t \qquad (2.74)$$

Collecting *sine* and *cosine* functions in Eq. (2.74) separately,

$$(-mA_1 \omega^2 + kA_1 - cA_2 \omega) \sin \omega t = Q_0 \sin \omega t \qquad (2.75a)$$
$$(-mA_2 \omega^2 + A_2 k + cA_1 \omega) \cos \omega t = 0 \qquad (2.75b)$$

From Eq. (2.75a),

$$A_1 \left(\frac{k}{m} - \omega^2 \right) - A_2 \left(\frac{c}{m} \omega \right) = \frac{Q_0}{m} \qquad (2.76)$$

and from Eq. (2.75b),

$$A_1 \left(\frac{c}{m} \omega \right) + A_2 \left(\frac{k}{m} - \omega^2 \right) = 0 \qquad (2.77)$$

Solution of Eqs. (2.76) and (2.77) will give the following relations for the constants A_1 and A_2:

$$A_1 = \frac{(k - m\omega^2)Q_0}{(k - m\omega^2)^2 + c^2\omega^2} \tag{2.78}$$

and

$$A_2 = \frac{-c\omega Q_0}{(k - m\omega^2)^2 + c^2\omega^2} \tag{2.79}$$

By substituting Eqs. (2.78) and (2.79) into Eq. (2.73) and simplifying, one can obtain

$$\boxed{z = Z\cos(\omega t + \alpha)} \tag{2.80}$$

where

$$\alpha = \tan^{-1}\left(-\frac{A_1}{A_2}\right) = \tan^{-1}\left(\frac{k - m\omega^2}{c\omega}\right) = \tan^{-1}\left[\frac{1 - (\omega^2/\omega_n^2)}{2D(\omega/\omega_n)}\right] \tag{2.81}$$

and

$$\boxed{Z = \sqrt{A_1^2 + A_2^2} = \frac{(Q_0/k)}{\sqrt{[1 - (\omega^2/\omega_n^2)]^2 + 4D^2(\omega^2/\omega_n^2)}}} \tag{2.82}$$

where $\omega_n = \sqrt{k/m}$ is the undamped natural frequency and D is the damping ratio.

Equation (2.82) can be plotted in a nondimensional form as $Z/(Q_0/k)$ against ω/ω_n. This is shown in Figure 2.9. In this figure, note that the maximum values of $Z/(Q_0/k)$ do not occur at $\omega = \omega_n$, as occurs in the case of forced vibration of a spring-mass system (Section 2.4). Mathematically, this can be shown as follows: From Eq. (2.82),

$$\frac{Z}{(Q_0/k)} = \frac{1}{\sqrt{[1 - (\omega^2/\omega_n^2)]^2 + 4D^2(\omega^2/\omega_n^2)}} \tag{2.83}$$

For maximum value of $Z/(Q_0/k)$,

$$\frac{\partial[Z/(Q_0/k)]}{\partial(\omega/\omega_n)} = 0 \tag{2.84}$$

From Eqs. (2.83) and (2.84),

$$\frac{\omega}{\omega_n}\left(1 - \frac{\omega^2}{\omega_n^2}\right) - 2D^2\left(\frac{\omega}{\omega_n}\right) = 0$$

or

$$\omega = \omega_n\sqrt{1 - 2D^2} \tag{2.85}$$

Hence,

$$f_m = f_n\sqrt{1 - 2D^2} \tag{2.86}$$

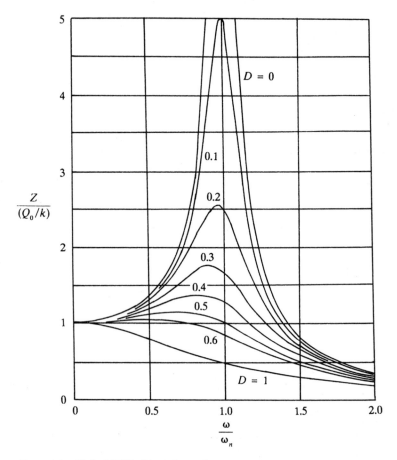

Figure 2.9　Plot of $Z/(Q_0/k)$ against ω/ω_n

where f_m is the frequency at *maximum amplitude* (the *resonant frequency for vibration with damping*) and f_n is the natural frequency = $(1/2\pi)\sqrt{k/m}$. Hence, the *amplitude of vibration at resonance* can be obtained by substituting Eq. (2.85) into Eq. (2.82):

$$Z_{res} = \frac{Q_0}{k} \frac{1}{\sqrt{[1 - (1 - 2D^2)]^2 + 4D^2(1 - 2D^2)}}$$

$$= \frac{Q_0}{k} \frac{1}{2D\sqrt{1 - D^2}} \tag{2.87}$$

Maximum Dynamic Force Transmitted to the Subgrade

For vibrating foundations, it is sometimes necessary to determine the dynamic force transmitted to the foundation. This can be given by summing the spring force and the damping force caused by relative motion between mass and dashpot; that is,

$$F_{dynam} = kz + c\dot{z} \tag{2.88a}$$

From Eq. (2.80),

$$z = Z\cos(\omega t + \alpha)$$

Therefore,

$$\dot{z} = -\omega Z \sin(\omega t + \alpha)$$

and

$$F_{dynam} = kZ\cos(\omega t + \alpha) - c\omega Z \sin(\omega t + \alpha) \tag{2.88b}$$

If one lets

$$kZ = A\cos\phi \quad \text{and} \quad c\omega Z = A\sin\phi,$$

then Eq. (2.88) can be written as

$$F_{dynam} = A\cos(\omega t + \phi + \alpha) \tag{2.89}$$

where

$$A = \sqrt{(A\cos\phi)^2 + (A\sin\phi)^2} = Z\sqrt{k^2 + (c\omega)^2} \tag{2.90}$$

Hence, the *magnitude* of *maximum dynamic force* will be equal to $Z\sqrt{k^2 + (c\omega)^2}$.

Example 2.6

A machine and its foundation weigh 140 kN. The spring constant and the damping ratio of the soil supporting the soil may be taken as 12×10^4 kN/m and 0.2, respectively. Forced vibration of the foundation is caused by a force that can be expressed as

$$Q(kN) = Q_0 \sin \omega t$$

$$Q_0 = 46 \text{ kN}, \qquad \omega = 157 \text{ rad/s}$$

Determine
a. the undamped natural frequency of the foundation,
b. amplitude of motion, and
c. maximum dynamic force transmitted to the subgrade.

Solution

a.

$$f_n = \frac{1}{2\pi}\sqrt{\frac{k}{m}} = \frac{1}{2\pi}\sqrt{\frac{12 \times 10^4}{140/9.81}} = \underline{14.59 \text{ cps}}$$

b. From Eq. (2.82),

$$Z = \frac{Q_0/k}{\sqrt{(1 - \omega^2/\omega_n^2)^2 + 4D^2(\omega^2/\omega_n^2)}}$$

$$\omega_n = 2\pi f_n = 2\pi(14.59) = 91.67 \text{ rad/s}$$

So

$$Z = \frac{46/(12 \times 10^4)}{\sqrt{[1 - (157/91.67)^2]^2 + 4(0.2)^2 \times (157/91.67)^2}}$$

$$= \frac{3.833 \times 10^{-4}}{\sqrt{3.737 + 0.469}} = 0.000187 \text{ m} = \underline{0.187 \text{ mm}}$$

c. From Eq. (2.90), the dynamic force transmitted to the subgrade

$$A = Z\sqrt{k^2 + (c\omega)^2}$$

From Eq. (2.47b),

$$c = 2D\sqrt{km} = 2(0.2)\sqrt{(12 \times 10^4)\left(\frac{140}{9.81}\right)} = 523.46 \text{ kN-s/m}$$

Thus,

$$F_{\text{dynam}} = 0.000187\sqrt{(12 \times 10^4)^2 + (523.46 \times 157)^2} = \underline{27.20 \text{ kN}} \qquad \blacksquare$$

2.7 Rotating-Mass-Type Excitation

In many cases of foundation equipment, vertical vibration of foundations is produced by counter-rotating masses as shown in Figure 2.10a. Since horizontal forces on the foundation at any instant cancel, the net vibrating force on the foundation can be determined to be equal to $2m_e e\omega^2 \sin \omega t$ (where m_e = mass of each counter-rotating element, e = eccentricity, and ω = angular frequency of the masses). In such cases, the equation of motion with viscous damping [Eq. (2.72)] can be modified to the form

$$m\ddot{z} + kz + c\dot{z} = Q_0 \sin \omega t \qquad (2.91)$$

$$Q_0 = 2m_e e\omega^2 = U\omega^2 \qquad (2.92)$$

$$U = 2m_e e \qquad (2.93)$$

and m is the mass of the foundation, including $2m_e$.

Equations (2.91)–(2.93) can be similarly solved by the procedure presented in Section 2.6.

The solution for displacement may be given as

$$\boxed{z = Z \cos(\omega t + \alpha)} \qquad (2.94)$$

where

$$\boxed{Z = \frac{(U/m)(\omega/\omega_n)^2}{\sqrt{(1 - \omega^2/\omega_n^2)^2 + 4D^2(\omega^2/\omega_n^2)}}} \qquad (2.95)$$

$$\alpha = \tan^{-1}\left[\frac{1 - (\omega^2/\omega_n^2)}{2D(\omega/\omega_n)}\right] \qquad (2.96)$$

In Section 2.6, a nondimensional plot for the amplitude of vibration was

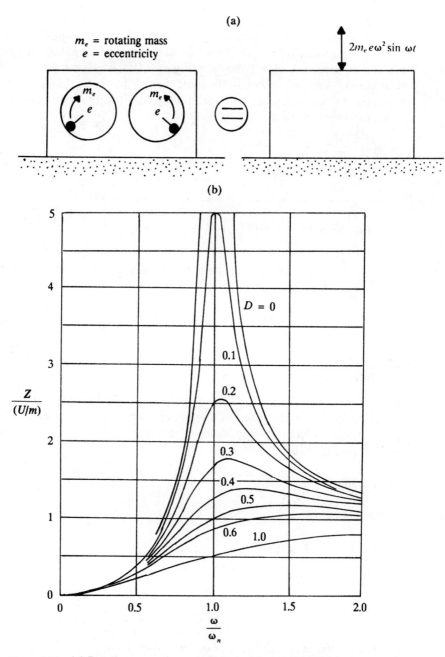

Figure 2.10   (a) Rotating mass-type excitation; (b) plot of $Z/(U/m)$ against ω/ω_n

given in Figure 2.9 [i.e., $Z(Q_0/k)$ versus ω/ω_n]. This was for a vibration produced by a sinusoidal forcing function ($Q_0 = $ const). For a rotating-mass type of excitation, a similar type of nondimensional plot for the amplitude of vibration can also be prepared. This is shown in Figure 2.10b, which is a plot of $Z/(U/m)$ versus ω/ω_n. Also proceeding in the same manner [as in Eq. (2.86) for the case where $Q_0 = $ const], the angular resonant frequency for rotating-mass-type excitation can be obtained as

$$\omega = \frac{\omega_n}{\sqrt{1 - 2D^2}} \tag{2.97}$$

or

$$\boxed{f_m = \text{damped resonant frequency} = \frac{f_n}{\sqrt{1 - 2D^2}}} \tag{2.98}$$

The amplitude at damped resonant frequency can be given [similar to Eq. (2.87)] as

$$Z_{res} = \frac{U/m}{2D\sqrt{1 - D^2}} \tag{2.99}$$

2.8   Determination of Damping Ratio

The damping ratio D can be determined from free and forced vibration tests on a system. In a *free vibration* test, the system is displaced from its equilibrium position, after which the amplitudes of displacement are recorded with time. Now, from Eq. (2.70)

$$\delta = \ln \frac{Z_n}{Z_{n+1}} = \frac{2\pi D}{\sqrt{1 - D^2}}$$

If D is small, then

$$\delta = \ln \frac{Z_n}{Z_{n+1}} = 2\pi D \tag{2.100}$$

It can also be shown that

$$n\delta = \ln \frac{Z_0}{Z_n} = 2\pi n D \tag{2.101}$$

where $Z_n = $ the peak amplitude of the nth cycle. Thus,

$$\boxed{D = \frac{1}{2\pi n} \ln \frac{Z_0}{Z_n}} \tag{2.102}$$

In a *forced vibration* test, the following procedure can be used to determine the damping ratio.

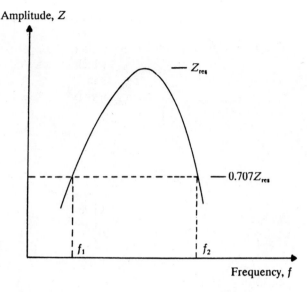

Figure 2.11   Bandwidth method of determination of damping ratio from forced vibration test

1. Vibrate the system with a constant force type of excitation and obtain a plot of amplitude (Z) with frequency (f), as shown in Figure 2.11.

2. Determine Z_{res} from Figure 2.11.

3. Calculate $0.707Z_{res}$. Obtain the frequencies f_1 and f_2 that correspond to $0.707Z_{res}$.

4. From Eq. (2.87)

$$Z_{res} = \left(\frac{Q_0}{k}\right)\left(\frac{1}{2D\sqrt{1 - D^2}}\right)$$

However, if D is small,

$$Z_{res} = \left(\frac{Q_0}{k}\right)\left(\frac{1}{2D}\right) \tag{2.103}$$

Again, from Eq. (2.83)

$$Z = 0.707Z_{res} = \frac{Q_0/k}{\sqrt{[1 - (f/f_n)^2]^2 + 4D^2(f/f_n)^2}} \tag{2.104}$$

Combining Eqs. (2.103) and (2.104),

$$\frac{0.707}{2D} = \frac{1}{\sqrt{[1 - (f/f_n)^2]^2 + 4D^2(f/f_n)^2}}$$

$$\left(\frac{f}{f_n}\right)^2 - 2\left(\frac{f}{f_n}\right)^2(1 - 2D^2) + (1 - 8D^2) = 0$$

$$\left(\frac{f}{f_n}\right)^2_{1,2} = (1 - 2D^2) \pm 2D\sqrt{1 + D^2}$$

or

$$\left(\frac{f_2}{f_n}\right)^2 - \left(\frac{f_1}{f_n}\right)^2 = 4D\sqrt{1+D^2} \approx 4D \qquad (2.105)$$

However,

$$\left(\frac{f_2}{f_n}\right)^2 - \left(\frac{f_1}{f_n}\right)^2 = \left(\frac{f_2-f_1}{f_n}\right)\left(\frac{f_2+f_1}{f_n}\right)$$

But

$$\frac{f_2+f_1}{f_n} \approx 2$$

So

$$\left(\frac{f_2}{f_n}\right)^2 - \left(\frac{f_1}{f_n}\right)^2 \approx 2\left(\frac{f_2-f_1}{f_n}\right) \qquad (2.106)$$

Now, combining Eqs. (2.105) and (2.106)

$$4D = 2\left(\frac{f_2-f_1}{f_n}\right)$$

or

$$\boxed{D = \frac{1}{2}\left(\frac{f_2-f_1}{f_n}\right)} \qquad (2.107)$$

Knowing the resonant frequency to be approximately equal to f_n, the magnitude of D can be calculated. This is referred to as the *bandwidth method*.

2.9   Vibration-Measuring Instrument

Based on the theories of vibration presented in the preceding sections, it is now possible to study the principles of a vibration-measuring instrument, as shown in Figure 2.12. The instrument consists of a spring-mass-dashpot system. It is mounted on a vibrating base. The relative motion of the mass m with respect to the vibrating base is monitored.

Let the motion of the base be given as

$$z' = Z' \sin \omega t \qquad (2.108)$$

Neglecting the transients, let the absolute motion of the mass be given as

$$z'' = Z'' \sin \omega t \qquad (2.109)$$

So, the equation of motion for the mass can be written as

$$m\ddot{z}'' + k(z'' - z') + c(\dot{z}'' - \dot{z}') = 0$$

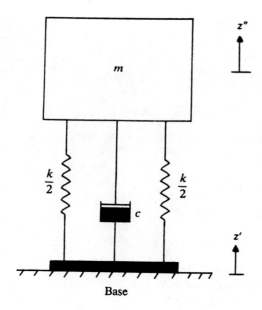

Figure 2.12   Principles of vibration-measuring instrument

Now let

$$z'' - z' = z$$

and

$$\dot{z}'' - \dot{z}' = \dot{z}$$

So

$$m\ddot{z} + kz + c\dot{z} = m\omega^2 Z' \sin \omega t \qquad (2.110)$$

The solution to the Eq. (2.110) can be given as [similar to Eqs. (2.80), (2.81), and (2.82)]

$$z = Z \cos(\omega t + \alpha) \qquad (2.111)$$

where

$$Z = \frac{m\omega^2 Z'}{\sqrt{(k - m\omega^2)^2 + (c\omega)^2}} \qquad (2.112)$$

and

$$\alpha = \tan^{-1}\left(\frac{k - m\omega^2}{c\omega}\right) \qquad (2.113)$$

Again, from Eq. (2.112),

$$\frac{Z}{Z'} = \frac{(\omega/\omega_n)^2}{\sqrt{[1 - (\omega/\omega_n)^2]^2 + 4D^2(\omega/\omega_n)^2}} \qquad (2.114)$$

If the natural frequency of the instrument ω_n is small and ω/ω_n is large, then for practically all values of D, the magnitude of Z/Z' is about 1. Hence the instrument works as a *velocity pickup*.

Also, from Eq. (2.114) one can write that

$$\frac{Z}{\omega^2 Z'} = \frac{1}{\omega_n^2 \sqrt{[1 - (\omega/\omega_n)^2]^2 + 4D^2(\omega/\omega_n)^2}} \tag{2.115}$$

If $D = 0.69$ and ω/ω_n is varied from zero to 0.4 (Prakash, 1981), then Eq. (2.115) will result in

$$\frac{Z}{\omega^2 Z'} \approx \frac{1}{\omega_n^2} = \text{const}$$

So

$$Z \propto \omega^2 Z'$$

However, $\omega^2 Z'$ is the absolute acceleration of the vibrating base. For this condition, the instrument works as an *acceleration pickup*. Note that, for this case, the natural frequency of the instrument and, thus, ω_n are large, and hence the ratio ω/ω_n is small.

SYSTEM WITH TWO DEGREES OF FREEDOM

2.10   Vibration of a Mass-Spring System

A mass-spring system with two degrees of freedom is shown in Figure 2.13a. The system may be excited into vibration in several ways. Two cases of practical interest are

a. sinusoidal force applied on mass m_1 resulting in forced vibration of the system, and

b. the vibration of the system triggered by an impact on mass m_2.

The procedure for calculating the natural frequencies of the system shown in Figure 2.13 is described first, followed by a method for calculating amplitudes of masses m_1 and m_2 for the two cases of excitation mentioned here.

A. Calculation of Natural Frequency

The free body diagrams for the vibration of the masses m_1 and m_2 are shown in Figure 2.13b. The equations of motion may be written as

$$m_1 \ddot{z}_1 + k_1 z_1 + k_2 (z_1 - z_2) = 0 \tag{2.116}$$

$$m_2 \ddot{z}_2 + k_2 (z_2 - z_1) = 0 \tag{2.117}$$

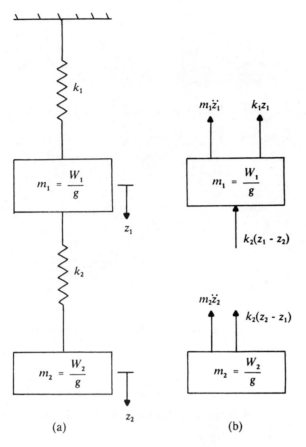

Figure 2.13   Mass-spring system with two degrees of freedom

where

k_1, k_2 = spring constants

z_1, z_2 = displacement of masses m_1 and m_2, respectively

Now, let

$$z_1 = A \sin \omega_n t \qquad\qquad\qquad (2.118a)$$

$$z_2 = B \sin \omega_n t \qquad\qquad\qquad (2.118b)$$

where ω_n = natural frequency of the system. Substitution of Eqs. (2.118a) and (2.118b) into Eqs. (2.116) and (2.117) yields

$$A(k_1 + k_2 - m_1\omega_n^2) - k_2 B = 0 \qquad\qquad (2.119a)$$

and

$$-Ak_2 + (k_2 - m_2\omega_n^2)B = 0 \qquad\qquad (2.119b)$$

For the nontrivial solution

$$\begin{vmatrix} k_1 + k_2 - m_1\omega_n^2 & -k_2 \\ -k_2 & k_2 - m_2\omega_n^2 \end{vmatrix} = 0$$

or

$$(k_1 + k_2 - m_1\omega_n^2)(k_2 - m_2\omega_n^2) = k_2^2$$

$$\omega_n^4 - \left(\frac{k_1 m_2 + k_2 m_2 + k_2 m_1}{m_1 m_2}\right)\omega_n^2 + \frac{k_1 k_2}{m_1 m_2} = 0 \tag{2.120}$$

Let

$$\eta = \frac{m_2}{m_1} \tag{2.121a}$$

$$\omega_{nl_1} = \sqrt{\frac{k_1}{m_1 + m_2}} \tag{2.121b}$$

$$\omega_{nl_2} = \sqrt{\frac{k_2}{m_2}} \tag{2.121c}$$

Substituting Eqs. (2.121a), (2.121b), and (2.121c) into Eq. (2.120) and simplifying, one obtains

$$\boxed{\omega_n^4 - (1 + \eta)(\omega_{nl_1}^2 + \omega_{nl_2}^2)\omega_n^2 + (1 + \eta)(\omega_{nl_1}^2)(\omega_{nl_2}^2) = 0} \tag{2.122}$$

Equation (2.122) represents the frequency equation for a two-degree system.

B. Amplitude of Vibration of Masses m_1 and m_2

Vibration Induced by a Force Acting on Mass m_1: Figure 2.14 shows the case where a force $Q = Q_0 \sin \omega t$ is acting on a mass m_1. The equations of motion may be written as

$$m_1 z_1 + k_1 z_1 + k_2(z_1 - z_2) = Q_0 \sin \omega t \tag{2.123a}$$

$$m_2 z_2 + k_2(z_2 - z_1) = 0 \tag{2.123b}$$

Let

$$z_1 = A_1 \sin \omega t \tag{2.124a}$$

$$z_2 = A_2 \sin \omega t \tag{2.124b}$$

Substitution of Eqs. (2.124a) and (2.124b) into Eqs. (2.123a) and (2.123b) yields

$$A_1(-m_1\omega^2 + k_1 + k_2) - A_2 k_2 = Q_0 \tag{2.125a}$$

$$A_2(k_2 - m_2\omega^2) - A_1 k_2 = 0 \tag{2.125b}$$

From Eq. (2.125b)

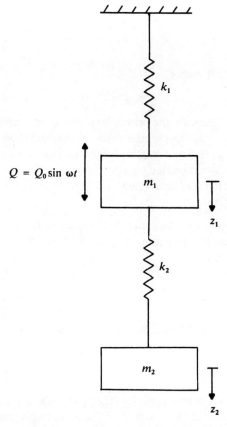

Figure 2.14   Vibration induced by a force on a mass-spring system with two degrees of freedom

$$A_2 = \frac{A_1 k_2}{(k_2 - m_2 \omega^2)} \tag{2.126}$$

Combining Eqs. (2.125a) and (2.126),

$$A_1(-m_1 \omega^2 + k_1 + k_2) - \frac{A_1 k_2^2}{(k_2 - m_2 \omega^2)} = 0$$

or

$$A_1 = \frac{Q_0(\omega_{nl_2}^2 - \omega^2)}{m_1 \Delta(\omega^2)} \tag{2.127}$$

where $\Delta(\omega^2) = \omega^4 - (1 + \eta)(\omega_{nl_1}^2 + \omega_{nl_2}^2)\omega^2 + (1 + \eta)(\omega_{nl_1}^2)(\omega_{nl_2}^2).$ (2.128)

Similarly, it can be shown that

$$A_2 = \frac{Q_0 \omega_{nl_2}^2}{m_1 \Delta(\omega^2)}$$ (2.129)

It may be observed from Eq. (2.127) that $A_1 = 0$ if

$$\omega_{nl_2} = \omega$$ (2.130)

Equations (2.127) and (2.130) illustrate the principle of *vibration absorber*. In a practical situation, the system k_1, m_1 represents a *main system*, and the system k_2, m_2 represents an *auxiliary system*. The vibration of the main system can, in principle, be reduced or even totally eliminated by attaching an auxiliary system to the main mass, designed in such a way that its natural frequency ω_{nl_2} is equal to the operating frequency ω.

Vibration Induced by an Impact on Mass m_2: A practical solution to this case is obtained by assuming that the vibration is being induced by an initial velocity v_0 to mass m_2. For this case, let

$$z_1 = C_1 \sin \omega_{n_1} t + C_2 \sin \omega_{n_2} t$$ (2.131a)

$$z_2 = D_1 \sin \omega_{n_1} t + D_2 \sin \omega_{n_2} t$$ (2.131b)

The initial conditions of vibration are defined as follows. At time $t = 0$:

$$z_1 = z_2 = 0$$ (2.132a)

$$\dot{z}_1 = 0 \quad \text{and} \quad \dot{z}_2 = v_0$$ (2.132b)

Substituting Eqs. (2.132a) and (2.132b) into Eqs. (2.116) and (2.117), applying the initial conditions as defined in Eqs. (2.132a) and (2.132b), and simplifying, one obtains

$$z_1 = \frac{(\omega_{nl_2}^2 - \omega_{n_1}^2)(\omega_{nl_2}^2 - \omega_{n_2}^2)}{\omega_{nl_2}^2(\omega_{n_1}^2 - \omega_{n_2}^2)} \left(\frac{\sin \omega_{n_1} t}{\omega_{n_1}} - \frac{\sin \omega_{n_2} t}{\omega_{n_2}} \right) v_0$$ (2.133a)

and

$$z_2 = \frac{1}{(\omega_{n_1}^2 - \omega_{n_2}^2)} \left[\frac{(\omega_{nl_2}^2 - \omega_{n_2}^2) \sin \omega_{n_1} t}{\omega_{n_1}} - \frac{(\omega_{nl_2}^2 - \omega_{n_1}^2) \sin \omega_{n_2} t}{\omega_{n_2}} \right] v_0$$ (2.133b)

The preceding relationships can be further simplified to determine the amplitudes Z_1 and Z_2 of masses m_1 and m_2, respectively:

$$Z_1 = \frac{(\omega_{nl_2}^2 - \omega_{n_1}^2)(\omega_{nl_2}^2 - \omega_{n_2}^2)}{\omega_{nl_2}^2(\omega_{n_1}^2 - \omega_{n_2}^2)\omega_{n_2}} v_0$$ (2.134a)

and

$$Z_2 = \frac{(\omega_{nl_2}^2 - \omega_{n_1}^2)v_0}{(\omega_{n_1}^2 - \omega_{n_2}^2)\omega_{n_2}}$$ (2.134b)

Example 2.7

Refer to Figure 2.13a. Calculate the natural frequencies of the system. Given:

$$\text{Weight:} \quad W_1 = 25 \text{ lb}; \ W_2 = 5 \text{ lb}$$

$$\text{Spring constant:} \quad k_1 = 100 \text{ lb/in.}; \ k_2 = 50 \text{ lb/in.}$$

Solution

From Eqs. (2.121a), (2.121b), and (2.121c)

$$\eta = \frac{m_2}{m_1} = \frac{W_2}{W_1} = \frac{5}{25} = 0.2$$

$$\omega_{nl_1} = \sqrt{\frac{k_1}{m_1 + m_2}} = \sqrt{\frac{(100)(32.2 \times 12)}{(25 + 5)}} = 35.78 \text{ rad/s}$$

$$\omega_{nl_2} = \sqrt{\frac{k_2}{m_2}} = \sqrt{\frac{(50)(32.2 \times 12)}{5}} = 61.96 \text{ rad/s}$$

From Eq. (2.122)

$$\omega_n^4 - (1 + \eta)(\omega_{nl_1}^2 + \omega_{nl_2}^2)\omega_n^2 + (1 + \eta)(\omega_{nl_1}^2)(\omega_{nl_2}^2) = 0$$

$$\omega_n^4 - (1 + 0.2)[(35.78)^2 + (61.96)^2]\omega_n^2 + (1 + 0.2)(35.78)^2(61.96)^2 = 0$$

$$\omega_n^4 - 6143.1\omega_n^2 + 5{,}897{,}727.9 = 0$$

$$\omega_{n_{1,2}}^2 = \frac{6143.1 \pm \sqrt{(6143.1)^2 - (4)(5{,}897{,}727.9)}}{2}$$

$$\omega_{n_1}^2 = 1190.15; \qquad \omega_{n_2}^2 = 4952.2$$

So

$$\underline{\omega_{n_1} = 34.5 \text{ rad/s}}; \qquad \underline{\omega_{n_2} = 70.37 \text{ rad/s}} \qquad \blacksquare$$

Example 2.8

Refer to Example 2.7. If a sinusoidally varying force $Q = 10 \sin \omega t$ lb is applied to the mass m_1 (Figure 2.13a), what would be the amplitudes of vibration given $\omega = 78.54$ rad/s?

Solution

From Eq. (2.128),

$$\Delta(\omega^2) = \omega^4 - (1 + \eta)(\omega_{nl_1}^2 + \omega_{nl_2}^2)\omega^2 + (1 + \eta)(\omega_{nl}^2)(\omega_{nl_2}^2)$$

$$= (78.54)^4 - (1 + 0.2)[(35.78)^2 + (61.96)^2](78.54)^2$$

$$+ (1 + 0.2)(35.78)^2(61.96)^2$$

$$= 6{,}054{,}603.6$$

Again, using Eqs. (2.127) and (2.129),

$$A_1 = \frac{Q_0(\omega_{nl_2}^2 - \omega^2)}{m_1 \Delta(\omega^2)} = \frac{(10)[(61.96)^2 - (78.54)^2]}{\left(\dfrac{25}{32.2 \times 12}\right)(6,054,603.6)}$$

$$= -0.059 \text{ in.}$$

So, the magnitude of A_1 is $\underline{0.059 \text{ in.}}$

$$A_2 = \frac{Q_0 \omega_{nl_2}^2}{m_1 \Delta(\omega^2)} = \frac{(10)(61.96)^2}{\left(\dfrac{25}{32.2 \times 12}\right)(6,054,603.6)} = \underline{0.097 \text{ in.}} \qquad \blacksquare$$

2.11   Coupled Translation and Rotation of a Mass-Spring System (Free Vibration)

Figure 2.15 shows a mass-spring system that will undergo translation and rotation. The equations of motion of the mass m can be given as

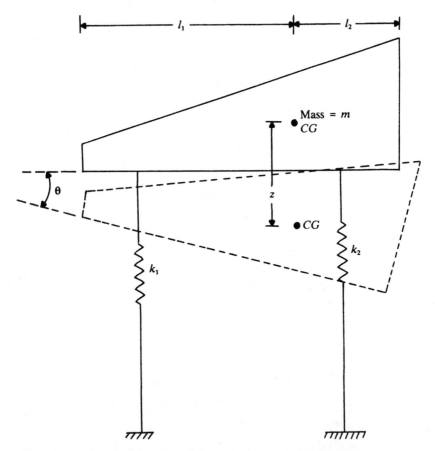

Figure 2.15   Coupled translation and rotation of a mass-spring system—free vibration

$$m\ddot{z} + k_1(z - l_1\theta) + k_2(z + l_2\theta) = 0 \tag{2.135}$$

$$mr^2\ddot{\theta} - l_1k_1(z - l_1\theta) + l_2k_2(z + l_2\theta) = 0 \tag{2.136}$$

where

θ = angle of rotation of the mass m

$$\ddot{\theta} = \frac{d^2\theta}{dt^2}$$

r = radius of gyration of the body about the center of gravity (*Note:* $mr^2 = J$ = mass moment of inertia about the center of gravity)

k_1, k_2 = spring constants

z = distance of translation of the center of gravity of the body

Now, let

$$k_1 + k_2 = k_z \tag{2.137}$$

and

$$l_1^2k_1 + l_2^2k_2 = k_\theta \tag{2.138}$$

So, the equations of motion can be written as

$$m\ddot{z} + k_z z + (l_2k_2 - l_1k_1)\theta = 0 \tag{2.139}$$

$$mr^2\ddot{\theta} + k_\theta\theta + (l_2k_2 - l_1k_1)z = 0 \tag{2.140}$$

If $l_1k_1 = l_2k_2$, Eq. (2.139) is independent of θ and Eq. (2.140) is independent of z. This means that the two motions (i.e., translation and rotation) can exist independently of each one another (uncoupled motion); that is,

$$m\ddot{z} + k_z z = 0 \tag{2.141}$$

and

$$mr^2\ddot{\theta} + k_\theta\theta = 0 \tag{2.142}$$

The natural circular frequency ω_{nz} of translation can be obtained by

$$\omega_{nz} = \sqrt{\frac{k_z}{m}} \tag{2.143}$$

Similarly, the natural circular frequency of rotation $\omega_{n\theta}$ can be given by

$$\omega_{n\theta} = \sqrt{\frac{k_\theta}{mr^2}} \tag{2.144}$$

However, if l_1k_1 is not equal to l_2k_2, the equations of motion (coupled motion) can be solved as follows: Let

$$\frac{k_z}{m} = E_1 \tag{2.145}$$

$$\frac{l_2 k_2 - l_1 k_1}{m} = E_2 \tag{2.146}$$

$$\frac{k_\theta}{m} = E_3 \tag{2.147}$$

Combining Eqs. (2.139), (2.140), (2.145)–(2.147),

$$\ddot{z} + E_1 z + E_2 \theta = 0 \tag{2.148}$$

$$\ddot{\theta} + \left(\frac{E_3}{r^2}\right)\theta + \left(\frac{E_2}{r^2}\right)z = 0 \tag{2.149}$$

For solution of these equations, let

$$z = Z \cos \omega_n t \tag{2.150}$$

and

$$\theta = \Theta \cos \omega_n t \tag{2.151}$$

Substitution of Eqs. (2.150) and (2.151) into Eqs. (2.148) and (2.149) results in

$$(E_1 - \omega_n^2)Z + E_2 \Theta = 0 \tag{2.152}$$

and

$$\left(\frac{E_3}{r^2} - \omega_n^2\right)\Theta + \left(\frac{E_2}{r^2}\right)Z = 0 \tag{2.153}$$

For nontrivial solutions of Eqs. (2.152) and (2.153),

$$\begin{vmatrix} E_1 - \omega_n^2 & E_2 \\ \dfrac{E_2}{r^2} & \dfrac{E_3}{r^2} - \omega_n^2 \end{vmatrix} = 0 \tag{2.154}$$

or

$$\omega_n^4 - \left(\frac{E_3}{r^2} + E_1\right)\omega_n^2 + \frac{E_1 E_3 - E_2^2}{r^2} = 0 \tag{2.155}$$

The natural frequencies ω_{n_1}, ω_{n_2} of the system can be determined from Eq. (2.155) as

$$\begin{matrix} \omega_{n_1} \\ \omega_{n_2} \end{matrix} = \frac{1}{\sqrt{2}}\left\{\left(\frac{E_3}{r^2} + E_1\right) \mp \left[\left(\frac{E_3}{r^2} - E_1\right)^2 + 4\frac{E_2^2}{r^2}\right]^{1/2}\right\}^{1/2} \tag{2.156}$$

Hence, the general equations of motion can be given as

$$z = Z_1 \cos \omega_{n_1} t + Z_2 \cos \omega_{n_2} t \tag{2.157}$$

and

$$\theta = \Theta_1 \cos \omega_{n_1} t + \Theta_2 \cos \omega_{n_2} t \tag{2.158}$$

The amplitude ratios can also be obtained from Eqs. (2.152) and (2.153) as

$$\frac{Z_1}{\Theta_1} = -\frac{E_2}{E_1 - \omega_{n_1}^2} = \frac{-(E_3/r^2 - \omega_{n_1}^2)}{E_2/r^2} \tag{2.159}$$

and

$$\frac{Z_2}{\Theta_2} = -\frac{E_2}{E_1 - \omega_{n_2}^2} = \frac{-(E_3/r^2 - \omega_{n_2}^2)}{E_2/r^2} \tag{2.160}$$

PROBLEMS

2.1 Define the following terms:
a. Spring constant
b. Coefficient of subgrade reaction
c. Undamped natural circular frequency
d. Undamped natural frequency
e. Period
f. Resonance
g. Critical damping coefficient
h. Damping ratio
i. Damped natural frequency

2.2 A machine foundation can be idealized to a mass-spring system, as shown in Figure 2.4. Given

$$\text{Weight of machine} + \text{foundation} = 400 \text{ kN}$$

$$\text{Spring constant} = 100,000 \text{ kN/m}$$

Determine the natural frequency of undamped free vibration of this foundation and the natural period.

2.3 Refer to Problem 2.2. What would be the static deflection z_s of this foundation?

2.4 Refer to Example 2.3. For this foundation let time $t = 0$, $z = z_0 = 0$, $\dot{z} = v_0 = 0$.
a. Determine the natural period T of the foundation.
b. Plot the dynamic force on the subgrade of the foundation due to the forced part of the response for time $t = 0$ to $t = 2T$.
c. Plot the dynamic force on the subgrade of the foundation due to the free part of the response for $t = 0$ to $t = 2T$.
d. Plot the total dynamic force on the subgrade [that is, the algebraic sum of (b) and (c)].
[*Hint*: Refer to Eq. (2.33).]

$$\text{Force due to forced part} = k \left(\frac{Q_0/k}{1 - \omega^2/\omega_n^2} \right) \sin \omega t$$

$$\text{Force due to free part} = k \left(\frac{Q_0/k}{1 - \omega^2/\omega_n^2} \right) \left(-\frac{\omega}{\omega_n} \sin \omega_n t \right)$$

2.5   A foundation of mass m is supported by two springs attached in series. (See Figure P2.5). Determine the natural frequency of the undamped free vibration.

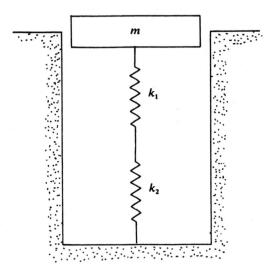

Figure P2.5

2.6   A foundation of mass m is supported by two springs attached in parallel (Figure P2.6). Determine the natural frequency of the undamped free vibration.

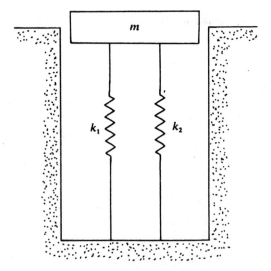

Figure P2.6

2.7   For the system shown in Figure P2.7, calculate the natural frequency and period given $k_1 = 100$ N/mm, $k_2 = 200$ N/mm, $k_3 = 150$ N/mm, $k_4 = 100$ N/mm, $k_5 = 150$ N/mm, and $m = 100$ kg.

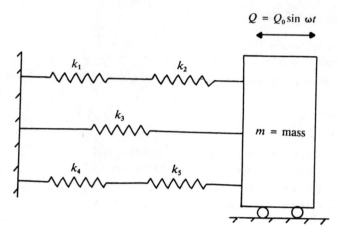

Figure P2.7

2.8   Refer to Problem 2.7. If a sinusoidally varying force $Q = 50 \sin \omega t$ (N) is applied to the mass as shown, what would be the amplitude of vibration given $\omega = 47$ rad/s?

2.9   A body weighs 30 lb. A *spring* and a *dashpot* are attached to the body in the manner shown in Figure 2.2b. The spring constant is 15 lb/in. The dashpot has a resistance of 0.15 lb at a velocity of 2.5 in./s. Determine the following for free vibration:
a. Damped natural frequency of the system
b. Damping ratio
c. Ratio of successive amplitudes of the body (Z_n/Z_{n+1})
d. Amplitude of the body 5 cycles after it is disturbed, assuming that at time $t = 0$, $z = 1$ in.

2.10   A machine foundation can be identified as a mass-spring system. This is subjected to a forced vibration. The vibrating force is expressed as

$$Q = Q_0 \sin \omega t \qquad Q_0 = 15,000 \text{ lb} \qquad \omega = 3100 \text{ rad/min}$$

Given

   Weight of machine + foundation = 65,000 lb

   Spring constant = 5000 kip/in.

Determine the maximum and minimum force transmitted to the subgrade.

2.11   Repeat Problem 2.10 if

$$Q_0 = 200 \text{ kN} \qquad \omega = 6000 \text{ rad/min}$$

   Weight of machine + foundation = 400 kN

   Spring constant = 120,000 kN/m

2.12   A foundation weighs 800 kN. The foundation and the soil can be approximated as a mass-spring-dashpot system as shown in Figure 2.2b. Given

Spring constant = 200,000 kN/m

Dashpot coefficient = 2340 kN-s/m

Determine the following:
a. Critical damping coefficient c_c
b. Damping ratio
c. Logarithmic decrement
d. Damped natural frequency

2.13 The foundation given in Problem 2.12 is subjected to a vertical force $Q = Q_0 \sin \omega t$ in which

$$Q_0 = 25 \text{ kN} \qquad \omega = 100 \text{ rad/s}$$

Determine
a. the amplitude of the vertical vibration of the foundation, and
b. the maximum dynamic force transmitted to the subgrade.

2.14 A mass-spring system with two degrees of freedom is shown in Figure P2.14. Determine the natural frequencies ω_{n_1} and ω_{n_2} as a function of k_1, k_2, k_3, m_1, and m_2.

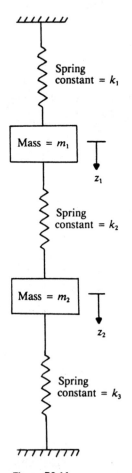

Figure P2.14

2.15 A spring-mass system consists of a spring k_1 and a mass m_1, as shown in Figure P2.15. An auxiliary spring k_2 and mass m_2 are attached as shown. What should be the value of k_2 so that the auxiliary spring-mass system acts as a vibration absorber for the main system (k_1, m_1)? Given $Q_0 = 100$ N and $\omega = 31$ rad/s.

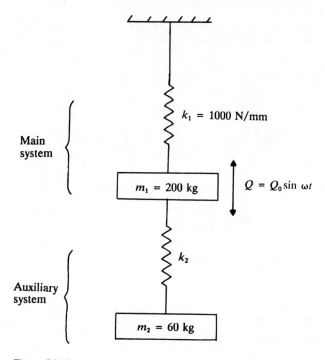

Figure P2.15

REFERENCE

Prakash, S. (1981). *Soil Dynamics*. McGraw-Hill, New York.

Chapter 3

WAVES IN ELASTIC MEDIUM

3.1   Introduction

If a stress is suddenly applied to a body, the part of the body closest to the source of disturbance will be affected first. The deformation of the body due to the load will gradually spread throughout the body via *stress waves*. The nature of propagation of stress waves in an elastic medium is the subject of discussion in this chapter. Stress wave propagation is of extreme importance in geotechnical engineering, since it allows determination of soil properties such as modulus of elasticity and shear modulus and also helps in the development of the design parameters of earthquake-resistant structures. The problem of stress wave propagation can be divided into three major categories:

a. Elastic stress waves in a bar

b. Stress waves in an infinite elastic medium

c. Stress waves in an elastic half-space

However, before the relationships for the stress waves can be developed, it is essential to have some knowledge of the fundamental definitions of stress, strain, and other related parameters that are generally encountered in an elastic medium. These definitions are given in Sections 3.2 and 3.3.

3.2   Stress and Strain

Notations for Stress

Figure 3.1 shows an element in an elastic medium whose sides measure dx, dy, and dz. The normal stresses acting on the plane normal to the x, y, and z axes are σ_x, σ_y, and σ_z, respectively. The shear stresses are τ_{xy}, τ_{yx}, τ_{yz}, τ_{zy}, τ_{zx}, and τ_{xz}. The notations for the shear stresses are as follows.

If τ_{ij} is a shear stress, it means that it is acting on a plane normal to the i axis, and its direction is parallel to the j axis. For equilibrium purposes, by taking moments, it may be seen that

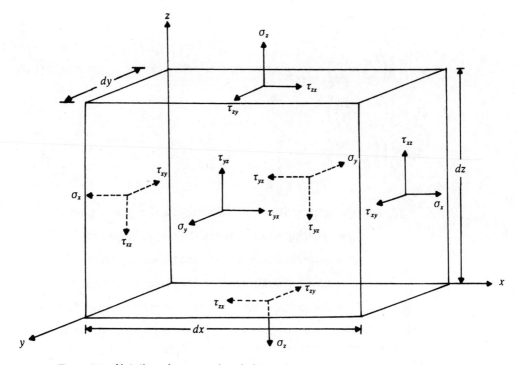

Figure 3.1   Notations for normal and shear stresses

$$\tau_{xy} = \tau_{yx} \tag{3.1}$$

$$\tau_{xz} = \tau_{zx} \tag{3.2}$$

$$\tau_{yz} = \tau_{zy} \tag{3.3}$$

Strain

Due to a given stress condition, let the displacements in the x, y, and z directions (Figure 3.1) be, respectively, u, v, and w. Then the equations for strains and rotations of elastic and isotropic materials in terms of displacements are as follows:

$$\varepsilon_x = \frac{\partial u}{\partial x} \tag{3.4}$$

$$\varepsilon_y = \frac{\partial v}{\partial y} \tag{3.5}$$

$$\varepsilon_z = \frac{\partial w}{\partial z} \tag{3.6}$$

$$\gamma'_{xy} = \frac{\partial v}{\partial x} + \frac{\partial u}{\partial y} \tag{3.7}$$

$$\gamma'_{yz} = \frac{\partial w}{\partial y} + \frac{\partial v}{\partial z} \tag{3.8}$$

$$\gamma'_{zx} = \frac{\partial u}{\partial z} + \frac{\partial w}{\partial x} \tag{3.9}$$

$$\overline{\omega}_x = \frac{1}{2}\left(\frac{\partial w}{\partial y} - \frac{\partial v}{\partial z}\right) \tag{3.10}$$

$$\overline{\omega}_y = \frac{1}{2}\left(\frac{\partial u}{\partial z} - \frac{\partial w}{\partial x}\right) \tag{3.11}$$

$$\overline{\omega}_z = \frac{1}{2}\left(\frac{\partial v}{\partial x} - \frac{\partial u}{\partial y}\right) \tag{3.12}$$

where

ε_x, ε_y, and ε_z = normal strains in the direction of x, y, and z, respectively

γ'_{xy} = shearing strain between the planes xz and yz

γ'_{yz} = shearing strain between the planes yx and zx

γ'_{zx} = shearing strain between the planes zy and xy

$\overline{\omega}_x$, $\overline{\omega}_y$, and $\overline{\omega}_z$ = the components of rotation about the x, y, and z axes.

These derivations are given in most of the textbooks on the theory of elasticity (e.g., Timoshenko and Goodier, 1970). Hence, they are not covered here in detail.

3.3   Hooke's Law

For an elastic, isotropic material, the normal strains and normal stresses can be related by the following equations:

$$\varepsilon_x = \frac{1}{E}[\sigma_x - \mu(\sigma_y + \sigma_z)] \tag{3.13}$$

$$\varepsilon_y = \frac{1}{E}[\sigma_y - \mu(\sigma_x + \sigma_z)] \tag{3.14}$$

$$\varepsilon_z = \frac{1}{E}[\sigma_z - \mu(\sigma_x + \sigma_y)] \tag{3.15}$$

where ε_x, ε_y, and ε_z are the respective normal strains in the directions of x, y, and z, E is Young's modulus, and μ is Poisson's ratio.

The shear stresses and the shear strains can be related by the following equations:

$$\tau_{xy} = G\gamma'_{xy} \tag{3.16}$$

$$\tau_{yz} = G\gamma'_{yz} \tag{3.17}$$

$$\tau_{zx} = G\gamma'_{zx} \tag{3.18}$$

where the shear modulus is

$$G = \frac{E}{2(1 + \mu)} \tag{3.19}$$

and γ'_{xy}, γ'_{yz}, and γ'_{zx} are the shear strains.

Equations (3.13)–(3.15) can be solved to express normal stresses in terms of normal strains as

$$\sigma_x = \lambda\bar{\varepsilon} + 2G\varepsilon_x \tag{3.20}$$

$$\sigma_y = \lambda\bar{\varepsilon} + 2G\varepsilon_y \tag{3.21}$$

$$\sigma_z = \lambda\bar{\varepsilon} + 2G\varepsilon_z \tag{3.22}$$

where

$$\lambda = \frac{\mu E}{(1 + \mu)(1 - 2\mu)} \tag{3.23}$$

$$\bar{\varepsilon} = \varepsilon_x + \varepsilon_y + \varepsilon_z \tag{3.24}$$

From Eqs. (3.19) and (2.23), it is easy to see that

$$\mu = \frac{\lambda}{2(\lambda + G)} \tag{3.25}$$

ELASTIC STRESS WAVES IN A BAR

3.4   Longitudinal Elastic Waves in a Bar

Figure 3.2 shows a rod, the cross-sectional area of which is equal to A. Let the Young's modulus and the unit weight of the material that constitutes the rod be equal to E and γ, respectively. Now, let the stress along section a–a of the rod increase by σ. The stress increase along the section b–b can then be given by $\sigma + (\partial\sigma/\partial x)\Delta x$. Based on Newton's second law,

$$\sum \text{force} = (\text{mass})\,(\text{acceleration})$$

Thus, summing the forces in the x direction,

$$-\sigma A + \left(\sigma + \frac{\partial\sigma}{\partial x}\Delta x\right)A = \frac{(A\Delta x\gamma)}{g}\frac{\partial^2 u}{\partial t^2} \tag{3.26}$$

where $A\Delta x\gamma$ = weight of the rod of length Δx, g is the acceleration due to gravity, u is displacement in the x direction, and t is time.

Equation (3.26) is based on the assumptions that (1) the stress is uniform over the entire cross-sectional area and (2) the cross section remains plane during the motion. Simplification of Eq. (3.26) gives

$$\frac{\partial\sigma}{\partial x} = \rho\left(\frac{\partial^2 u}{\partial t^2}\right) \tag{3.27}$$

where $\rho = \gamma/g$ is the density of the material of the bar. However,

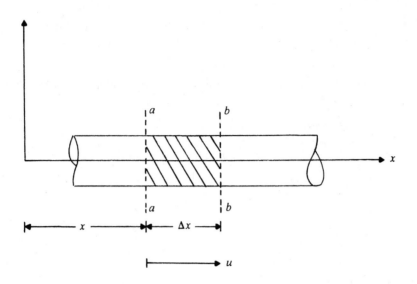

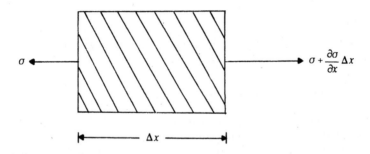

Figure 3.2   Longitudinal elastic wave in a bar

$$\sigma = (\text{strain})\,(\text{Young's modulus}) = \left(\frac{\partial u}{\partial x}\right)E \qquad (3.28)$$

Substitution of Eq. (3.28) into (3.27) yields

$$\frac{\partial^2 u}{\partial t^2} = \left(\frac{E}{\rho}\right)\left(\frac{\partial^2 u}{\partial x^2}\right)$$

or

$$\boxed{\frac{\partial^2 u}{\partial t^2} = v_c^2\,\frac{\partial^2 u}{\partial x^2}} \qquad (3.29)$$

where

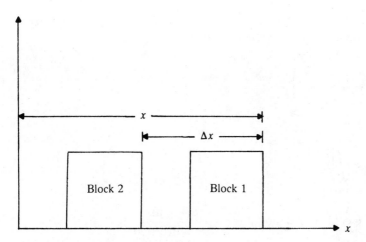

Figure 3.3   Motion of longitudinal elastic wave in a bar

$$v_c = \sqrt{\frac{E}{\rho}} \tag{3.30}$$

The term v_c is the velocity of the *longitudinal* stress wave propagation. This fact can be demonstrated as follows. The solution to Eq. (3.29) can be written in the form

$$u = F(v_c t + x) + G(v_c t - x) \tag{3.31}$$

where $F(v_c t + x)$ and $G(v_c t - x)$ represent some functions of $(v_c t + x)$ and $(v_c t - x)$, respectively. At a given time t, let the function $F(v_c t + x)$ be represented by block 1 in Figure 3.3, and

$$u_t = F(v_c t + x)$$

At time $t + \Delta t$, the function will be represented by block 2 in Figure 3.3. Thus,

$$u_{t+\Delta t} = F[v_c(t + \Delta t) + (x - \Delta x)] \tag{3.32}$$

If the block moves unchanged in shape from position 1 to position 2,

$$u_t = u_{t+\Delta t}$$

or

$$F(v_c t + x) = F[v_c(t + \Delta t) + (x - \Delta x)]$$

or

$$v_c \Delta t = \Delta x \tag{3.33}$$

Thus the velocity of the longitudinal stress wave propagation is equal to $\Delta x / \Delta t = v_c$. In a similar manner, it can be shown that the function $G(v_c t - x)$ represents a wave traveling in the positive direction of x.

If the bar described above is confined, so that no lateral expansion is possible, then the above equation can be modified as

$$\frac{\partial^2 u}{\partial t^2} = v_c'^2 \frac{\partial^2 u}{\partial x^2} \qquad (3.34)$$

where

$$v_c' = \sqrt{\frac{M}{\rho}} \qquad (3.35)$$

M = constrained modulus = $\dfrac{E(1 - \mu)}{(1 - 2\mu)(1 + \mu)}$

μ = Poisson's ratio

3.5   Velocity of Particles in the Stressed Zone

It is important for readers to differentiate between the velocity of the longitudinal wave propagation (v_c) and the velocity of the particles in the stressed zone. In order to distinguish them, consider a compressive stress pulse of intensity σ_x and duration t' (Figure 3.4a) be applied to the end of a rod (shown in Figure 3.4b). When this stress pulse is applied initially, a small zone of the rod will undergo compression. With time this compression will be transmitted to successive zones. During a time interval Δt the stress will travel through a distance

$$\Delta x = v_c \Delta t$$

At any time $t > t'$, a segment of the rod of length $\bar{x}$ will constitute the compressed zone. Note that

$$\bar{x} = v_c t'$$

The elastic shortening of the rod then is

$$u = \left(\frac{\sigma_x}{E}\right)(\bar{x}) = \left(\frac{\sigma_x}{E}\right)(v_c t')$$

Note that u is the displacement of the end of the rod. Now, the velocity of the end of the rod and, thus, the particle velocity is

$$\dot{u} = \frac{u}{t'} = \frac{\sigma_x v_c}{E}$$

Also, it is important to note the following:

1. Particle velocity $\dot{u}$ is a function of the intensity of stress σ_x. However, the longitudinal wave propagation velocity is a function of material property only.

2. The wave propagation velocity and the particle velocity are in the same direction when a compressive stress is applied. However when a tensile stress is

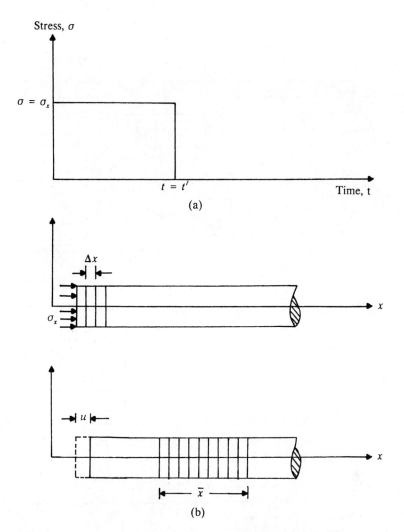

Figure 3.4   Velocity of wave propagation and velocity of particles

applied, the wave propagation velocity and the particle velocity are in opposite directions.

3.6   Reflection of Elastic Stress Waves at the End of a Bar

Bars must terminate at some point. One needs to consider the case of what happens when one of these disturbances, $F(v_c t - x)$ or $G(v_c t + x)$ [Eq. (3.31)], meets the end of the bar.

Figure 3.5a shows a compression wave moving along a bar in the positive x direction. Additionally, a tension wave of the same length is moving along the negative direction of x. When the two waves meet each other (at section a–a), the compression and tension cancel each other, resulting in zero stress; however,

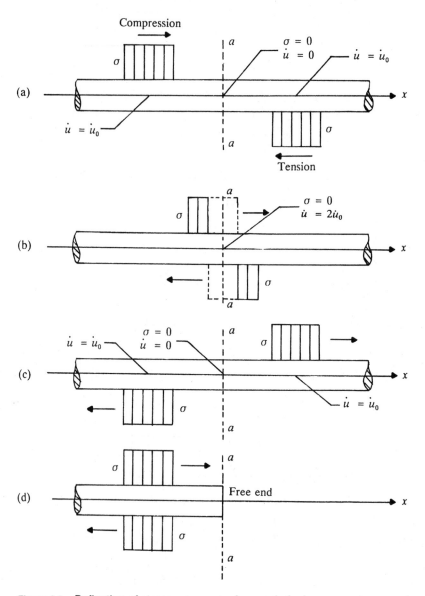

Figure 3.5   Reflection of stress waves at a free end of a bar

the particle velocity is doubled (Figure 3.5b). This is because the particle velocity for a compression wave is in the direction of the motion, and in the tension wave, the particle velocity is opposed to the direction of motion. After the two waves pass each other, the stress and the particle velocity again return to zero at section a–a (Figure 3.5c). The section a–a corresponds to having the stress condition that a free end of a bar would have. Figure 3.5d shows the portion of the rod located to the left of section a–a, and the section can be considered as a free end. By observation it can be seen that, at the free end of a bar, a *compression wave*

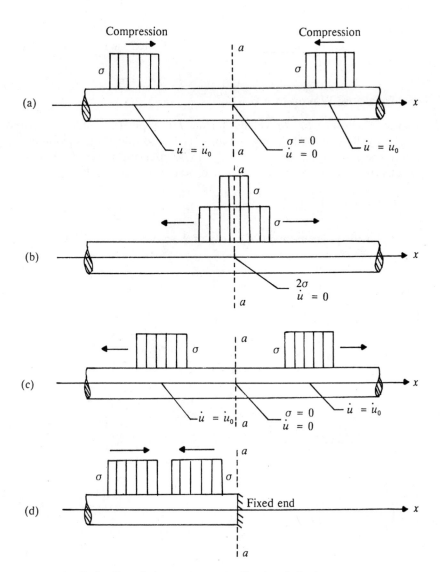

Figure 3.6   Reflection of stress waves at a fixed end of a bar

is reflected back as a tension wave having the same magnitude and shape. In a similar manner, a tension wave is reflected back as a compression wave at the free end of a bar.

Figure 3.6a shows a bar in which two identical compression waves are traveling in opposite directions. When the two waves cross each other at section a–a, the magnitude of the stress will be doubled. However, the particle velocity $\dot{u}$ will be equal to zero (Figure 3.6b). After the two waves pass each other, the stress and the particle velocity return to zero at section a–a (Figure 3.6c). Section a–a remains stationary and behaves as a fixed end of a rod. By observation it can be seen (Figure 3.6d) that a compression wave is reflected back as a compression wave of the same magnitude and shape, but the stress is doubled at the fixed end. In a similar manner, a tension wave is reflected back as a tension wave at the fixed end of a bar.

3.7   Torsional Waves in a Bar

Figure 3.7 shows a rod to which a torque T is applied at a distance x, and the end at x will be rotated through an angle θ. The torque at the section located at a distance $x + \Delta x$ can be given by $T + (\partial T/\partial x)\Delta x$ and the corresponding rotation by $\theta + (\partial \theta/\partial x)\Delta x$. Applying Newton's second law of motion,

$$-T + \left(T + \frac{\partial T}{\partial x}\Delta x\right) = \rho J \Delta x \frac{\partial^2 \theta}{\partial t^2} \tag{3.36}$$

where J is the polar moment of inertia of the cross section of the bar.

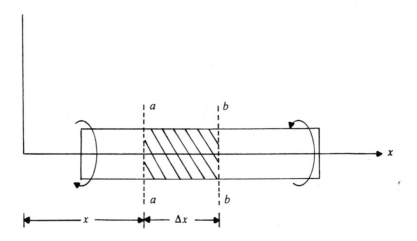

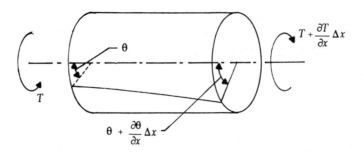

Figure 3.7   Torsional waves in a bar

However, torque T can be expressed by the relation

$$T = JG \frac{\partial \theta}{\partial x} \qquad (3.37)$$

Substitution of Eq. (3.37) into Eq. (3.36) results in

$$\frac{\partial^2 \theta}{\partial t^2} = \frac{G}{\rho} \frac{\partial^2 \theta}{\partial x^2} \qquad (3.38)$$

or

$$\frac{\partial^2 \theta}{\partial t^2} = v_s^2 \frac{\partial^2 \theta}{\partial x^2} \qquad (3.39)$$

where

$$v_s = \sqrt{\frac{G}{\rho}} \qquad (3.40)$$

is the velocity of torsional waves. Note that Eqs. (3.39) and (3.29) are of similar form.

3.8   Longitudinal Vibration of Short Bars

The solution to the wave equations for short bars vibrating in a natural mode can be written in the general form as

$$u(x, t) = U(x)(A_1 \sin \omega_n t + A_2 \cos \omega_n t) \qquad (3.41)$$

where A_1 and A_2 are constants, ω_n is the natural circular frequency of vibration, and $U(x)$ is the amplitude of displacement along the length of the rod and is independent of time.

For longitudinal vibration of uniform bars, if Eq. (3.41) is substituted into Eq. (3.29), it yields

$$\frac{\partial^2 u(x, t)}{\partial x^2} - \frac{\rho}{E} \frac{\partial^2 u(x, t)}{\partial t^2} = 0$$

or

$$\frac{\partial^2 U(x)}{\partial x^2} + \frac{\rho}{E}(\omega_n^2) \cdot U(x) = 0 \qquad (3.42)$$

The solution to Eq. (3.42) may be expressed in the form

$$U(x) = B_1 \sin \left(\frac{\omega_n x}{v_c} \right) + B_2 \cos \left(\frac{\omega_n x}{v_c} \right) \qquad (3.43)$$

where B_1 and B_2 are constants. These constants may be determined by the end condition to which a rod may be subjected.

A. End Condition: Free–Free

For the free–free condition, the stress and thus the strain at the ends are zero. So at $x = 0$, $dU(x)/dx = 0$; and at $x = L$, $dU(x)/dx = 0$, where L is the length of the bar. Differentiating Eq. (3.43) with respect to x,

$$\frac{dU(x)}{dx} = \frac{B_1 \omega_n}{v_c} \cos\left(\frac{\omega_n x}{v_c}\right) - \frac{B_2 \omega_n}{v_c} \sin\left(\frac{\omega_n x}{v_c}\right) \tag{3.44}$$

Substitution of the first boundary condition into Eq. (3.44) results in

$$0 = \frac{B_1 \omega_n}{v_c}; \quad \text{i.e.,} \quad B_1 = 0 \tag{3.45}$$

Again, from the second boundary condition and Eq. (3.44),

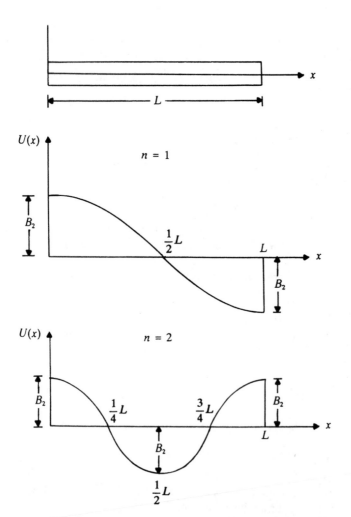

Figure 3.8 Longitudinal vibration of a short bar—free–free end condition

$$0 = -\left(\frac{B_2 \omega_n}{v_c}\right) \sin\left(\frac{\omega_n L}{v_c}\right)$$

Since B_2 is not equal to zero,

$$\frac{\omega_n L}{v_c} = n\pi \tag{3.46}$$

or

$$\boxed{\omega_n = \frac{n\pi v_c}{L}} \tag{3.47}$$

where $n = 1, 2, 3, \ldots$. Thus,

$$\boxed{v_c = \frac{\omega_n L}{n\pi}} \tag{3.48}$$

The equation for the amplitude of displacement for this case can be given by combining Eqs. (3.43), (3.45), and (3.48), or

$$U(x) = B_2 \cos\left(\frac{n\pi x}{L}\right) \tag{3.49}$$

The variation of the nature of $U(x)$ for the first two harmonics (i.e., $n = 1$ and 2) is shown in Figure 3.8. The equation for $u(x, t)$ for all modes of vibration can also be given by combining Eqs. (3.49) and (3.41).

B. End Condition: Fixed–Fixed

For a fixed–fixed end condition, at $x = 0$, $U(x) = 0$ (i.e., displacement is zero); and at $x = L$, $U(x) = 0$.

Substituting the first boundary condition into Eq. (3.43) results in

$$0 = B_2 \tag{3.50}$$

Again, combining the second boundary condition and Eq. (3.43),

$$0 = B_1 \sin\left(\frac{\omega_n L}{v_c}\right)$$

Since $B_1 \neq 0$,

$$\frac{\omega_n L}{v_c} = n\pi \tag{3.51}$$

where $n = 1, 2, 3, \ldots$; or

$$\boxed{\omega_n = \frac{n\pi v_c}{L}} \tag{3.52}$$

or

$$v_c = \frac{\omega_n L}{n\pi} \tag{3.53}$$

The displacement amplitude equation can now be given by combining Eqs. (3.43), (3.50), and (3.52) as

$$U(x) = B_1 \sin\left(\frac{n\pi x}{L}\right) \tag{3.54}$$

Figure 3.9 shows the variation of $U(x)$ for the first two harmonics ($n = 1$ and 2).

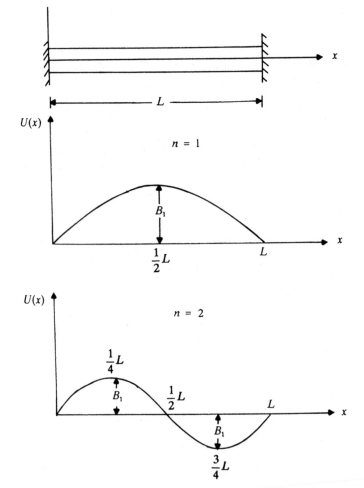

Figure 3.9 Longitudinal vibration of a short bar—fixed–fixed end condition

C. End Condition: Fixed–Free

The boundary conditions for the fixed–free case can be given as follows:

At $x = 0$ (fixed end), $U(x) = 0$

At $x = L$ (free end), $\dfrac{dU(x)}{dx} = 0$

From the first boundary condition and Eq. (3.43),

$$U(x) = 0 = B_2 \tag{3.55}$$

Again, from the second boundary condition and Eq. (3.43)

$$\frac{dU(x)}{dx} = 0 = \frac{B_1 \omega_n}{v_c} \cos\left(\frac{\omega_n L}{v_c}\right) \tag{3.56}$$

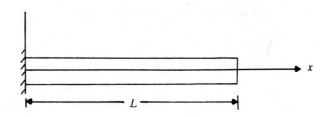

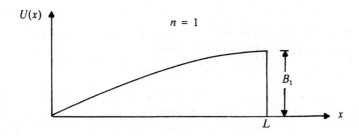

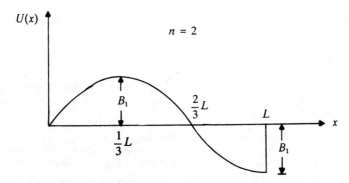

Figure 3.10 Longitudinal vibration of a short bar—fixed–free end condition

or

$$\frac{\omega_n L}{v_c} = (2n - 1)\frac{1}{2}\pi$$

where $n = 1, 2, 3, \ldots$; so

$$\boxed{\omega_n = \frac{1}{2}(2n - 1)\pi\left(\frac{v_c}{L}\right)} \tag{3.57}$$

The displacement amplitude equation can now be written by combining Eqs. (3.43), (3.55), and (3.57) as

$$U(x) = B_1 \sin\left[\frac{\frac{1}{2}(2n - 1)\pi x}{L}\right] \tag{3.58}$$

Figure 3.10 shows the variation of $U(x)$ for the first two harmonics.

3.9   Torsional Vibration of Short Bars

The torsional vibration of short bars can be treated in a manner similar to the longitudinal vibration given in Section 3.8 by writing the equation for natural modes of vibration as

$$\theta(x, t) = \Theta(x)(A_1 \sin \omega_n t + A_2 \cos \omega_n t) \tag{3.59}$$

where Θ = amplitude of angular distortion and A_1 and A_2 are constants.
Solution of Eqs. (3.39) and (3.59) results in

$$\omega_n = \frac{n\pi v_s}{L} \tag{3.60}$$

for the free–free end and fixed–fixed end conditions and

$$\omega_n = \frac{\frac{1}{2}(2n - 1)\pi v_s}{L} \tag{3.61}$$

for the fixed–free end condition, where L is the length of the bar and $n = 1, 2, 3, \ldots$.

STRESS WAVES IN AN INFINITE ELASTIC MEDIUM

3.10   Equation of Motion in an Elastic Medium

Figure 3.11 shows the stresses acting on an element of elastic medium with sides measuring dx, dy, and dz. For obtaining the differential equations of motion, one needs to sum the forces in the x, y, and z directions. Along the x direction,

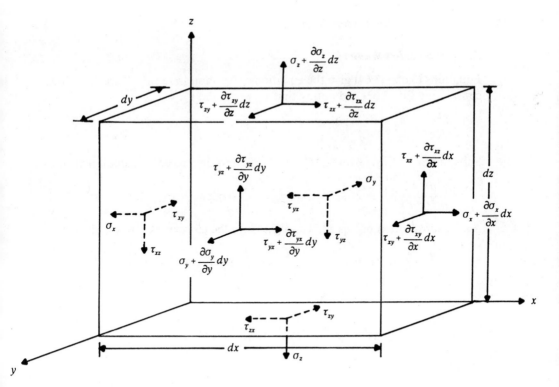

Figure 3.11   Derivation of the equation of motion in an elastic medium

$$\left[\left(\sigma_x + \frac{\partial \sigma_x}{\partial x}dx\right) - \sigma_x\right](dy)(dz) + \left[\left(\tau_{zx} + \frac{\partial \tau_{zx}}{\partial z}dz\right) - \tau_{zx}\right](dx)(dy)$$

$$+ \left[\left(\tau_{yx} + \frac{\partial \tau_{yx}}{\partial y}dy\right) - \tau_{yx}\right](dx)(dz) = \rho(dx)(dy)(dz)\frac{\partial^2 u}{\partial t^2}$$

where ρ is the density of the medium and u is the displacement component along the x direction. Alternatively,

$$\frac{\partial \sigma_x}{\partial x} + \frac{\partial \tau_{yx}}{\partial y} + \frac{\partial \tau_{zx}}{\partial z} = \rho \frac{\partial^2 u}{\partial t^2} \tag{3.62}$$

Similarly, summing forces on the element in the y and z directions

$$\frac{\partial \sigma_y}{\partial y} + \frac{\partial \tau_{xy}}{\partial x} + \frac{\partial \tau_{zy}}{\partial z} = \rho \frac{\partial^2 v}{\partial t^2} \tag{3.63}$$

and

$$\frac{\partial \sigma_z}{\partial z} + \frac{\partial \tau_{xz}}{\partial x} + \frac{\partial \tau_{yz}}{\partial y} = \rho \frac{\partial^2 w}{\partial t^2} \tag{3.64}$$

where v and w are the components of displacement in the y and z directions, respectively.

3.11   Equations for Stress Waves

A. Compression Waves

Equations (3.62)–(3.64) give the equations of motion in terms of stresses. Now, considering Eq. (3.62) and noting that $\tau_{xy} = \tau_{yx}$ and $\tau_{xz} = \tau_{zx}$,

$$\rho \frac{\partial^2 u}{\partial t^2} = \frac{\partial \sigma_x}{\partial x} + \frac{\partial \tau_{xy}}{\partial y} + \frac{\partial \tau_{xz}}{\partial z}$$

Substitution of Eqs. (3.16), (3.18), and (3.20) into the preceding equation yields

$$\rho \frac{\partial^2 u}{\partial t^2} = \frac{\partial}{\partial x}(\lambda \bar{\varepsilon} + 2G\varepsilon_x) + \frac{\partial}{\partial y}(G\gamma'_{xy}) + \frac{\partial}{\partial z}(G\gamma'_{xz})$$

Again, substitution of Eqs. (3.7) and (3.9) into the last expression will yield

$$\rho \frac{\partial^2 u}{\partial t^2} = \frac{\partial}{\partial x}(\lambda \bar{\varepsilon} + 2G\varepsilon_x) + G\frac{\partial}{\partial y}\left(\frac{\partial v}{\partial x} + \frac{\partial u}{\partial y}\right) + G\frac{\partial}{\partial z}\left(\frac{\partial u}{\partial z} + \frac{\partial w}{\partial x}\right)$$

or

$$\rho \frac{\partial^2 u}{\partial t^2} = \lambda \frac{\partial \bar{\varepsilon}}{\partial x} + G\left(\frac{\partial^2 u}{\partial x^2} + \frac{\partial^2 v}{\partial x\,\partial y} + \frac{\partial^2 w}{\partial x\,\partial y} + \frac{\partial^2 u}{\partial x^2} + \frac{\partial^2 u}{\partial y^2} + \frac{\partial^2 u}{\partial z^2}\right) \qquad (3.65)$$

But

$$\frac{\partial^2 u}{\partial x^2} + \frac{\partial^2 v}{\partial x\,\partial y} + \frac{\partial^2 w}{\partial x\,\partial z} = \frac{\partial \bar{\varepsilon}}{\partial x} \qquad (3.66)$$

So

$$\rho \frac{\partial^2 u}{\partial t^2} = (\lambda + G)\frac{\partial \bar{\varepsilon}}{\partial x} + G\nabla^2 u \qquad (3.67)$$

where

$$\nabla^2 = \frac{\partial^2}{\partial x^2} + \frac{\partial^2}{\partial y^2} + \frac{\partial^2}{\partial z^2} \qquad (3.68)$$

Similarly, by proper substitution in Eqs. (3.63) and (3.64), the following relations can be obtained:

$$\rho \frac{\partial^2 v}{\partial t^2} = (\lambda + G)\frac{\partial \bar{\varepsilon}}{\partial y} + G\nabla^2 v \qquad (3.69)$$

and

$$\rho \frac{\partial^2 w}{\partial t^2} = (\lambda + G)\frac{\partial \bar{\varepsilon}}{\partial z} + G\nabla^2 w \qquad (3.70)$$

Now, differentiating Eqs. (3.67), (3.69), and (3.70) with respect to x, y, and z, respectively, and adding,

$$\rho \frac{\partial^2}{\partial t^2}\left(\frac{\partial u}{\partial x} + \frac{\partial v}{\partial y} + \frac{\partial w}{\partial z}\right) = (\lambda + G)\left(\frac{\partial^2 \bar{\varepsilon}}{\partial x^2} + \frac{\partial^2 \bar{\varepsilon}}{\partial y^2} + \frac{\partial^2 \bar{\varepsilon}}{\partial z^2}\right)$$

$$+ G\nabla^2\left(\frac{\partial u}{\partial x} + \frac{\partial v}{\partial y} + \frac{\partial w}{\partial z}\right)$$

or

$$\rho \frac{\partial^2 \bar{\varepsilon}}{\partial t^2} = (\lambda + G)(\nabla^2 \bar{\varepsilon}) + G(\nabla^2 \bar{\varepsilon}) = (\lambda + 2G)\nabla^2 \bar{\varepsilon} \tag{3.71}$$

Therefore,

$$\boxed{\frac{\partial^2 \bar{\varepsilon}}{\partial t^2} = \frac{\lambda + 2G}{\rho}\nabla^2 \bar{\varepsilon} = v_p^2 \nabla^2 \bar{\varepsilon}} \tag{3.72}$$

where

$$\boxed{v_p = \sqrt{\frac{\lambda + 2G}{\rho}}} \tag{3.73}$$

Equation (3.73) is in the same form as the wave equation given in Eq. (3.29). Also note that $\bar{\varepsilon}$ is the volumetric strain and v_p is the *velocity of the dilatational waves.* This is also referred to as the *primary wave, P-wave,* or *compression wave.* Also another fact that needs to be pointed out here is that the expression for v_c was given as $v_c = \sqrt{E/\rho}$. Comparing the expressions for v_c and v_p, one can see that the velocity of compression waves is faster than v_c.

B. Distortional Waves or Shear Waves

Differentiating Eq. (3.69) with respect to z and Eq. (3.70) with respect to y,

$$\rho \frac{\partial^2}{\partial t^2}\left(\frac{\partial v}{\partial z}\right) = (\lambda + G)\frac{\partial^2 \bar{\varepsilon}}{(\partial y)(\partial z)} + G\nabla^2 \frac{\partial v}{\partial z} \tag{3.74}$$

and

$$\rho \frac{\partial^2}{\partial t^2}\left(\frac{\partial w}{\partial y}\right) = (\lambda + G)\frac{\partial^2 \bar{\varepsilon}}{(\partial y)(\partial z)} + G\nabla^2 \frac{\partial w}{\partial y} \tag{3.75}$$

Subtracting Eq. (3.74) from (3.75) yields

$$\rho \frac{\partial}{\partial t^2}\left(\frac{\partial w}{\partial y} - \frac{\partial v}{\partial z}\right) = G\nabla^2\left(\frac{\partial w}{\partial y} - \frac{\partial v}{\partial z}\right)$$

However, $\partial w/\partial y - \partial v/\partial z = 2\bar{\omega}_x$ [Eq. (3.10)]; thus,

$$\rho \frac{\partial^2 \bar{\omega}_x}{\partial t^2} = G\nabla^2 \bar{\omega}_x \tag{3.76}$$

or

$$\boxed{\frac{\partial^2 \overline{\omega}_x}{\partial t^2} = \frac{G}{\rho} \nabla^2 \overline{\omega}_x = v_s^2 \nabla^2 \overline{\omega}_x}$$ (3.77)

where $v_s = \sqrt{G/\rho}$.

Equation (3.77) represents the equation for distortional waves and the *velocity* of propagation is v_s. This is also referred to as the *shear wave*, or *S-wave*. Comparison of the shear wave velocity given above with that in a rod [Eq. (3.40)] shows that they are the same. Using the process of similar manipulation, one can also obtain two more equations similar to Eq. (3.77):

$$\boxed{\frac{\partial^2 \overline{\omega}_y}{\partial t^2} = v_s^2 \nabla^2 \overline{\omega}_y}$$ (3.78)

and

$$\boxed{\frac{\partial^2 \overline{\omega}_z}{\partial t^2} = v_s^2 \nabla^2 \overline{\omega}_z}$$ (3.79)

3.12   General Comments

Based on the derivations for the velocities of compression waves and shear waves as derived in the preceding section, the following general observations can be made.

1. There are two types of stress waves that can propagate through an infinite elastic medium; however, they travel at different velocities.

2. From Eq. (3.73),

$$v_p = \sqrt{\frac{\lambda + 2G}{\rho}}$$

However

$$\lambda = \frac{\mu E}{(1 + \mu)(1 - \mu)}$$

and

$$G = \frac{E}{2(1 + \mu)}$$

Substitution of the preceding two relationships into the expression of v_p yields

$$v_p = \sqrt{\frac{E(1 - \mu)}{\rho(1 + \mu)(1 - 2\mu)}}$$ (3.80)

Similarly,

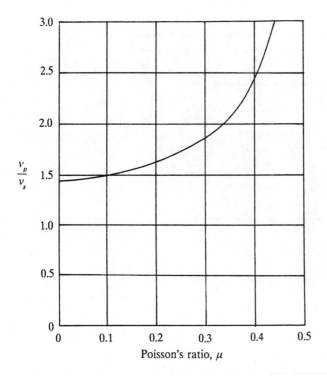

Figure 3.12   Variation of v_p/v_s with μ [Eq. (3.82)]

$$v_s = \sqrt{\frac{G}{\rho}} = \sqrt{\frac{E}{2(1 + \mu)\rho}} \tag{3.81}$$

Combining Eqs. (3.80) and (3.81),

$$\frac{v_p}{v_s} = \sqrt{\frac{2(1 - \mu)}{(1 - 2\mu)}} \tag{3.82}$$

Figure 3.12 shows a plot of v_p/v_s versus μ based on Eq. (3.82). It can be seen from the plot that for all values of μ, v_p/v_s is greater than 1.

3. Table 3.1 gives some typical values of v_p and v_s as encountered through various types of soils and rocks. Techniques for field determination of the velocities of compression waves and shear waves traveling through various soil media are described in Chapter 4.

4. Wave propagation through saturated soils involves the soil skeleton and water in the void spaces. A comprehensive theoretical study of this problem is given by Biot (1956). This study shows that there are two compressive waves and one shear wave through the saturated medium. Some investigators have referred to the two compressive waves as the *fluid wave* (transmitted through the fluid) and the *frame wave* (transmitted through the soil structure), although there is coupled motion of the fluid and the frame waves. As far as the shear wave is concerned, the pore water has no rigidity to shear. Hence, the shear wave in the soil is dependent only on the properties of the soil skeleton.

Table 3.1 Typical Values of v_p and v_s

Soil type	Compressive wave velocity, v_p (ft/s)	Shear wave velocity, v_s (ft/s)
Fine sand	1,000	300–500
Dense sand	1,500	750
Gravel	2,500	600–750
Moist clay	4,000–4,500	500
Granite	13,000–18,000	7,000–11,000
Sandstone	4,500–14,000	2,000–7,000

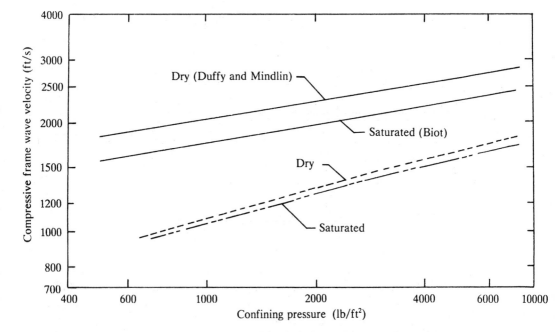

Figure 3.13 Comparison of experimental and theoretical results for compressive frame wave velocities in dry and saturated Ottawa sand (after Hardin and Richart, 1963)

Figure 3.13 shows the theoretical variation of the *compressive frame wave* velocities in dry and saturated sands, based on Biot's theory, using the values of the constants representative for a quartz sand (Hardin and Richart, 1963). Along with that, for comparison purposes, are shown the experimental *longitudinal wave* velocities [v_c from Eq. (3.30)] for dry and saturated Ottawa sands. For a given confining pressure, the difference of wave velocities between dry and saturated specimens is negligible and may be accounted for by the difference in the unit weight of the soil.

The velocity of *compression* waves (v_w) through water can be expressed as

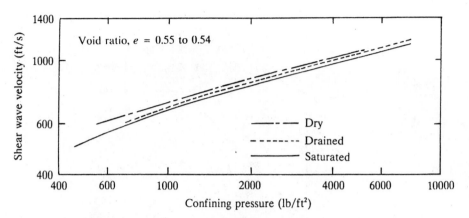

Figure 3.14 Variation of shear wave velocity with confining pressure for Ottawa sand (after Hardin and Richart, 1963)

$$v_w = \sqrt{\frac{B_w}{\rho_w}} \qquad (3.83)$$

where B_w is the bulk modulus of water and ρ_w is the density of water. Usually the value of v_w is of the order of 4800 ft/s (1463 m/s).

Figure 3.14 shows the variation of the experimental shear wave velocity for dry, drained, and saturated Ottawa sand. It may be noted that for a given confining pressure, the range of variation of v_s is very small.

STRESS WAVES IN AN ELASTIC HALF-SPACE

3.13 Rayleigh Waves

Equations derived in Section 3.11 are for stress waves in the body of an infinite, elastic, and isotropic medium. Another type of wave, called a *Rayleigh wave*, also exists near the boundary of an elastic half-space. This type of wave was first investigated by Lord Rayleigh (1885). In order to study this, consider a plane wave through an elastic medium with a plane boundary as shown in Figure 3.15. Note that the plane $x-y$ is the boundary of the elastic half-space and z is positive downward. Let u and w represent the displacements in the directions x and z, respectively, and be independent of y. Therefore,

$$u = \frac{\partial \phi}{\partial x} + \frac{\partial \psi}{\partial z} \qquad (3.84)$$

and

$$w = \frac{\partial \phi}{\partial z} - \frac{\partial \psi}{\partial x} \qquad (3.85)$$

where ϕ and ψ are two potential functions. The dilation $\bar{\varepsilon}$ can be defined as

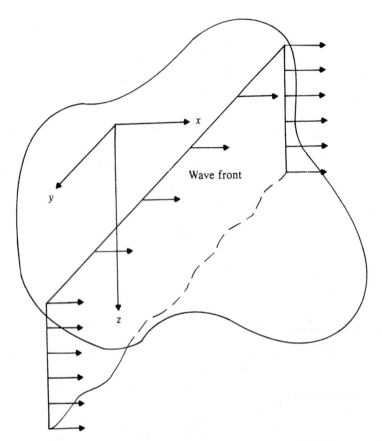

Figure 3.15 Plane wave through an elastic medium with a plane boundary

$$\bar{\varepsilon} = \varepsilon_x + \varepsilon_y + \varepsilon_z = \frac{\partial u}{\partial x} + \frac{\partial v}{\partial y} + \frac{\partial w}{\partial z}$$

$$= \left(\frac{\partial^2 \phi}{\partial x^2} + \frac{\partial^2 \psi}{\partial x \, \partial z}\right) + (0) + \left(\frac{\partial^2 \phi}{\partial z^2} - \frac{\partial^2 \psi}{\partial x \, \partial z}\right) = \frac{\partial^2 \phi}{\partial x^2} + \frac{\partial^2 \phi}{\partial z^2} = \nabla^2 \phi \qquad (3.86)$$

Similarly, the rotation in the x–z plane can be given by

$$2\bar{\omega}_y = \frac{\partial u}{\partial z} - \frac{\partial w}{\partial x} = \frac{\partial^2 \psi}{\partial x^2} + \frac{\partial^2 \psi}{\partial z^2} = \nabla^2 \psi \qquad (3.87)$$

Substituting Eqs. (3.84) and (3.86) into Eq. (3.67) yields

$$\rho \frac{\partial^2}{\partial t^2}\left(\frac{\partial \phi}{\partial x} + \frac{\partial \psi}{\partial z}\right) = (\lambda + G)\frac{\partial}{\partial x}(\nabla^2 \phi) + G\nabla^2\left(\frac{\partial \phi}{\partial x} + \frac{\partial \psi}{\partial z}\right)$$

or

$$\rho \frac{\partial}{\partial x}\left(\frac{\partial^2 \phi}{\partial t^2}\right) + \rho \frac{\partial}{\partial z}\left(\frac{\partial^2 \psi}{\partial t^2}\right) = (\lambda + 2G)\frac{\partial}{\partial x}(\nabla^2 \phi) + G\frac{\partial}{\partial z}(\nabla^2 \psi) \qquad (3.88)$$

In a similar manner, substituting Eqs. (3.85) and (3.86) into Eqs. (3.70), we get

$$\rho\frac{\partial}{\partial z}\left(\frac{\partial^2\phi}{\partial t^2}\right) - \rho\frac{\partial}{\partial x}\left(\frac{\partial^2\psi}{\partial t^2}\right) = (\lambda + 2G)\frac{\partial}{\partial z}(\nabla^2\phi) - G\frac{\partial}{\partial x}(\nabla^2\psi) \tag{3.89}$$

Equations (3.88) and (3.89) will be satisfied if (1) $\rho(\partial^2\phi/\partial t^2) = (\lambda + 2G)\nabla^2\phi$ or

$$\frac{\partial^2\phi}{\partial t^2} = \left(\frac{\lambda + 2G}{\rho}\right)\nabla^2\phi = v_p^2\nabla^2\phi \tag{3.90}$$

and (2) $\rho(\partial^2\psi/\partial t^2) = G\nabla^2\psi$ or

$$\frac{\partial^2\psi}{\partial t^2} = \frac{G}{\rho}\nabla^2\psi = v_s^2\nabla^2\psi \tag{3.91}$$

Now, consider a sinusoidal wave traveling in the positive x direction. Let the solutions of ϕ and ψ be expressed as

$$\phi = F(z)\exp[i(\omega t - fx)] \tag{3.92}$$

and

$$\psi = G(z)\exp[i(\omega t - fx)] \tag{3.93}$$

where $F(z)$ and $G(z)$ are functions of depth

$$f = \frac{2\pi}{\text{wavelength}} \tag{3.94}$$

$$i = \sqrt{-1} \tag{3.95}$$

Substituting Eq. (3.92) into Eq. (3.90), we get

$$\left(\frac{\partial^2}{\partial t^2}\right)\{F(z)\exp[i(\omega t - fx)]\} = v_p^2\nabla^2\{F(z)\exp[i(\omega t - fx)]\}$$

or

$$-\omega^2 F(z) = v_p^2[F''(z) - f^2 F(z)] \tag{3.96}$$

Similarly, substituting Eq. (3.93) into Eq. (3.91) results in

$$-\omega^2 G(z) = v_s^2[G''(z) - f^2 G(z)] \tag{3.97}$$

where

$$F''(z) = \frac{\partial^2 F(z)}{\partial z^2} \tag{3.98}$$

$$G''(z) = \frac{\partial^2 G(z)}{\partial z^2} \tag{3.99}$$

Equations (3.96) and (3.97) can be rearranged to the form

$$F''(z) - q^2 F(z) = 0 \tag{3.100}$$

and

$$G''(z) - s^2 G(z) = 0 \tag{3.101}$$

where

$$q^2 = f^2 - \frac{\omega^2}{v_p^2} \tag{3.102}$$

$$s^2 = f^2 - \frac{\omega^2}{v_s^2} \tag{3.103}$$

Solutions to Eqs. (3.100) and (3.101) can be given as

$$F(z) = A_1 e^{-qz} + A_2 e^{qz} \tag{3.104}$$

and

$$G(z) = B_1 e^{-sz} + B_2 e^{sz} \tag{3.105}$$

where A_1, A_2, B_1, and B_2 are constants.

From Eqs. (3.104) and (3.105), it can be seen that A_2 and B_2 must equal zero; otherwise $F(z)$ and $G(z)$ will approach infinity with depth, which is not the type of wave that is considered here. With A_2 and B_2 equal zero,

$$F(z) = A_1 e^{-qz} \tag{3.106}$$

$$G(z) = B_1 e^{-sz} \tag{3.107}$$

Combining Eqs. (3.92) and (3.106) and Eqs. (3.93) and (3.107),

$$\phi = (A_1 e^{-qz})[e^{i(\omega t - fx)}] \tag{3.108}$$

and

$$\psi = (B_1 e^{-sz})[e^{i(\omega t - fx)}] \tag{3.109}$$

The boundary conditions for the two preceding equations are at $z = 0$, $\sigma_z = 0$, $\tau_{zx} = 0$, and $\tau_{zy} = 0$. From Eq. (3.22),

$$\sigma_{z(z=0)} = \lambda \bar{\varepsilon} + 2G\varepsilon_z = \lambda \bar{\varepsilon} + 2G\left(\frac{\partial w}{\partial z}\right) = 0 \tag{3.110}$$

Combining Eqs. (3.85), (3.86), and (3.108)–(3.110), one obtains

$$A_1[(\lambda + 2G)q^2 - \lambda f^2] - 2iB_1 Gfs = 0 \tag{3.110}$$

or

$$\frac{A_1}{B_1} = \frac{2iGfs}{(\lambda + 2G)q^2 - \lambda f^2} \tag{3.112}$$

Similarly,

$$\tau_{zx(z=0)} = G\gamma_{zx} = G\left(\frac{\partial w}{\partial x} + \frac{\partial u}{\partial z}\right) = 0 \tag{3.113}$$

Again, combining Eqs. (3.84), (3.85), (3.108), (3.109), and (3.113),

$$2iA_1 fq + (s^2 + f^2)B_1 = 0$$

or

$$\frac{A_1}{B_1} = \frac{-(s^2 + f^2)}{2ifq} \tag{3.114}$$

Equating the right-hand sides of Eqs. (3.112) and (3.114),

$$\frac{2iGfs}{(\lambda + 2G)q^2 - \lambda f^2} = -\frac{(s^2 + f^2)}{2ifq}$$

$$4Gf^2 sq = (s^2 + f^2)[(\lambda + 2G)q^2 - \lambda f^2]$$

or

$$16G^2 f^4 s^2 q^2 = (s^2 + f^2)^2 [(\lambda + 2G)q^2 - \lambda f^2]^2 \tag{3.115}$$

Substituting for q and s and then dividing both sides of Eq. (3.115) by $G^2 f^8$, we get

$$16\left(1 - \frac{\omega^2}{v_p^2 f^2}\right)\left(1 - \frac{\omega^2}{v_s^2 f^2}\right) = \left[2 - \left(\frac{\lambda + 2G}{G}\right)\frac{\omega^2}{v_p^2 f^2}\right]^2 \left(2 - \frac{\omega^2}{v_s^2 f^2}\right)^2 \tag{3.116}$$

From Eq. (3.94)

$$\text{Wavelength} = \frac{2\pi}{f} \tag{3.117}$$

However,

$$\text{Wavelength} = \frac{\text{velocity of wave}}{(\omega/2\pi)} = \frac{v_r}{(\omega/2\pi)} \tag{3.118}$$

where v_r is the *Rayleigh* wave velocity. Thus, from Eqs. (3.117) and (3.118), $2\pi/f = 2\pi v_r/\omega$, or

$$f = \frac{\omega}{v_r} \tag{3.119}$$

So,

$$\frac{\omega^2}{v_p^2 f^2} = \frac{\omega^2}{v_p^2(\omega^2/v_r^2)} = \frac{v_r^2}{v_p^2} = \alpha^2 V^2 \tag{3.120}$$

Similarly,

$$\frac{\omega^2}{v_s^2 f^2} = \frac{\omega^2}{v_s^2(\omega^2/v_r^2)} = \frac{v_r^2}{v_s^2} = V^2 \tag{3.121}$$

where

$$\alpha^2 = \frac{v_s^2}{v_p^2} \tag{3.122}$$

However $v_p^2 = (\lambda + 2G)/\rho$ and $v_s^2 = G/\rho$. Thus

$$\alpha^2 = \frac{v_s^2}{v_p^2} = \frac{G}{\lambda + 2G} \tag{3.123}$$

Table 3.2 Values of V [Eq. (3.126)]

μ	$V = v_r/v_s$
0.25	0.919
0.29	0.926
0.33	0.933
0.4	0.943
0.5	0.955

The term α^2 can also be expressed in terms of Poisson's ratio. From the relations given in Eq. (3.25),

$$\lambda = \frac{2\mu G}{1 - 2\mu} \tag{3.124}$$

Substitution of this relation in Eq. (3.123) yields

$$\alpha^2 = \frac{G}{2\mu G/(1 - 2\mu) + 2G} = \frac{(1 - 2\mu)G}{2\mu G + 2G - 4\mu G} = \frac{(1 - 2\mu)}{(2 - 2\mu)} \tag{3.125}$$

Again, substituting Eqs. (3.120), (3.121), and (3.123) into Eq. (3.116),

$$16(1 - \alpha^2 V^2)(1 - V^2) = (2 - V^2)^2(2 - V^2)^2$$

or

$$V^6 - 8V^4 - (16\alpha^2 - 24)V^2 - 16(1 - \alpha^2) = 0 \tag{3.126}$$

Equation (3.126) is a cubic equation in V^2. For a given value of Poisson's ratio, the proper value of V^2 can be found and, hence, so can the value of v_r in terms of v_p or v_s. An example of this is shown in Example 3.1. Table 3.2 gives some values of v_r/v_s ($= V$) for various values of Poisson's ratio.

3.14   Displacement of Rayleigh Waves

From Eqs. (3.84) and (3.85),

$$u = \frac{\partial \phi}{\partial x} + \frac{\partial \psi}{\partial z} \tag{3.84}$$

and

$$w = \frac{\partial \phi}{\partial z} - \frac{\partial \psi}{\partial x} \tag{3.85}$$

Substituting the relations developed for ϕ and ψ [Eqs. (3.108), (3.109)] in these equations, one obtains

$$u = -(if A_1 e^{-qz} + B_1 s e^{-sz})[e^{i(\omega t - f x)}] \tag{3.127}$$

$$w = -(A_1 q e^{-qz} - B_1 if e^{-sz})[e^{i(\omega t - f x)}] \tag{3.128}$$

However, from Eq. (3.114), $B_1 = -2iA_1 fq/(s^2 + f^2)$. Substituting this relation in Eqs. (3.127) and (3.128) gives

$$u = A_1 fi\left(-e^{-qz} + \frac{2qs}{s^2 + f^2}e^{-sz}\right)[e^{i(\omega t - fx)}] \tag{3.129}$$

and

$$w = A_1 q\left(-e^{-qz} + \frac{2f^2}{s^2 + f^2}e^{-sz}\right)[e^{i(\omega t - fx)}] \tag{3.130}$$

From the preceding two equations, it is obvious that the <u>rate of attenuation of the displacement along the x direction with depth z will depend on the factor</u> <u>U</u>, where

$$U = -e^{-qz} + \frac{2qs}{s^2 + f^2}e^{-sz} = -e^{-(q/f)(fz)} + \left[\frac{2(q/f)(s/f)}{s^2/f^2 + 1}\right]e^{-(s/f)(fz)} \tag{3.131}$$

Similarly, the rate of attenuation of the displacement along the z direction with depth will depend on

$$W = -e^{-qz} + \frac{2f^2}{s^2 + f^2}e^{-sz} = -e^{-(q/f)(fz)} + \frac{2}{s^2/f^2 + 1}e^{-(s/f)(zf)} \tag{3.132}$$

However,

$$q^2 = f^2 - \frac{\omega^2}{v_p^2} \tag{3.102}$$

or

$$\frac{q^2}{f^2} = 1 - \frac{\omega^2}{f^2 v_p^2} = 1 - \frac{v_r^2}{v_p^2} = 1 - \alpha^2 V^2 \tag{3.133}$$

Also,

$$s^2 = f^2 - \frac{\omega^2}{v_s^2} \tag{3.103}$$

$$\frac{s^2}{f^2} = 1 - \frac{\omega^2}{f^2 v_s^2} = 1 - \frac{v_r^2}{v_s^2} = 1 - V^2 \tag{3.134}$$

If the Poisson's ratio is known, one can determine the value of V from Eq. (3.126). Substituting the previously determined values of V in Eqs. (3.133) and (3.134), q/f and s/f can be determined; hence, U and W are determinable as functions of z and f. From Example 3.1, it can be seen that for $\mu = 0.25$, $V = 0.9194$. Thus,

$$\frac{q^2}{f^2} = 1 - \alpha^2 V^2 = 1 - \left(\frac{1 - 2\mu}{2 - 2\mu}\right)V^2 = 1 - \left(\frac{1 - 0.5}{2 - 0.5}\right)(0.9194)^2 = 0.7182$$

or

$$\frac{q}{f} = 0.8475$$

$$\frac{s^2}{f^2} = 1 - V^2 = 1 - (0.9194)^2 = 0.1547$$

or

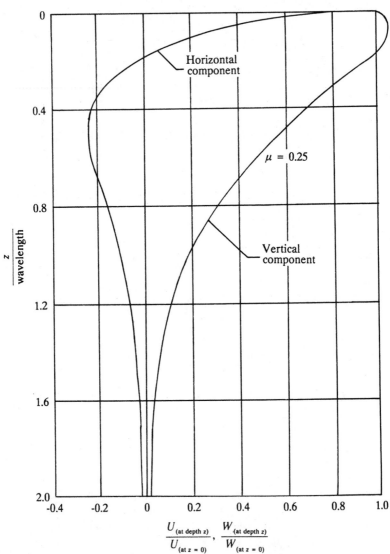

Figure 3.16 Variation of the amplitude of vibration of the horizontal and vertical components of Rayleigh waves with depth ($\mu = 0.25$)

$$\frac{s}{f} = 0.3933$$

Substituting these values of q/f and s/f into Eqs. (3.131) and (3.132),

$$U_{(\mu=0.25)} = -\exp(-0.8475fz) + 0.5773\exp(-0.3933fz) \tag{3.135}$$

$$W_{(\mu=0.25)} = -\exp(-0.8475fz) + 1.7321\exp(-0.3933fz) \tag{3.136}$$

Based on Eqs. (3.135) and (3.136), the following observations can be made:

1. The magnitude of U decreases rapidly with increasing value of fz. At $fz = 1.21$, U becomes equal to zero; so, at $z = 1.21/f$, there is no motion parallel to the surface. It has been shown in Eq. (3.94) that $f = 2\pi/(\text{wavelength})$. Thus, at $z = 1.21/f = 1.21(\text{wavelength})/2\pi = 0.1926(\text{wavelength})$, the value of U is zero. At greater depths, U becomes finite; however it is of the opposite sign, so the vibration takes place in opposite phase.

2. The magnitude of W first increases with fz, reaches a maximum value at $z = 0.076(\text{wavelength})$ (i.e., $fz = 0.4775$), and then decreases with depth.

Figure 3.16 shows a nondimensional plot of the variation of amplitude of vertical and horizontal components of Rayleigh waves with depth for $\mu = 0.25$. Equations (3.135) and (3.136) show that the path of a particle in the medium is an *ellipse* with its *major axis normal to the surface*.

Example 3.1

Given $\mu = 0.25$, determine the value of the Rayleigh wave velocity in terms of v_s.

Solution

From Eq. (3.126),

$$V^6 - 8V^4 - (16\alpha^2 - 24)V^2 - 16(1 - \alpha^2) = 0$$

For $\mu = 0.25$,

$$\alpha^2 = \frac{1 - 2\mu}{2 - 2\mu} = \frac{1 - 0.5}{2 - 0.5} = \frac{1}{3}$$

$$V^6 - 8V^4 - (\tfrac{16}{3} - 24)V^2 - 16(1 - \tfrac{1}{3}) = 0$$

$$3V^6 - 24V^4 + 56V^2 - 32 = 0$$

$$(V^2 - 4)(3V^4 - 12V^2 + 8) = 0$$

Therefore,

$$V^2 = 4, \qquad 2 + \frac{2}{\sqrt{3}}, \qquad 2 - \frac{2}{\sqrt{3}}$$

If $V^2 = 4$,

$$\frac{s^2}{f^2} = 1 - V^2 = 1 - 4 = -3$$

and s/f is imaginary. This is also the case for $V^2 = 2 + 2/\sqrt{3}$.

Keeping Eqs. (3.129), (3.131), and (3.130), (3.132) in mind, one can see that when q/f and s/f are imaginary, it does not yield the type of wave that is being discussed here. Thus,

$$V^2 = 2 - \frac{2}{\sqrt{3}} \qquad V = \frac{v_r}{v_s} = 0.9194$$

or

$$\underline{v_r = 0.9194 v_s}$$ ■

3.15 Attenuation of the Amplitude of Elastic Waves with Distance

If an impulse of short duration is created at the surface of an elastic half-space, the body waves travel into the medium with hemispherical wave fronts, as shown in Figure 3.17. The *Rayleigh waves* will propagate radially outward along a *cylindrical wave front.* At some distance from the point of disturbance, the displacement of the ground will be of the nature shown in Figure 3.18. Since *P*-waves are the fastest, they will arrive first, followed by *S*-waves and then the Rayleigh waves. As may be seen from Figure 3.18, the ground displacement due to the Rayleigh wave arrival is much greater than that for *P*- and *S*-waves. The amplitude of disturbance gradually decreases with distance.

Referring to Figure 3.18a and b, it can be seen that the particle motion due to Rayleigh waves starting at ① can be combined to give the lines of the surface particle motion as shown in Figure 3.18c. The part of the motion is a *retrograde ellipse.*

When *body waves* spread out along a hemispherical wave front, the energy is distributed over an area that increases with the square of the radius:

$$E' \propto \frac{1}{r^2} \tag{3.137}$$

where E' is the energy per unit area and r is the radius. However, the amplitude is proportional to the square root of the energy per unit area:

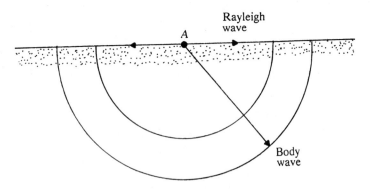

Note: Point A is the source of disturbance

Figure 3.17 Propagation of body waves and Rayleigh waves

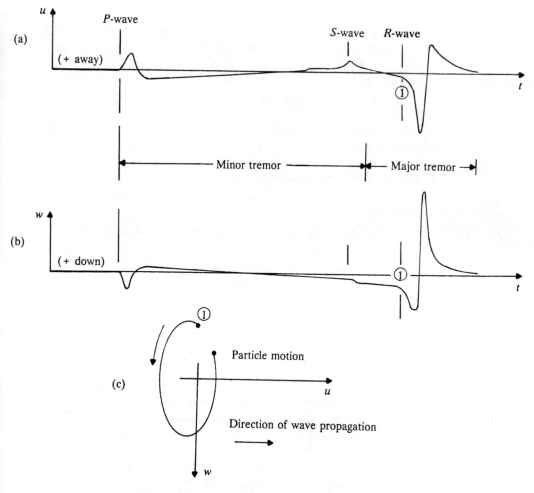

Figure 3.18 Wave system from surface point source in ideal medium (after Richart, Hall, and Woods, 1970)

$$\text{Amplitude} \propto \sqrt{E'} \propto \sqrt{\frac{1}{r^2}}$$

or

$$\text{Amplitude} \propto \frac{1}{r} \tag{3.138}$$

Along the surface of the half-space only, the amplitude of the body waves is proportional to $1/r^2$.

Similarly, the amplitude of the Rayleigh waves, which spread out in a cylindrical wave front, is proportional to $1/\sqrt{r}$. Thus the attenuation of the amplitude of the *Rayleigh* waves is *slower* than that for the body waves.

The loss of the amplitude of waves due to spreading out is called *geometrical damping*. In addition to the above damping, there is another type of loss—that from *absorption* in real earth material. This is called *material damping*. Thus, accounting for both types of damping, the vertical amplitude of Rayleigh waves can be given by the relation

$$\overline{w}_n = \overline{w}_1 \sqrt{\frac{r_1}{r_n}} \exp[-\beta(r_n - r_1)] \tag{3.139}$$

where $\overline{w}_n$ and $\tilde{w}_1$ are vertical amplitudes at distances r_n and r_1, and β is the absorption coefficient.

Equation (3.139) is given by Bornitz (1931). (See also Hall and Richart, 1963.) The magnitude of β depends on the type of soil.

REFERENCES

Biot, M. A. (1956). "Theory of Propagation of Elastic Waves in a Fluid Saturated Soil." *Journal of the Acoustical Society of America*, Vol. 28, pp. 168–178.

Bornitz, G. (1931). *Über die Ausbreitung der von Graszkolbenmaschinen erzeugten Bodenschwingungen in die Tiefe*, J. Springer, Berlin.

Duffy, J., and Mindlin, R. D. (1957). "Stress-Strain Relations of a Granular Medium," Transactions, ASME, pp. 585–593.

Hall, J. R., Jr., and Richart, F. E., Jr. (1963). "Dissipation of Elastic Wave Energy in Granular Soils," *Journal of the Soil Mechanics and Foundations Division*, ASCE, Vol. 89, No. SM6, pp. 27–56.

Hardin, B. O., and Richart, F. E., Jr. (1963). "Elastic Wave Velocities in Granular Soils." *Journal of the Soil Mechanics and Foundations Division*, ASCE, Vol. 89, No. SM1, pp. 33–65.

Rayleigh, Lord (1885). "On Wave Propagated Along the Plane Surface of Elastic Solid," *Proceedings*, London Mathematical Society, Vol. 17, pp. 4–11.

Richart, F. E., Jr., Hall, J. R., Jr., and Woods, R. D. (1970). *Vibrations of Soils and Foundations*, Prentice Hall, Inc., Englewood Cliffs, New Jersey.

Timoshenko, S. P., and Goodier, J. N. (1970). *Theory of Elasticity*, McGraw-Hill, New York.

Chapter 4

PROPERTIES OF DYNAMICALLY LOADED SOILS

4.1   Introduction

Many problems in civil engineering practice require the knowledge of the properties of soils subjected to dynamic load. Those problems include the dynamic bearing capacity of foundations, response of machine foundations subjected to cyclic loading, soil-structure interaction during the propagation of stress waves generated due to an earthquake, and earthquake resistance of dams and embankments. This chapter is devoted primarily to describing various laboratory and field test procedures available to predict the soil properties subjected to dynamic loading. It is divided into three major parts:

   a. Laboratory tests and results

   b. Field tests and measurements

   c. Empirical correlations for the shear modulus and damping ratio obtained from field and laboratory tests. These are the two most important parameters needed for most design work.

LABORATORY TESTS AND RESULTS

4.2   Shear Strength of Soils Under Rapid Loading Conditions

Saturated Clay

In most common soil test programs, the undrained shear strength of saturated cohesive soils is determined by conducting *unconsolidated-undrained triaxial tests.* The soil specimen for this type of test is initially subjected to a confining pressure σ_3 in a triaxial test chamber, as shown in Figure 4.1a. After that an axial

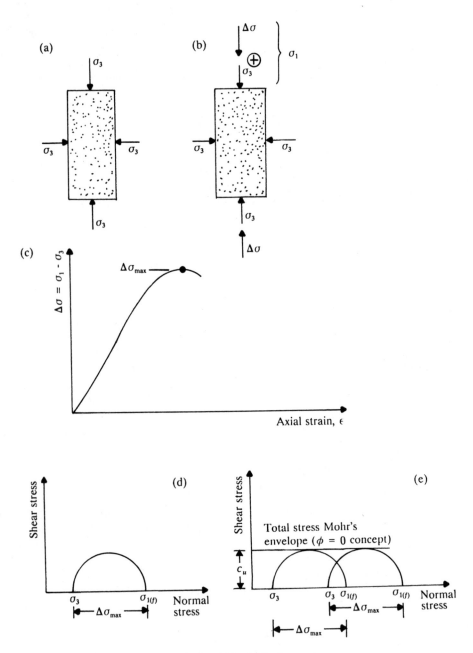

Figure 4.1   Unconsolidated-undrained triaxial test

stress $\Delta\sigma$ is applied to the specimen (Figure 4.1b). The axial stress $\Delta\sigma$ is gradually increased from zero to higher values at a constant rate of compressive strain. The strain rate $\dot{\varepsilon}$ is maintained at about 0.5% or less. The general nature of $\Delta\sigma$ versus axial strain ε diagram thus obtained is shown in Figure 4.1c. The *total* major and minor principal stresses at failure can now be given as

Major principal stress (total) $= \sigma_{1(f)} = \sigma_3 + \Delta\sigma_{max}$

Minor principal stress (total) $= \sigma_3$

The *total stress Mohr's circle* at failure is shown in Figure 4.1d. It can be shown (see Das, 1990) that for a given saturated clayey soil, the magnitude of $\Delta\sigma_{max}$ is practically independent of the confining pressure σ_3, as shown in Figure 4.1e. The total stress Mohr's envelope for this case is parallel to the normal stress axis and is referred to as the $\phi = 0$ *condition* (where $\phi =$ soil friction angle). The undrained shear strength c_u is expressed as

$$c_u = \frac{\Delta\sigma_{max}}{2} = \frac{\sigma_{1(f)} - \sigma_3}{2} \tag{4.1}$$

The undrained shear strength obtained by conducting tests at such low-axial

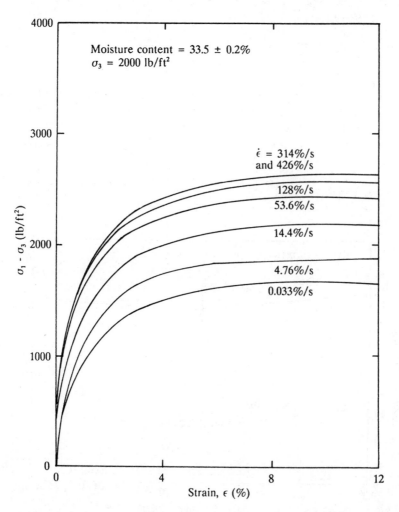

Figure 4.2   Unconsolidated-undrained triaxial test results on Buckshot clay (after Carroll, 1963)

strain rates is representative of the *static loading condition*, or

$$c_u = c_{u(static)}$$

Experimental results have shown that the magnitude of $\Delta\sigma_{max} = \sigma_{1(f)} - \sigma_3$ gradually increases with the increase of axial strain rate $\dot{\varepsilon}$. This conclusion can be seen from the laboratory test results on Buckshot clay (Figure 4.2). From Figure 4.2, it can be observed that $c_u = (\Delta\sigma_{max})/2 = [\sigma_{1(f)} - \sigma_3]/2$ obtained between strain rates of 50% to 425% are not too different and can be approximated to be a single value (Carroll, 1963). This value can be referred to as the *dynamic undrained shear strength*, or

$$c_u = c_{u(dynamic)}$$

Carroll suggested that for most practical cases, one can assume that

$$\frac{c_{u(dynamic)}}{c_{u(static)}} \approx 1.5 \qquad\qquad (4.2)$$

Sand

Several vacuum triaxial test results on different dry sands (that is, standard Ottawa sand, Fort Peck sand, and Camp Cooke sand) were reported by Whitman and Healy (1963). These tests were conducted with various effective confining pressures ($\bar{\sigma}_3$) and axial strain rates. The *compressive strength* $\Delta\sigma_{max}$ determined from these tests can be given as

$$\Delta\sigma_{max} = \bar{\sigma}_{1(f)} - \bar{\sigma}_3 \qquad\qquad (4.3)$$

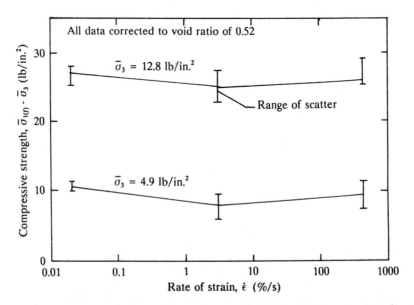

Figure 4.3   Strain-rate effect for dry Ottawa sand (after Whitman and Healy, 1963)

where

$\bar{\sigma}_3$ = effective minor principal stress

$\bar{\sigma}_{1(f)}$ = effective major principal stress at failure

An example of the effect of axial strain rate on dry Ottawa sand is shown in Figure 4.3. It can be seen that for a given $\bar{\sigma}_3$ the magnitude of $\Delta\sigma_{max}$ decreases initially with the increase of the strain rate to a minimum value and increases thereafter. From fundamentals of soil mechanics it is known that

$$\phi = \sin^{-1}\left(\frac{\bar{\sigma}_{1(f)} - \bar{\sigma}_3}{\bar{\sigma}_{1(f)} + \bar{\sigma}_3}\right) \qquad (4.4)$$

where ϕ = drained soil friction angle.

Based on Figure 4.3 and Eq. (4.4) it is obvious that the initial increase of the strain rate results in a decrease of the soil friction angle. The minimum dynamic friction angle may be given as (Vesic, 1973)

$$\phi_{\text{dynamic}} \approx \phi - 2° \qquad (4.5)$$

(obtained from static tests—that is, small strain rate of loading)

4.3   Strength and Deformation Characteristics of Soils under Transient Load

In many circumstances it may be necessary to know the strength and deformation characteristics of soils under *transient loading*. A typical example of tran-

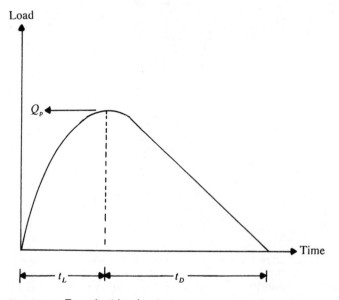

Figure 4.4   Transient load

sient loading is that occurring due to a blast. Figure 4.4 shows the nature of an idealized load versus time variation for such a case. In this figure, Q_P is the peak load, t_L is the time of loading, and t_D is the time of decay.

Casagrande and Shannon (1949) conducted some early investigations to study the stress-deformation and strength characteristics of Manchester sand and Cambridge clay soils. Undrained tests were conducted in three specially devised apparatuses—one falling-beam apparatus and two pendulum-loading apparatuses. In these specially devised pieces of equipment, the loading pattern on soil specimens was similar to that shown in Figure 4.4. Figure 4.5a shows the variation of stress and strain with time for an unconfined Cambridge clay specimen with $t_L = 0.02$ s. Similarly, Figure 4.5b compares the nature of varia-

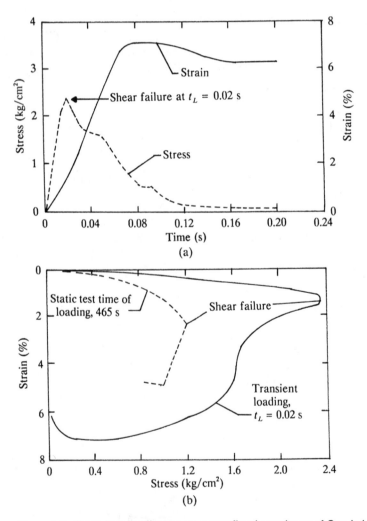

Figure 4.5 Transient loading on an unconfined specimen of Cambridge clay: (a) stress-strain-time variation for $t_L = 0.02$ s; (b) comparison of stress versus strain variation for static and transient loadings (after Casagrande and Shannon, 1949)

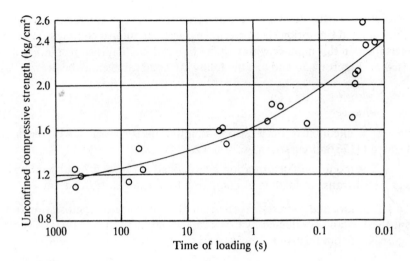

Figure 4.6   Unconfined compressive strength of Cambridge clay for varying time of loading (after Casagrande and Shannon, 1949)

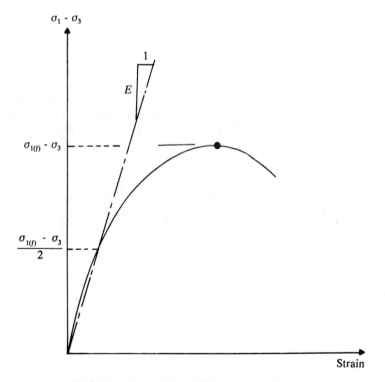

Figure 4.7   Definition of modulus of deformation, E

tion of strain versus stress for static and transient ($t_L = 0.02$ s) loading conditions on unconfined <u>Cambridge clay</u> specimens. The unconfined compressive strength determined in this manner with varying times of loading is shown in Figure 4.6. Based on Figures 4.5b and 4.6, the following conclusions may be drawn.

1. $\dfrac{q_{u(\text{transient})}}{q_{u(\text{static})}} \approx 1.5$ to 2

where q_u = unconfined compression strength. This is consistent with the findings of Carroll (1963) discussed in Section 4.2.

2. The modulus of deformation E as defined in Figure 4.7 is about two times as great for transient loading as compared to that for static loading.

The nature of the stress-versus-strain plot for confined compression tests on <u>Manchester sand</u> conducted by Casagrande and Shannon (1949) is as shown in Figure 4.8. From this study it was concluded that

1. $\dfrac{[\bar{\sigma}_{1(f)} - \bar{\sigma}_3]_{\text{transient}}}{[\bar{\sigma}_{1(f)} - \bar{\sigma}_3]_{\text{static}}} \approx 1.1$

and

2. The modulus of deformation as defined by Figure 4.7 is approximately the same for transient and static loading conditions.

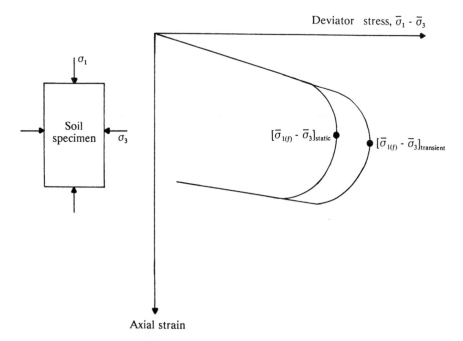

Figure 4.8 Confined-compression test on sand—stress-versus-strain behavior under static and transient loading

4.4   Travel-Time Test for Determination of Longitudinal and Shear Wave Velocities (v_c and v_s)

Using electronic equipment, the time t_c required for travel of elastic waves through a soil specimen of length L can be measured in the laboratory. For *longitudinal* waves

$$v_c = \frac{L}{t_c} \tag{4.6}$$

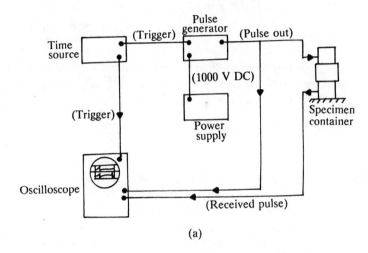

(a)

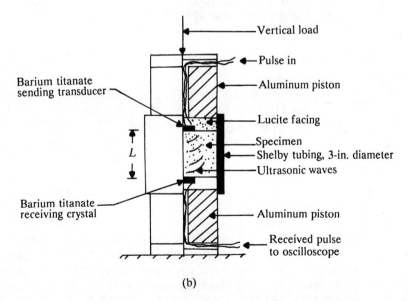

(b)

Figure 4.9   Travel-time method: (a) schematic diagram of the laboratory setup for measuring v_c'; (b) details of the soil specimen and container for the laboratory setup (after Whitman and Lawrence, 1963)

The modulus of elasticity E can then be calculated from Eq. (3.30) as

$$v_c = \sqrt{\frac{E}{\rho}}$$

or

$$E = \rho v_c^2 = \rho \frac{L^2}{t_c^2} \tag{4.7}$$

If the soil specimen is *confined laterally,* then the travel time will give the value of v_c' as shown in Eq. (3.35). Thus $v_c = L/t_c'$, and

$$M = \rho \frac{L^2}{t_c'^2} \tag{4.8}$$

where t_c' = time of travel of longitudinal waves in a laterally confined specimen.

Similarly, if the travel time t_s for *torsional waves* through a soil of length L is determined, the velocity v_s can be given as $v_s = L/t_c$, and

$$G = \rho v_s^2 = \rho \frac{L^2}{t_c^2} \tag{4.9}$$

Whitman and Lawrence (1963) have provided limited test results for v_c' in 20–30 Ottawa sand. The schematic diagram of the apparatus for measuring v_c' is shown in Figure 4.9a. The soil specimen was confined in 3-in. (76.2-mm) diameter Shelby tube (Figure 4.9b). Vertical load was applied by an aluminum piston. In this system, a pulse was sent from one piezo-electric crystal and received by a second one at the opposite end. The received signal was displayed on an oscilloscope, which allowed measurement of t_c'. It was found that the velocity v_c' increases with the increase of axial pressure.

4.5   Resonant Column Test

The resonant column test essentially consists of a soil column that is excited to vibrate in one of its natural modes. Once the frequency at *resonance* is known, the wave velocity can easily be determined. The soil column in the resonant column device can be excited longitudinally or torsionally, yielding velocities of v_c or v_s, respectively. The resonant column technique was first applied to testing of soils in Japan by Ishimato and Iida (1937) and Iida (1938, 1940). Since then it has been extensively used in many countries, with several modifications using different end conditions to constrain the specimen. One of the earlier types of resonant column device in the United States was used by Wilson and Dietrich (1960) for testing clay specimens.

Hardin and Richart (1963) reported the use of two types of resonant column devices—one for longitudinal vibration and the other for torsional vibration. The specimens were free at each end (*free–free end condition*). A schematic diagram of the laboratory experimental setup is shown in Figure 4.10. The power supply and amplifier No. 1 were used to amplify the sinusoidal output signal of the oscillator, which had a frequency range of 5 cps to 600,000 cps. The amplified signals were fed into the driver, producing the desired vibrations. Figure 4.11a shows the schematic diagram of the driver for torsional oscillation.

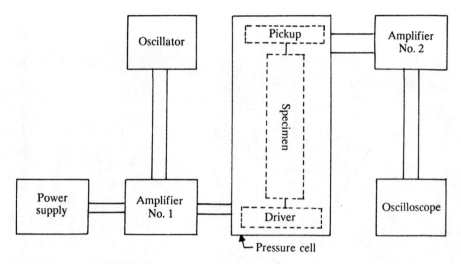

Figure 4.10 Schematic diagram of experimental setup for resonant column test of Hardin and Richart—free–free end condition

Similarly, the schematic diagram of the driver for longitudinal vibration is shown in Figure 4.11b. These devices will give results for *low-amplitude vibration conditions*. With *free–free end conditions*, for longitudinal vibrations at resonance

$$v_c = \frac{\omega_n L}{n\pi} \tag{3.53}$$

For $n = 1$ (that is, normal mode of vibration),

$$v_c = \frac{\omega_n L}{\pi} = \frac{2\pi f_n L}{\pi} = 2f_n L$$

or

$$v_c = \sqrt{\frac{E}{\rho}} = 2f_n L$$

or

$$\boxed{E = 4f_n^2 \rho L^2} \tag{4.10}$$

Similarly, for torsional vibration, at resonance (with $n = 1$)

$$v_s = 2f_n L$$

or

$$v_s = \sqrt{\frac{G}{\rho}} = 2f_n L$$

or

$$\boxed{G = 4f_n^2 \rho L^2} \tag{4.11}$$

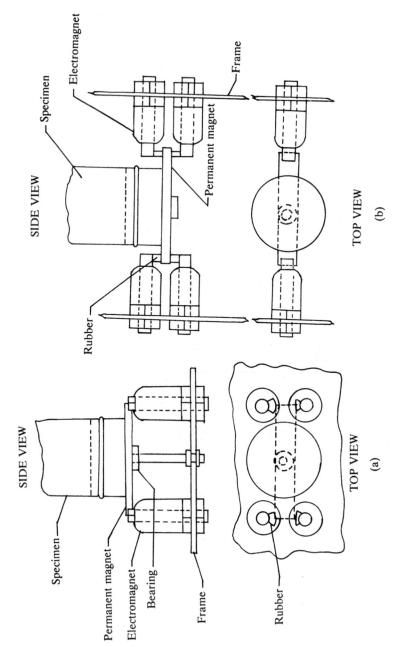

Figure 4.11 Drawings for steady-state vibration drivers in the resonant column device with free–free end conditions: (a) for torsional vibration; (b) for longitudinal vibration (after Hardin and Richart, 1963)

Once the magnitudes of E and G are known, the value of the Poisson's ratio can be obtained as

$$\mu = \frac{E}{2G} - 1 \qquad\qquad (4.12)$$

Hall and Richart (1963) also used two other types of resonant column

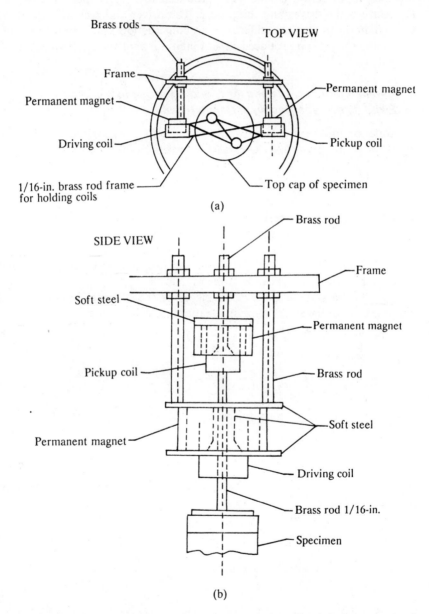

(a)

(b)

Figure 4.12   Driving and measuring components for a fixed–free resonant column device (after Hall and Richart, 1963)

devices (one for longitudinal vibration and the other for torsional vibration). The end conditions for these two types of devices were *fixed–free*—fixed at the bottom and free at the top of the specimen. The general layouts of the laboratory setup for this equipment were almost the same as shown in Figure 4.10, except for the fact that the driver and the pickup were located at the top of the specimen. This is shown in Figure 4.12. Since the driver and the pickup were located close together, a correction circuit was introduced to correct the inductive coupling between the driver and the pickup. The driver and pickup were attached to a common frame. The differences in construction and arrangement of the driver and the pickup produced either longitudinal or torsional vibration of the specimen.

A. Derivation of Expressions for v_c and E for Use in the Fixed–Free-Type Resonant Column Test

An equation for the circular natural frequency for the longitudinal vibration of short rods with *fixed–free end conditions* was derived in Eq. (3.57) as

$$\omega_n = \frac{(2n-1)\pi}{2}\frac{v_c}{L}$$

However, in a fixed–free-type resonant column test, the driving mechanism and also the motion-monitoring device have to be attached to the top of the specimen (Figure 4.13), in effect changing the boundary conditions assumed in deriving Eq. (3.57). So a modified equation for the circular natural frequency needs to be derived. This can be done as follows.

Let the mass of the attachments placed on the specimen be equal to m. For the vibration of the soil column in a natural mode,

$$u(x,t) = U(x)(A_1 \sin \omega_n t + A_2 \cos \omega_n t) \tag{3.41}$$

and

$$U(x) = B_1 \sin\left(\frac{\omega_n x}{v_c}\right) + B_2 \cos\left(\frac{\omega_n x}{v_c}\right) \tag{3.43}$$

At $x = 0$, $U(x) = 0$. So B_2 in Eq. (3.43) is zero. Thus

$$U(x) = B_1 \sin\left(\frac{\omega_n x}{v_c}\right) \tag{4.13}$$

At $x = L$, the inertia force of mass m is acting on the soil column, and this can be expressed as

$$F = -m\frac{\partial^2 u}{\partial t^2} \tag{4.14}$$

where F = inertia force. Also, the strain

$$\frac{\partial u}{\partial x} = \frac{F}{AE} \tag{4.15}$$

where A = cross-sectional area of the specimen and E = modulus of elasticity. Combining Eqs. (3.41), (4.13), and (4.15) we get

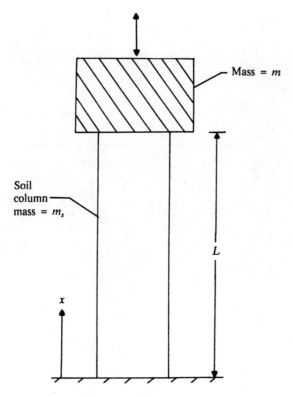

Figure 4.13   Derivation of Eq. (4.20)

$$\frac{F}{AE} = \frac{\partial u}{\partial x} = \left(\frac{\partial U}{\partial x}\right)(A_1 \sin \omega_n t + A_2 \cos \omega_n t)$$

$$= \frac{\partial}{\partial x}\left[B_1 \sin\left(\frac{\omega_n x}{v_c}\right)\right](A_1 \sin \omega_n t + A_2 \cos \omega_n t)$$

$$= \left(\frac{B_1 \omega_n}{v_c}\right)\left[\cos\left(\frac{\omega_n x}{v_c}\right)\right](A_1 \sin \omega_n t + A_2 \cos \omega_n t) \qquad (4.16)$$

Again, combining Eqs. (3.41), (4.13), and (4.14),

$$F = -m\frac{\partial^2 u}{\partial t^2} = -m\left[B_1 \sin\left(\frac{\omega_n x}{v_c}\right)\right]\left(\frac{\partial^2}{\partial t^2}\right)(A_1 \sin \omega_n t + A_2 \cos \omega_n t)$$

$$= m\omega_n^2 B_1 \sin\left(\frac{\omega_n x}{v_c}\right)(A_1 \sin \omega_n t + A_2 \cos \omega_n t) \qquad (4.17)$$

Now, from Eqs. (4.16) and (4.17),

$$\frac{AE}{v_c}\cos\left(\frac{\omega_n x}{v_c}\right) = m\omega_n \sin\left(\frac{\omega_n x}{v_c}\right) \qquad (4.18)$$

At $x = L$

$$AE = m\omega_n v_c \tan\left(\frac{\omega_n L}{v_c}\right) \tag{4.19}$$

However $v_c = \sqrt{E/\rho}$; or $E = v_c^2 \rho$. Substitution of this in Eq. (4.19) gives

$$Av_c^2 \rho = m\omega_n v_c \tan\left(\frac{\omega_n L}{v_c}\right)$$

$$\frac{A\rho}{m} = \frac{\omega_n}{v_c} \tan\left(\frac{\omega_n L}{v_c}\right)$$

$$\frac{AL\rho}{m} = \frac{\omega_n L}{v_c} \tan\left(\frac{\omega_n L}{v_c}\right)$$

or

$$\boxed{\frac{AL\gamma}{W} = \alpha \tan \alpha} \tag{4.20}$$

where

$\gamma = \rho g = $ unit weight of soil

$W = mg = $ weight of the attachments on top of the specimen

and

$$\alpha = \frac{\omega_n L}{v_c} \tag{4.21a}$$

The values of α corresponding to some values of $AL\gamma/W$ [Eq. (4.20)] are given in Table 4.1.

In any resonant column test, the ratio of $AL\gamma/W$ will be known. With a known value of $AL\gamma/W$, the value of α can be determined, and the natural frequency of vibration can be obtained from the test. Thus

$$\alpha = \frac{\omega_n L}{v_c} = \frac{2\pi f_n L}{v_c}$$

Table 4.1 Values of α and Corresponding $AL\gamma/W$ [Eq. (4.20)]

$\dfrac{AL\gamma}{W}$	α (rad)
0.1	0.32
0.3	0.53
0.5	0.66
0.7	0.75
1	0.86
2	1.08
4	1.27
10	1.43

or

$$v_c = \frac{2\pi f_n L}{\alpha}$$
(4.21b)

The modulus of elasticity of the soil can then be obtained as

$$E = \rho v_c^2 = \rho \left(\frac{2\pi f_n L}{\alpha}\right)^2 = 39.48 \left(\frac{f_n^2 L^2}{\alpha^2}\right)\rho$$
(4.22)

B. Derivation of Expressions for v_s and G for Use in the Fixed–Free-Type Resonant Column Test

In the resonant column tests where soil specimens are subjected to torsional vibration with fixed–free end conditions, the mass of the driving and motion-monitoring devices (Figure 4.14) can also be taken into account. For this condition, an equation similar to Eq. (4.20) can be derived that is of the form

$$\frac{J_s}{J_m} = \frac{\omega_n L}{v_s} \tan\left(\frac{\omega_n L}{v_s}\right) = \alpha \tan \alpha$$
(4.23)

where J_s = mass polar moment of inertia of the soil specimen and J_m = mass polar moment of inertia of the attachments with mass m. Thus

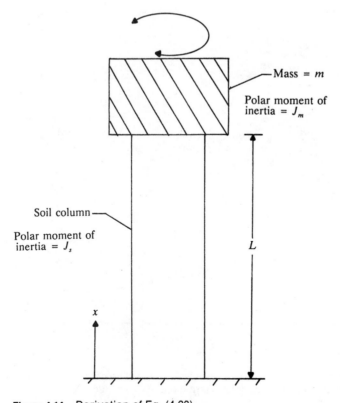

Mass = m

Polar moment of inertia = J_m

Soil column

Polar moment of inertia = J_s

L

x

Figure 4.14   Derivation of Eq. (4.23)

$$v_s = \frac{\omega_n L}{\alpha} = \frac{2\pi f_n L}{\alpha} \qquad (4.24)$$

and

$$G = \rho v_s^2 = 39.48 \left(\frac{f_n^2 L^2}{\alpha^2} \right) \rho \qquad (4.25)$$

C. Typical Laboratory Test Results from Resonant Column Tests

Most of the laboratory test results obtained from resonant column tests are for *low amplitudes* of vibration. By low amplitudes of vibration is meant strain amplitudes of the order of 10^{-4} or less.

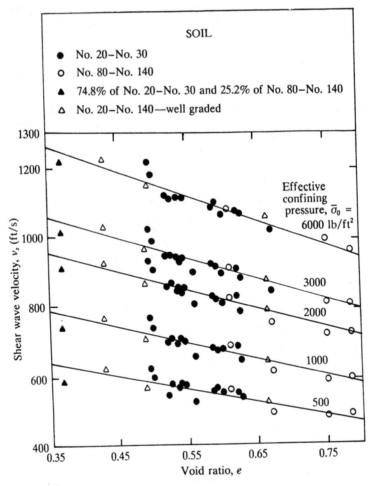

Figure 4.15 Variation of shear wave velocity with effective confining pressure $\bar{\sigma}_0$ for round-grained dry Ottawa sand (after Hardin and Richart, 1963)

Typical values of v_c and v_s with low amplitudes of vibration for No. 20–30 Ottawa sand compacted at a void ratio of about 0.55 are shown in Figures 3.13 and 3.14. These tests were conducted using the free–free and fixed–free types of resonant column device developed by Hardin and Richart (1963) and Hall and Richart (1963). Based on the results given in Figures 3.13 and 3.14, the following general conclusions can be drawn:

1. The values of v_c and v_s in soils increase with the increase of the effective average confining pressure $\bar{\sigma}_0$.

2. The values of v_c and v_s for saturated soils are slightly lower than those for dry soils. This can be accounted for by the increase of the unit weight of soil due to the presence of water in the void spaces.

Hardin and Richart (1963) also reported the results of several resonant column tests conducted in dry Ottawa sand. The shear wave velocities determined from these tests are shown in Figure 4.15. The peak-to-peak shear strain

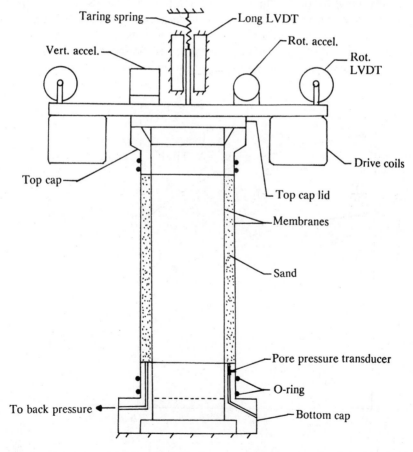

Figure 4.16   Schematic diagram of hollow-specimen resonant column device (after Drnevich, 1972)

amplitude for these tests was 10^{-3} rad. From Figure 4.15 it may be seen that the values of v_s are independent of the gradation, grain-size distribution, and also the relative density of compaction. However, v_s is dependent on the void ratio and the effective confining pressure.

D. Shear Modulus for Large Strain Amplitudes

For solid cylindrical specimens torsionally excited by resonant column devices, the shear strain varies from *zero at the center to a maximum at the periphery*, and it is difficult to evaluate a representative strain. For that reason, hollow cylindrical soil specimens in a resonant column device (Drnevich, Hall, and Richart, 1966, 1967) may be used to determine the shear modulus and damping at large strain amplitudes. Figure 4.16 shows a schematic diagram of this type of apparatus, in which the average shearing strain in the soil specimen is not greatly different from the maximum to the minimum. The variation of the shear modulus of dense C-190 Ottawa sand with the shear strain amplitude γ' is shown in Figure 4.17. Note that the value of G decreases with γ', but it decreases more rapidly for $\gamma' > 10^{-4}$. This is true for all soils. The reason for this can be explained by the use of Figure 4.18, which is a shear-stress-versus-strain diagram for a soil. The stress-strain relationships of soils are curvilinear. The shear modulus that is experimentally determined is the secant modulus obtained by joining the extreme points on the hysteresis loop. Note that when the amplitude

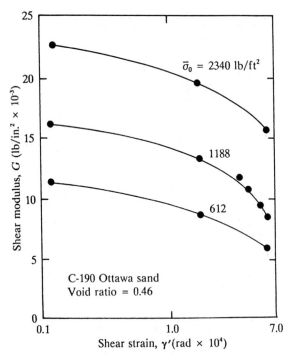

Figure 4.17   Effect of strain amplitude on shear modulus of sand (after Drnevich, Hall, and Richart, 1967)

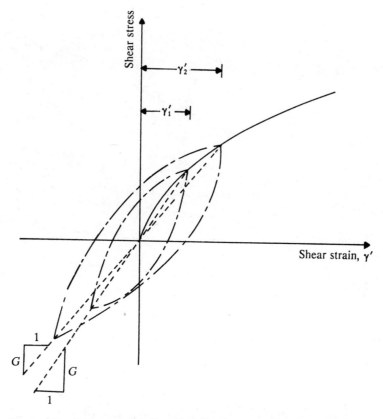

Figure 4.18 Nature of variation of shear stress versus shear strain

of strain is small (that is, $\gamma' = \gamma_1'$; Figure 4.18), the value of G is larger compared to that for the larger strain level (that is, $\gamma' = \gamma_2'$).

E. Effect of Prestraining on the Shear Modulus of Soils

The effect of shear modulus of soils due to prestraining was reported by Drnevich, Hall, and Richart (1967). These tests were conducted using C-190 Ottawa sand specimens. The specimens were first vibrated at a large amplitude for a certain number of cycles under a constant effective confining pressure ($\bar{\sigma}_0$). After that the shear modulii were determined by torsionally vibrating the specimens at small amplitudes (shearing strain $< 10^{-5}$). Figure 4.19 shows the results of six series of this type of test for dense sand (void ratio = 0.46). In general, the value of G increases with the increase of prestrain cycles.

F. Determination of Internal Damping

In Section 3.15, a distinction was made between *internal damping* and *material damping*. The internal damping of a soil specimen can be determined by resonant column tests.

In Chapter 2, the derivation of the expression for the logarithmic decrement

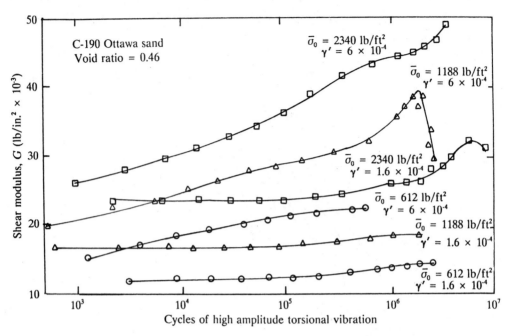

Figure 4.19 Effect of number of cycles of high-amplitude vibration on shear modulus determined at low amplitude (after Drnevich, Hall, and Richart, 1967)

was given as

$$\delta = \ln \frac{X_n}{X_{n+1}} = \frac{2\pi D}{1 - D^2} \tag{2.70}$$

where δ = logarithmic decrement and D = damping ratio.

The preceding equation is for the case of free vibration of a mass-spring-dashpot system. The damping ratio is given by the expression

$$D = \frac{c}{c_{cr}} = \frac{c}{2\sqrt{km_s}} \tag{1.47b}$$

where m_s = mass of the soil specimen (in this case).

For soils, the value of D is small and Eq. (2.70) can be approximated as

$$\delta = \ln \frac{X_n}{X_{n+1}} = 2\pi D \tag{4.26}$$

Now, combining Eqs. (1.47b) and (4.26)

$$\delta = \frac{\pi c}{\sqrt{km_s}} \tag{4.27}$$

The logarithmic decrement of a soil specimen (and hence the damping ratio D) can easily be measured by using a fixed–free-type resonant column device.

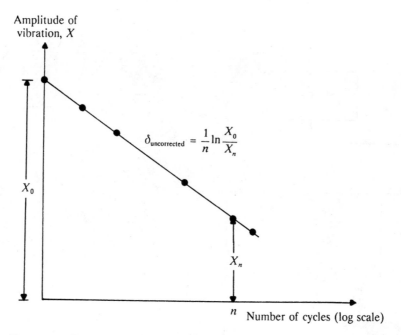

Figure 4.20 Plot of the amplitude of vibration against the corresponding number of cycles for determination of logarithmic decrement

The soil specimen is first set into steady-state forced vibration. The driving power is then shut off and the decay of the amplitude of vibration is plotted against the corresponding number of cycles. This plots as a straight line on a semilogarithmic graph paper, as shown in Figure 4.20. The logarithmic decrement can then be evaluated as

$$\delta_{uncorrected} = \left(\frac{1}{n}\right)\left(\ln\frac{X_0}{X_n}\right) \tag{4.28}$$

However, in a fixed–free type of resonant column device, the driving and the motion-monitoring equipment is placed on the top of the specimen. Hence, for determination of the true logarithmic decrement of the soil specimen, a correction to Eq. (4.28) is necessary. This has been discussed by Hall and Richart (1963). Consider the case of longitudinal vibration of a soil column, as shown in Figure 4.21, in which m = mass of the attachments on the top of the soil specimen and m_s = mass of the soil specimen. With the addition of mass m, Eq. (4.27) can be modified as

$$\delta_{uncorrected} = \frac{\pi c}{\sqrt{k(m_s + m)}} \tag{4.29}$$

From Eqs. (4.27) and (4.29),

$$\frac{\delta}{\delta_{uncorrected}} = \sqrt{\frac{m_s + m}{m_s}} = \sqrt{1 + \frac{m}{m_s}} \tag{4.30}$$

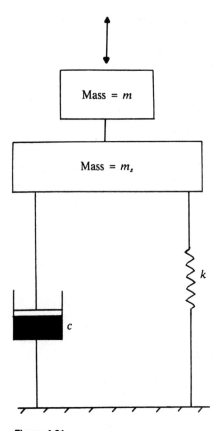

Figure 4.21

In order to use Eq. (4.30), it will be required to convert the mass m_s into an equivalent concentrated mass. The equivalent concentrated mass can be shown to be equal to $0.405m_s$. Thus, replacing m_s in Equation (4.30) by $0.405m_s$,

$$\delta = \delta_{\text{uncorrected}}\sqrt{1 + \frac{m}{0.405m_s}} \tag{4.31}$$

A similar correction may be used for specimens subjected to torsional vibration, which will be of the form

$$\delta = \delta_{\text{uncorrected}}\sqrt{1 + \frac{J_m}{0.405J_s}}$$

Hardin (1965) suggested a relation for δ of *dry sand* in low amplitude torsional vibration as

$$\delta = 9\pi(\gamma')^{0.2}(\bar{\sigma}_0)^{-0.5} \tag{4.32}$$

Equation (4.32) is valid for $\gamma' = 10^{-6}$ to 10^{-4} and $\bar{\sigma}_0 = 500$ lb/ft² to 3000 lb/ft² (24 to 144 kN/m²).

4.6   Cyclic Simple Shear Test

A cyclic simple shear test is a convenient method for determining the shear modulus and damping ratio of soils. It is also a convenient device for studying the liquefaction parameters of saturated cohesionless soils (Chapter 10). In cyclic simple shear tests a soil specimen, usually 20–30 mm high with a side length (or diameter) of 60–80 mm, is subjected to a vertical effective stress $\bar{\sigma}_v$ and a cyclic shear stress τ, as shown in Figure 4.22. The horizontal load necessary to deform the specimen is measured by a load cell, and the shear deformation of the specimen is measured by a linear variable differential transformer.

The shear modulus of a soil in the cyclic simple shear test can be determined as

$$G = \frac{\text{amplitude of cyclic shear stress, } \tau}{\text{amplitude of cyclic shear strain, } \gamma'} \tag{4.33}$$

The damping ratio at a given shear strain amplitude can be obtained from the hysteretic stress-strain properties. Referring to Figure 4.23 (also see Figure 4.18), the damping ratio can be given as

$$D = \frac{1}{2\pi} \frac{\text{area of the hysteresis loop}}{\text{area of triangle } OAB \text{ and } OA'B'} \tag{4.34}$$

Figure 4.24 shows a plot of shear modulus G with cyclic shear strain γ' for two values of $\bar{\sigma}_v$ (Silver and Seed, 1971) obtained from cyclic simple shear tests on a medium dense sand (relative density, $R_D = 60\%$). From the results of this study, the following can be stated:

1. For a given value of γ' and $\bar{\sigma}_v$, the shear modulus increases with the number of cycles of shear stress application. Most of the increase in G takes place in the first ten cycles, after which the rate of increase is relatively small.

2. For a given value of $\bar{\sigma}_v$ and number of cycles of stress application, the magnitude of G decreases with the amplitude of shear strain γ'. (*Note:* Similar results are shown in Figure 4.17.)

3. For a given value of γ' and number of cycles, the magnitude of G increases with the increase of $\bar{\sigma}_v$.

The nature of the shear-stress-versus-shear-strain behavior of a dense sand under cyclic loading is shown in Figure 4.25 (p. 114). Using the hysteresis loops of this type and Eq. (4.34), the damping ratios obtained from a cyclic simple shear test for a medium dense sand are shown in Figure 4.26. Note the following:

1. For a given value of $\bar{\sigma}_v$ and amplitude of shear strain γ', the damping ratio decreases with the number of cycles. Since, in most seismic events, the number of significant cycles is likely to be less than 20 (Chapter 7), the values determined at 5 cycles are likely to provide reasonable values for all practical purposes.

2. For a given number of cycles and $\bar{\sigma}_v$, the magnitude of D decreases with the decrease of γ'.

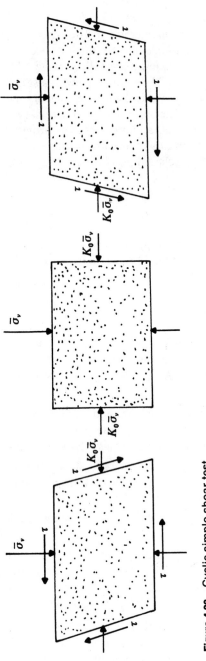

Figure 4.22 Cyclic simple shear test

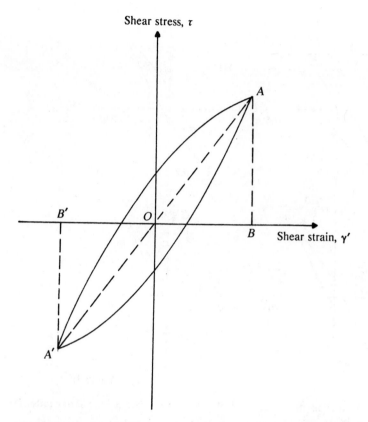

Figure 4.23 Determination of damping ratio from hysteresis loop [Eq. (4.34)]

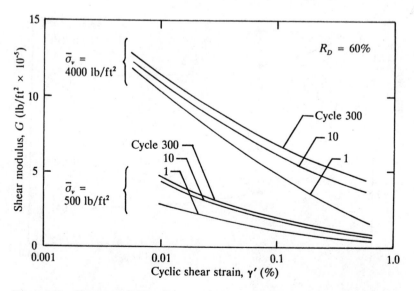

Figure 4.24 Shear modulus–shear strain relationship for medium dense sand (after Silver and Seed, 1971)

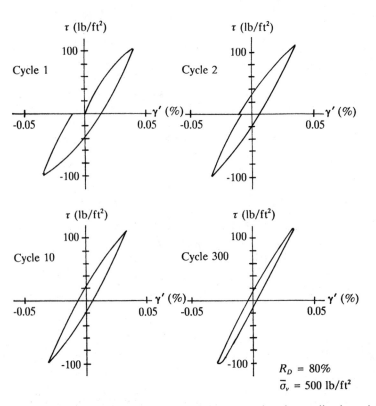

Figure 4.25   Stress-strain behavior of dense sand under cyclic shear (after Silver and Seed, 1971)

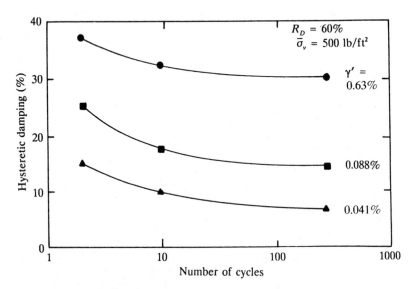

Figure 4.26   Effect of number of stress cycles on hysteretic damping for medium dense sand (after Silver and Seed, 1971)

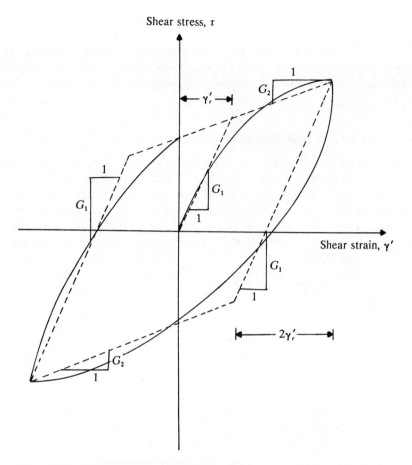

Figure 4.27   Bilinear idealization of shear-stress-versus-shear-strain plots

Other parameters remaining the same (that is, R_D, number of cycles, and amplitude of shear strain), a vertical stress increase will decrease the damping ratio. In many seismic analysis studies, it is convenient to represent the nonlinear shear-stress-versus-shear-strain relationship in the form of a bilinear model (also see Figure 7.10), as shown in Figure 4.27 (Thiers and Seed, 1968). In this figure G_1 is the shear modulus up to a limiting strain of γ_r', and G_2 is the modulus for strain beyond γ_r'.

Advantages of the Cyclic Simple Shear Test

There are several advantages in conducting cyclic simple shear tests. They are more representative of the field conditions, since the specimens can be consolidated in a K_0 state. Solid soil specimens used in resonant column tests can provide good results up to a shear strain amplitude of about $10^{-3}\%$. Similarly, the hollow samples used in resonant column studies provide results within a strain amplitude range of $10^{-3}\%$ to about 1%. However, cyclic simple shear tests can be conducted for a wider range of strain amplitude (that is, $10^{-2}\%$ to about

5%). This range is the general range of strain encountered in the ground motion during seismic activities.

The pore water pressure developed during the vibration of saturated soil specimens by a resonant column device is not usually measured. However, in cyclic simple shear tests, the pore water pressure can be measured at the boundary (see Section 10.10 and Figure 10.20).

4.7   Cyclic Torsional Simple Shear Test

Another technique used to study the behavior of soils subjected to cyclic loading involves a torsional simple shear device. The torsional simple shear device accommodates a "doughnutlike" specimen, as shown in Figure 4.28 (Ishibashi and Sherif, 1974). The specimen has inside and outside radii of $r_1 = 2$ in. (101.6 mm) and $r_2 = 1$ in. (50.8 mm). The inside and outside heights of the specimen are $h_1 = 1$ in. (25.4 mm) and $h_2 = 0.5$ in. (12.7 mm). The soil is initially subjected to a vertical effective stress $\bar{\sigma}_v$, an outside and inside horizontal effective stress of $\bar{\sigma}_h$, and a cyclic shear stress of τ (Figure 4.29). When a shear stress τ is applied, line AB moves to the position of $A'B'$ (Figure 4.29). So, the shearing strain is

$$\gamma'_A = \frac{r_1 \theta}{h_1}$$

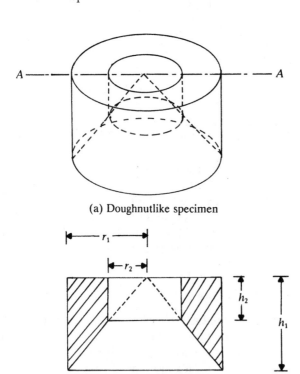

(a) Doughnutlike specimen

(b) Section at A-A

Figure 4.28   Soil specimen for torsional simple shear test

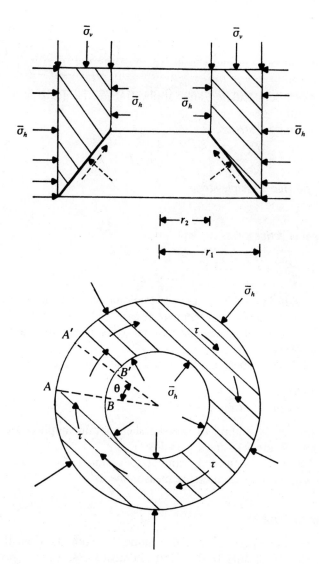

Figure 4.29 Applied stresses on a torsional simple shear test specimen

and

$$\gamma_B' = \frac{r_2 \theta}{h_2}$$

For uniform shear strain throughout the sample,

$$\gamma_A' = \gamma_B'$$

or

$$\frac{r_1 \theta}{h_1} = \frac{r_2 \theta}{h_2}$$

So

$$\boxed{\frac{r_1}{r_2} = \frac{h_1}{h_2}}$$ (4.35)

The following can be calculated after application of the horizontal shear stress on the specimen.

Major effective principal stress:

$$\bar{\sigma}_1 = \frac{\bar{\sigma}_v + \bar{\sigma}_h}{2} + \sqrt{\tau_h^2 + \left(\frac{\bar{\sigma}_v - \bar{\sigma}_h}{2}\right)^2}$$ (4.36a)

Intermediate effective principle stress:

$$\bar{\sigma}_2 = \bar{\sigma}_h$$ (4.36b)

Minor principal effective stress:

$$\bar{\sigma}_3 = \frac{\bar{\sigma}_v + \bar{\sigma}_h}{2} - \sqrt{\tau_h^2 + \left(\frac{\bar{\sigma}_v - \bar{\sigma}_h}{2}\right)^2}$$ (4.36c)

With proper design [Eq. (4.35)], a cyclic torsional shear device can apply nearly uniform shear strain on the specimen. It can apply shear strains up to about 1%. It also eliminates any sidewall frictional stresses that are encountered in cyclic simple shear tests.

The shear modulus of a specimen tested can be determined as

$$G = \frac{\text{amplitude of shear stress, } \tau}{\text{amplitude of shear strain, } \gamma'}$$

The damping ratio corresponding to a given shear strain amplitude can be determined by using Figure 4.18 and Eq. (4.34).

Liquefaction studies on saturated granular soils can also be conducted by this device along with pore water pressure measurement.

4.8   Cyclic Triaxial Test

Cyclic triaxial tests can be performed to determine the modulus of elasticity E and the damping ratio D of soils. In these tests, in most cases, the soil specimen is subjected to a confining pressure $\sigma_0 = \sigma_3$. After that, an axial cyclic stress $\Delta\sigma_d$ is applied to the specimen, as shown in Figure 4.30. The tests conducted for the evaluation of the modulus of elasticity and damping ratio are *strain-controlled* tests. A servo-system is used to apply cycles of controlled deformation.

Figure 4.31 shows the nature of a hysteresis loop obtained from a dynamic triaxial test. From this,

$$E = \frac{\Delta\sigma_d}{\varepsilon}$$

Once the magnitude of E is determined, the value of shear modulus can be calculated by assuming a representative value of Poisson's ratio, or

$$G = \frac{E}{2(1 + \mu)}$$

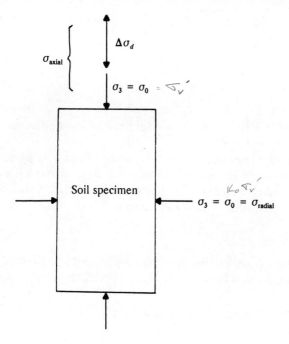

Figure 4.30 Cyclic triaxial test

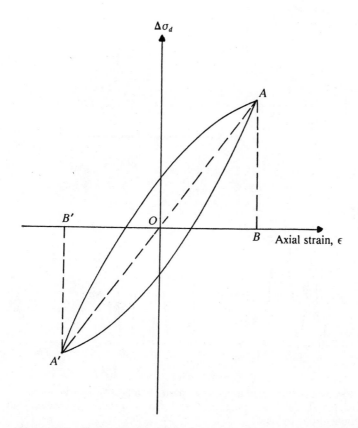

Figure 4.31 Determination of damping ratio from cyclic triaxial test

Again referring to Figure 4.31, the damping ratio can be calculated as

$$D = \frac{1}{2\pi} \frac{\text{area of the hysteresis loop}}{\text{area of triangle } OAB \text{ and } OA'B'}$$

Stress-controlled dynamic triaxial tests are used for liquefaction studies on saturated granular soils (see Chapter 10).

A more elaborate type of dynamic test device has also been used by several investigators to study the cyclic stress-strain history and shear characteristics of soils. Matsui, O-Hara, and Ito (1980) used a dynamic triaxial system that could generate sinusoidally varying axial and radial stresses.

A. Cyclic Strength of Clay

During earthquakes, the soil underlying building foundations and in structures such as earth embankments is subjected to a series of vibratory stress applica-

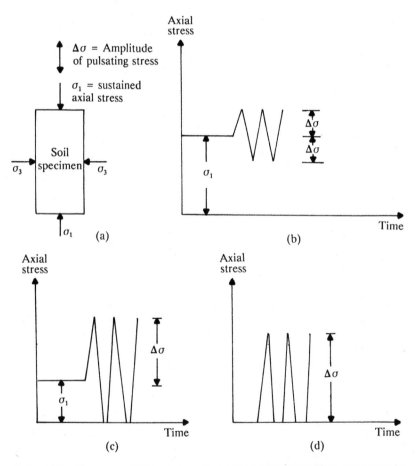

Figure 4.32 Stress conditions on a soil specimen [*Note*: (b) One-directional loading with symmetrical stress pulses; (c) and (d) one-directional loading with nonsymmetrical stress pulses]

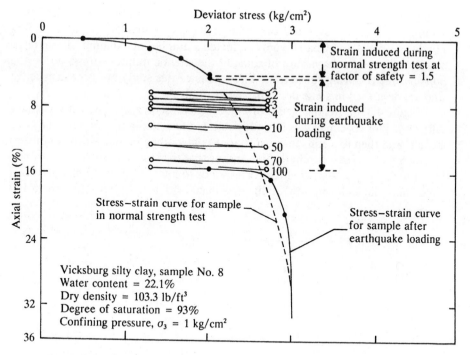

Figure 4.33   Stress-versus-strain relationship for Vicksburg silty clay under sustained and axial pulsating stress (after Seed and Chan, 1966)

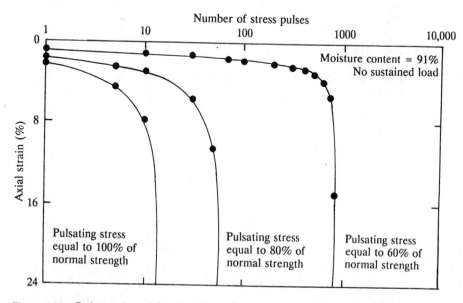

Figure 4.34   Deformation of San Francisco Bay mud specimens subjected to pulsating stress (after Seed and Chan, 1966)

tions. These vibratory stresses may induce large deformation in soil and thus failure. In order to evaluate the strength of clay under earthquake loading conditions, Seed and Chan (1966) conducted a number of dynamic triaxial tests. Figure 4.32 shows the nature of some of the stress conditions imposed on the soil specimens during those tests. The results of this study are very instructive and are described in some detail in this section.

Figure 4.33 shows the results of a laboratory test on a specimen of Vicksburg silty clay. This specimen was initially subjected to a confining pressure of $\sigma_3 = 1$ kg/cm^2 and then to a conventional (static) axial loading in undrained conditions up to 66% of its static strength. This means that the *sustained stress* $\sigma_1 - \sigma_3$ was equal to $0.66[\sigma_{1(f)} - \sigma_3]$, which corresponds to a factor of safety of 1.5. At this time the axial deformation of the specimen was about 5%. After that, 100

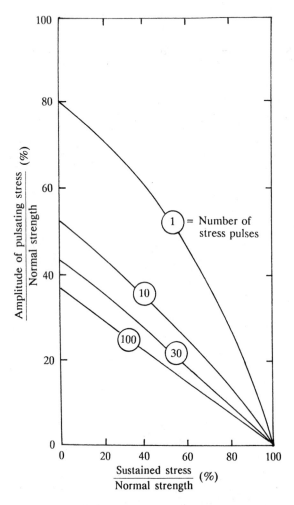

Figure 4.35 Combinations of sustained and pulsating stress intensities causing failure—San Francisco Bay mud (after Seed and Chan, 1966)

transient stress pulses were applied to the specimen. (*Note*: Loading type is similar to that shown in Figure 4.32b.) These stress pulses induced an additional axial stress of about 11%, although the static strength was never exceeded.

Figure 4.34 shows the nature of soil deformation on three soil specimens of San Francisco Bay mud subjected to pulsating stress levels of 100%, 80%, and 60% of normal strength (that is, static strength). For these tests, *no sustained stress* was applied. (*Note*: Loading type is similar to that shown in Figure 4.32d.) It can be seen that, for each level of pulsating stress, the specimen ultimately failed. Figure 4.35 is a plot of the pulsating stress level (as a percent of normal strength) versus sustained stress level (as a percent of normal strength) causing failure of San Francisco Bay mud at various numbers of transient stress pulses. Similar plots could be developed for various soils to help in the design procedure of various structures.

4.9   **Summary of Cyclic Tests**

In the preceding sections, various types of laboratory test methods were presented, from which the fundamental soil properties such as the shear modulus, modulus of elasticity, and damping ratio are determined. These parameters are used in the design and evaluation of the behavior of earthen, earth-supported, and earth-retaining structures. As was discussed in the preceding sections, the magnitudes of G and D are functions of the shear strain amplitude γ'. Hence, while selecting the values of G and D for a certain design work, it is essential to

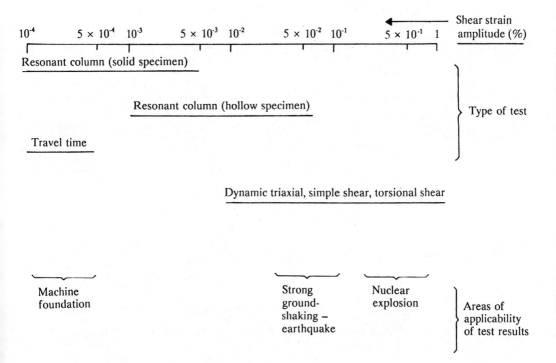

Figure 4.36   Range and applicability of dynamic laboratory tests

know the following:

a. Type of test from which the parameters were obtained

b. Magnitude of the shear strain amplitude

Figure 4.36 provides a guide that is useful for geotechnical engineers, as it gives the amplitude of shear strain levels, type of applicable dynamic tests, and the areas of applicability of the test results. Table 4.2 gives a comparison of the relative quantities of various laboratory techniques for measuring dynamic soil properties. Similarly, Table 4.3 gives a summary of the parameters measured in dynamic or cyclic laboratory tests.

Table 4.2 Relative Quality of Laboratory Techniques for Measuring Dynamic Soil Properties[a]

	Relative Quality of Test Results				
	G Shear modulus	E Young's modulus	D Material damping	Effect of number of cycles	Attenuation
Resonant column	Good	Good	Good	Good	—
with adaptation	—	—	—	—	Fair
Ultrasonic pulse	Fair	Fair	—	—	Poor
Cyclic triaxial	—	Good	Good	Good	—
Cyclic simple shear	Good	—	Good	Good	—
Cyclic torsional shear	Good	—	Good	Good	—

[a] After Silver (1981).

Table 4.3 Parameters Measured in Dynamic or Cyclic Laboratory Tests[a]

	Resonant column	Cyclic triaxial	Cyclic simple shear	Torsional shear
Load	Resonant frequency	Axial force	Horizontal force	Torque
Deformation				
Axial	Vertical displacement	Vertical displacement	Vertical displacement	Vertical displacement
Shear	Acceleration	Not measured	Horizontal displacement	Rotation
Lateral	Not usually measured	Not usually measured	Often controlled	Not usually measured
Volumetric	None for undrained tests Volume of fluid moving into or out of the sample for drained tests			
Pore water pressure	Not usually measured	Measured at boundary	Measured at boundary	Measured at boundary

[a] After Silver (1981).

FIELD TEST MEASUREMENTS

4.10 Reflection and Refraction of Elastic Body Waves—Fundamental Concepts

When an elastic stress wave impinges on the boundary of two layers, the wave is reflected and refracted. As has already been discussed in Chapter 3, there are two types of body waves—that is, compression waves (or P-waves) and shear waves (or S-waves). In the case of P-waves, the direction of the movement of the particles coincides with the direction of propagation. This is shown by the arrows in Figure 4.37a. The shear waves can be separated into two components:

a. *SV-waves*, in which the motion of the particles is in the plane of propagation as shown by the arrows in Figure 4.37b

b. *SH-waves*, in which the motion of the particles is perpendicular to the plane of propagation, as shown by a dark dot in Figure 4.37c

If a P-wave impinges on the boundary between two layers, as shown in Figure 4.38a, there will be two reflected waves and two refracted waves. The reflected waves consist of (1) a P-wave shown as P_1 in layer 1 and (2) an SV-wave shown as SV_1 in layer 1.

The refracted waves will consist of (1) a P-wave, shown as P_2 in layer 2, and

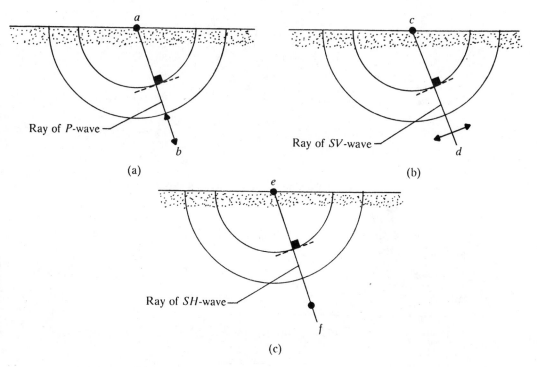

(a)   (b)

(c)

Figure 4.37 *P*-wave, *SV*-wave, and *SH*-wave

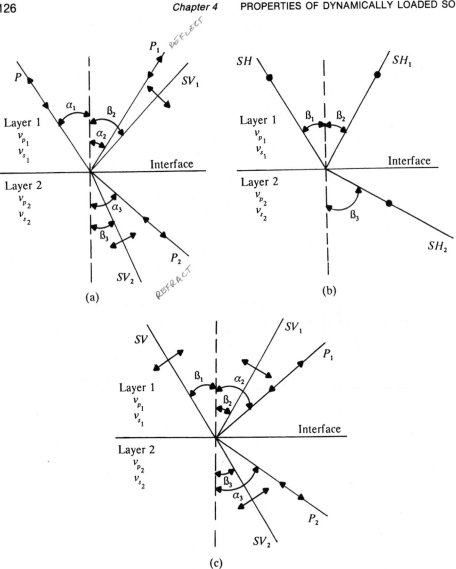

Figure 4.38 Reflection and refraction for (a) an incident P-ray; (b) an incident SH-ray; and (c) an incident SV-ray

(2) an SV-wave, shown as SV_2 in layer 2. Referring to the angles in Figure 4.38a, it can be shown that

$$\alpha_1 = \alpha_2 \tag{4.37}$$

and

$$\frac{\sin \alpha_1}{v_{p_1}} = \frac{\sin \alpha_2}{v_{p_1}} = \frac{\sin \beta_2}{v_{s_1}} = \frac{\sin \alpha_3}{v_{p_2}} = \frac{\sin \beta_3}{v_{s_2}} \tag{4.38}$$

where

v_{p_1} and v_{p_2} = the velocities of the P-wave front in layers 1 and 2, respectively

v_{s_1} and v_{s_2} = the velocities of the S-wave front in layers 1 and 2, respectively

If an *SH*-wave impinges the boundary between two layers, as shown in Figure 4.38b, there will be one reflected *SH*-wave (shown as SH_1) and one refracted *SH*-wave (shown as SH_2). For this case

$$\beta_1 = \beta_2 \tag{4.39}$$

and

$$\boxed{\frac{\sin \beta_1}{v_{s_1}} = \frac{\sin \beta_3}{v_{s_2}}} \tag{4.40}$$

Lastly, if an *SV*-wave impinges the boundary between two layers, as shown in Figure 4.38c, there will be two reflected waves and two refracted waves. The reflected waves are (1) a P-wave, shown as P_1 in layer 1, and (2) an *SV*-wave, shown as SV_1 in layer 1. The refracted waves are (a) a P-wave, shown as P_2 in layer 2, and (b) an *SV*-wave, shown as SV_2 in layer 2. For this case, $\beta_1 = \beta_2$:

$$\boxed{\frac{\sin \beta_1}{v_{s_1}} = \frac{\sin \alpha_2}{v_{p_1}} = \frac{\sin \beta_2}{v_{s_1}} = \frac{\sin \beta_3}{v_{s_2}} = \frac{\sin \alpha_3}{v_{p_2}}} \tag{4.41}$$

The mathematical derivations of these facts will not be shown here. For further details the reader is referred to Kolsky (1963, pp. 24–38).

4.11   Seismic Refraction Survey (Horizontal Layering)

Seismic refraction surveys are sometimes used to determine the wave propagation velocities through various soil layers in the field and to obtain thicknesses of each layer. Consider the case where there are two layers of soil, as shown in Figure 4.39a. Let the velocities of P-waves in layers 1 and 2 be v_{p_1} and v_{p_2}, respectively, and let $v_{p_1} < v_{p_2}$. A is a source of impulsive energy. If seismic waves are generated at A, the energy from that point will travel in hemispherical wave fronts. Consider the case of P-waves, since they are the fastest. If a detecting device is placed at point B, which is located at a *small* distance x from A, the P-wave that travels through the upper medium will reach it first before any other wave. The travel time for this first arrival may be given as

$$t = \frac{x}{v_{p_1}} \tag{4.42}$$

where $\overline{AB} = x$.

Again, consider the first arrival time of a P-wave at a point G, which is

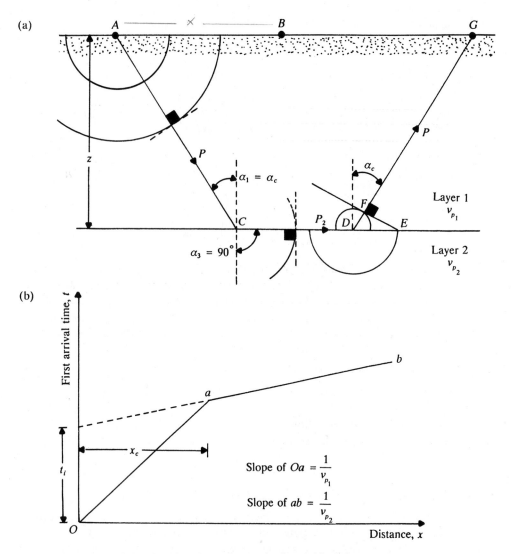

Figure 4.39 Seismic refraction survey—horizontal layering

located at a greater distance from A. In order to understand this, one considers a spherical P-wave front that originates at A striking the interface of the two layers. At some point C, the refracted P-wave front in the lower medium will be such that the tangent to the sphere will be perpendicular to the interface. In that case, the refracted P-ray (shown as P_2 in Figure 4.39a) will be parallel to the boundary and will travel with a velocity v_{p_2}. Note that because $v_{p_1} < v_{p_2}$, this wave front will travel faster than those described previously. From Eq. (4.38)

$$\frac{\sin \alpha_1}{v_{p_1}} = \frac{\sin \alpha_3}{v_{p_2}}$$

Since $\alpha_3 = 90°$, $\sin \alpha_3 = 1$, and

$$\alpha_1 = \sin^{-1}\left(\frac{v_{p_1}}{v_{p_2}}\right) = \alpha_c \tag{4.43}$$

where α_c = critical angle of incidence.

The wave front just described traveling with a velocity v_{p_2} will create vibrating stresses at the interface, and this will generate wave fronts that will spread out into the upper medium. These P-waves will spread with a velocity of v_{p_1}. The spherical wave front traveling downward from D in layer 2 will have a radius equal to DE after a time Δt. At the same time Δt, the spherical wave front traveling upward from point D will have a radius equal to DF. The resultant wave front in the upper layer will follow a line EF. It can be seen from the diagram that

$$\frac{v_{p_1}\Delta t}{v_{p_2}\Delta t} = \frac{DF}{DE} = \sin i_c \tag{4.44}$$

So ray DFG will make an angle i_c with the vertical. It can be mathematically shown that for x greater than a *critical value* x_c, the P-wave that travels the path $ACDG$ will be the *first to arrive* at point G. Let the time of travel for the P-wave along the path $ACDG$ be equal to t. Thus, $t = t_{AC} + t_{CD} + t_{DG}$, or

$$t = \left(\frac{z}{\cos i_c}\right)\left(\frac{1}{v_{p_1}}\right) + \frac{x - 2z\tan i_c}{v_{p_2}} + \left(\frac{z}{\cos i_c}\right)\left(\frac{1}{v_{p_1}}\right)$$

$$= \frac{x}{v_{p_2}} - \frac{2z\sin i_c}{v_{p_2}\cos i_c} + \frac{2z}{v_{p_1}\cos i_c}$$

where $x = \overline{AG}$. But $v_{p_2} = v_{p_1}/\sin i_c$ [from Eq. (4.44)]; thus

$$t = \frac{x}{v_{p_2}} - \frac{2z\sin^2 i_c}{v_{p_1}\cos i_c} + \frac{2z}{v_{p_1}\cos i_c} = \frac{x}{v_{p_2}} + \frac{2z}{v_{p_1}}\left(\frac{1 - \sin^2 i_c}{\cos i_c}\right)$$

$$= \frac{x}{v_{p_2}} + \frac{2z}{v_{p_1}}\cos i_c \tag{4.45}$$

Since $\sin i_c = v_{p_1}/v_{p_2}$

$$\cos i_c = \sqrt{1 - \sin^2 i_c} = \sqrt{1 - \left(\frac{v_{p_1}}{v_{p_2}}\right)^2} \tag{4.46}$$

Substituting Eq. (4.46) into Eq. (4.45), one obtains

$$t = \frac{x}{v_{p_2}} + \frac{2z\sqrt{v_{p_2}^2 - v_{p_1}^2}}{(v_{p_1})(v_{p_2})} \tag{4.47}$$

If detecting instruments are placed at various distances from the source of disturbance to obtain first arrival times and the results are plotted in graphical form, the graph will be like that shown in Figure 4.39b. The line Oa represents the data that follow Eq. (4.42). The slope of this line will give $1/v_{p_1}$. The line ab represents the data that follow Eq. (4.47). The slope of this line is $1/v_{p_2}$. Thus the velocities of v_{p_1} and v_{p_2} can now be obtained.

If line ab is projected back to $x = 0$, one obtains

$$t = t_i = \frac{2z\sqrt{v_{p_2}^2 - v_{p_1}^2}}{(v_{p_1})(v_{p_2})}$$

or

$$z = \frac{(t_i)(v_{p_1})(v_{p_2})}{2\sqrt{v_{p_2}^2 - v_{p_1}^2}} = \frac{t_i v_{p_1}}{2\cos i_c} \qquad (4.48)$$

where t_i is the intercept time. Hence, the thickness of layer 1 can be easily obtained.

The *critical distance* x_c (Figure 4.39b) beyond which the wave refracted at the interface arrives at the detector before the direct wave can be obtained by equating the right-hand sides of Equations (4.42) and (4.47):

$$\frac{x_c}{v_{p_1}} = \frac{x_c}{v_{p_2}} + \frac{2z\sqrt{v_{p_2}^2 - v_{p_1}^2}}{v_{p_1}v_{p_2}}$$

or

$$x_c = 2z\frac{\sqrt{v_{p_2}^2 - v_{p_1}^2}}{v_{p_1}v_{p_2}}\frac{v_{p_1}v_{p_2}}{v_{p_2} - v_{p_1}} = 2z\sqrt{\frac{v_{p_2} + v_{p_1}}{v_{p_2} - v_{p_1}}} \qquad (4.49)$$

The depth of the first layer can be calculated from Eq. (4.49) as

$$z = \frac{x_c}{2}\sqrt{\frac{v_{p_2} - v_{p_1}}{v_{p_2} + v_{p_1}}} \qquad (4.50)$$

A. Refraction Survey in a Three-Layered Soil Medium

Figure 4.40 considers the case of a refraction survey through a three-layered soil medium. Let v_{p_1}, v_{p_2}, and v_{p_3} be the P-wave velocities in layers 1, 2, and 3, respectively, as shown in Figure 4.40a ($v_{p_1} < v_{p_2} < v_{p_3}$). If A in Figure 4.40a is a source of disturbance, the P-wave traveling through layer 1 will arrive first at B, which is located a small distance away from A. The travel time for this can be given by Eq. (4.42) as $t = x/v_{p_1}$. At a greater distance x, the first arrival will correspond to the wave taking the path $ACDE$. The travel time for this can be given by Eq. (4.47) as

$$t = \frac{x}{v_{p_2}} + \frac{2z_1\sqrt{v_{p_2}^2 - v_{p_1}^2}}{(v_{p_1})(v_{p_2})}$$

where z_1 = thickness of top layer.

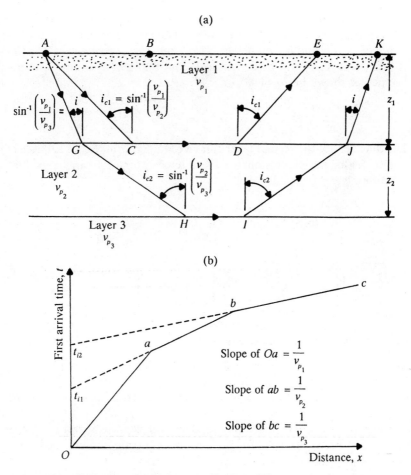

Figure 4.40   Refraction survey in a three-layer soil

At a still larger distance, the first arrival will correspond to the path *AGHIJK*. Note that the refracted ray *H-I* will travel with a velocity of v_{p_3}. The angle i_{c2} is the critical angle for layer 3.

$$i_{c2} = \sin^{-1}\left(\frac{v_{p_2}}{v_{p_3}}\right) \tag{4.51}$$

For this path (*AGHIJK*) the total travel time can be derived as

$$t = \frac{x}{v_{p_3}} + \frac{2z_1\sqrt{v_{p_3}^2 - v_{p_1}^2}}{(v_{p_3})(v_{p_1})} + \frac{2z_2\sqrt{v_{p_3}^2 - v_{p_2}^2}}{(v_{p_3})(v_{p_2})} \tag{4.52}$$

where z_2 = thickness of layer 2.

So, if detecting instruments are placed at various distances from the source of disturbance to obtain first arrival times, they can be plotted in a *t*-versus-*x* graph. This graph will appear as shown in Figure 4.40b. The line Oa corresponds to Eq. (4.42), ab corresponds to Eq. (4.47), and bc corresponds to Eq. (4.52). The slopes of Oa, ab, and bc will be $1/v_{p_1}$, $1/v_{p_2}$ and $1/v_{p_3}$, respectively. The thickness of the first layer z_1 can be determined from the intercept time t_{i1} in a similar manner, as shown in Eq. (4.48), or

$$z_1 = \frac{(t_{i1})(v_{p_1})(v_{p_2})}{2\sqrt{v_{p_2}^2 - v_{p_1}^2}}$$

The thickness of the second layer can be obtained from Eq. (4.52). Referring to Figure 4.40b, the expression for the intercept time t_{i2} can be evaluated by substituting $x = 0$ into Eq. (4.52):

$$t = t_{i2} = \frac{2z_1\sqrt{v_{p_3}^2 - v_{p_1}^2}}{(v_{p_3})(v_{p_1})} + \frac{2z_2\sqrt{v_{p_3}^2 - v_{p_2}^2}}{(v_{p_3})(v_{p_2})}$$

or

$$z_2 = \frac{1}{2}\left[t_{i2} - \frac{2z_1\sqrt{v_{p_3}^2 - v_{p_1}^2}}{(v_{p_3})(v_{p_1})}\right]\frac{(v_{p_3})(v_{p_2})}{\sqrt{v_{p_3}^2 - v_{p_2}^2}} \qquad (4.53)$$

B. Refraction Survey for Multilayer Soil

In general, if there are n layers, the first arrival time at various distances from the source of disturbance will plot as shown in Figure 4.41. There will be n segments in the t-versus-x plot. The slope of the nth segment will give the value $1/v_{p_n}$ $(n = 1, 2, \ldots)$.

The value of P-wave velocity in a natural deposit of soil will depend on several factors, such as confining pressure, moisture content, and void ratio. Some typical values of v_p are given in Table 3.1.

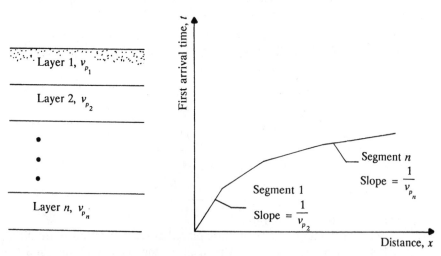

Figure 4.41   Refraction survey for multilayer soil

Example 4.1

Following are the results of a refraction survey (horizontal layering of soil). Determine the *P*-wave velocities of the soil layers and their thicknesses.

Distance (m)	Time of first arrival (ms)
2.5	5.5
5	11.1
7.5	16.1
15	24.0
25	30.8
35	38.2
45	46.1
55	51.3
60	52.8

Solution

The time-distance plot is given in Figure 4.42. From the plot,

$$v_{p_1} = \frac{5}{10.6 \times 10^{-3}} = \underline{472 \text{ m/s}}$$

$$v_{p_2} = \frac{10}{7.2 \times 10^{-3}} = \underline{1389 \text{ m/s}}$$

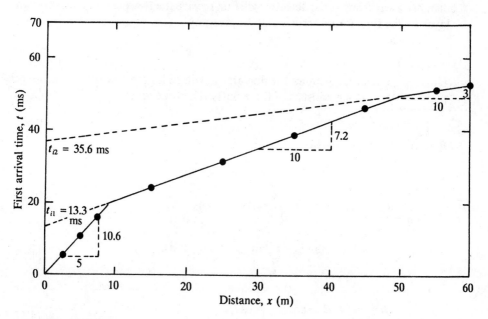

Figure 4.42

$$v_{p_3} = \frac{10}{3 \times 10^{-3}} = \underline{3333 \text{ m/s}}$$

$t_{i1} = 13.3 \times 10^{-3}$ s; $t_{i2} = 35.6 \times 10^{-3}$ s. From Eq. (4.48)

$$z_1 = \frac{(t_{i1})(v_{p_1})(v_{p_2})}{2\sqrt{v_{p_2}^2 - v_{p_1}^2}} = \frac{(13.3 \times 10^{-3})(472)(1389)}{2\sqrt{(1389)^2 - (472)^2}}$$

$$= \underline{3.39 \text{ m}}$$

From Eq. (4.53)

$$z_2 = \frac{1}{2}\left[t_{i2} - \frac{2z_1\sqrt{v_{p_3}^2 - v_{p_1}^2}}{(v_{p_3})(v_{p_1})} \right] \frac{(v_{p_3})(v_{p_2})}{\sqrt{v_{p_3}^2 - v_{p_2}^2}}$$

$$= \frac{1}{2}\left[35.6 \times 10^{-3} - \frac{(2)(3.39)\sqrt{(3333)^2 - (472)^2}}{(3333)(472)} \right]$$

$$\times \frac{(3333)(1389)}{\sqrt{(3333)^2 - (1389)^2}}$$

$$= \frac{1}{2}(0.02138)(1528) = \underline{16.33 \text{ m}}$$ ■

4.12   Refraction Survey in Soils with Inclined Layering

Figure 4.43a shows two soil layers. The interface of soil layers 1 and 2 is inclined at an angle β with respect to the horizontal. Let the *P*-wave velocities in layers 1 and 2 be v_{p_1} and v_{p_2}, respectively ($v_{p_1} < v_{p_2}$).

If a disturbance is created at *A* and a detector is placed at *B*, which is a small distance away from *A*, the detector will first receive the *P*-wave traveling through layer 1. The time for its arrival may be given by

$$t_d = \frac{x}{v_{p_1}}$$

However, at a larger distance the first arrival will be for the *P*-wave following the path *ACDE*—which consists of three parts. The time taken can be written as

$$t_d = t_{AC} + t_{CD} + t_{DE} \tag{4.54}$$

Referring to Figure 4.43a,

$$t_{AC} = \frac{z'}{v_{p_1} \cos i_c} \tag{4.55}$$

$$t_{CD} = \frac{CD}{v_{p_2}} = \frac{AA_4 - AA_1 - A_2A_3 - A_3A_4}{v_{p_2}}$$

$$= \frac{x \cos \beta - z' \tan i_c - z' \tan i_c - x \sin \beta \tan i_c}{v_{p_2}} \tag{4.56}$$

$$t_{DE} = \frac{DA_3 + A_3E}{v_{p_1}} = \frac{\dfrac{z'}{\cos i_c} + \dfrac{x \sin \beta}{\cos i_c}}{v_{p_1}} \tag{4.57}$$

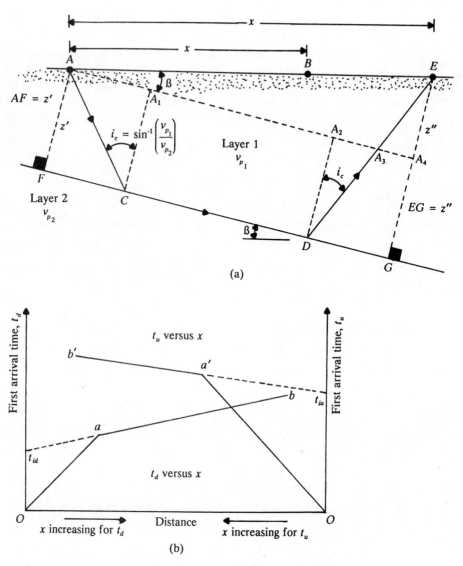

Figure 4.43   Refraction survey in soils with inclined layering

Substitution of Eqs. (4.55), (4.56), and (4.57) into Eq. (4.54) and simplification yields

$$t_d = \frac{2z' \cos i_c}{v_{p_1}} + \frac{x}{v_{p_1}} \sin(i_c + \beta) \tag{4.58}$$

Now, if the source of disturbance is E and the detector is placed at A, the first arrival time along the refracted ray path may be given by

$$t_u = \frac{2z'' \cos i_c}{v_{p_1}} + \frac{x}{v_{p_1}} \sin(i_c - \beta) \tag{4.59}$$

In the actual survey, one can have a source of disturbance such as A and observe the first arrival time at several points to the right of A and have a source of disturbance such as E and observe the first arrival time at several points to the left of E. These results can be plotted in a graphical form, as shown in Figure 4.43b (time-versus-x plot). From Figure 4.43b note that the slopes of Oa and $O'a'$ are both $1/v_{p_1}$. The slope of the branch ab will be $[\sin(i_c + \beta)]/v_{p_1}$, as can be seen from Eq. (4.58). Similarly, the slope of the branch $a'b'$ will be $[\sin(i_c - \beta)]/v_{p_1}$ [see Eq. (4.59)]. Let

$$m_d = \frac{\sin(i_c + \beta)}{v_{p_1}} \tag{4.60}$$

and

$$m_u = \frac{\sin(i_c - \beta)}{v_{p_1}} \tag{4.61}$$

From Eq. (4.60),

$$i_c = \sin^{-1}(m_d v_{p_1}) - \beta \tag{4.62}$$

Again, from Eq. (4.61)

$$i_c = \sin^{-1}(m_u v_{p_1}) + \beta \tag{4.63}$$

Solving the two preceding equations,

$$\boxed{i_c = \tfrac{1}{2}[\sin^{-1}(v_{p_1} m_d) + \sin^{-1}(v_{p_1} m_u)]} \tag{4.64}$$

and

$$\boxed{\beta = \tfrac{1}{2}[\sin^{-1}(v_{p_1} m_d) - \sin^{-1}(v_{p_1} m_u)]} \tag{4.65}$$

Once i_c is determined, the value of v_{p_2} can be obtained as

$$\boxed{v_{p_2} = \frac{v_{p_1}}{\sin i_c}} \tag{4.66}$$

Again referring to Figure 4.43b, if the ab and $a'b'$ branches are projected back, they will intercept the time axes at t_{id} and t_{iu}, respectively. From Eqs. (4.58) and (4.59), it can be seen that

$$t_{id} = \frac{2z' \cos i_c}{v_{p_1}}$$

or

$$\boxed{z' = \frac{(t_{id}) v_{p_1}}{2 \cos i_c}} \tag{4.67}$$

and

$$t_{iu} = \frac{2z'' \cos i_c}{v_{p_1}}$$

or

$$z'' = \frac{(t_{iu})v_{p_1}}{2 \cos i_c} \qquad (4.68)$$

Since i_c and v_{p_1} are known and t_{id} and t_{iu} can be determined from a graph, one can obtain the values of z' and z''.

Example 4.2

Referring to Figure 4.43a, the results of a refraction survey are as follows. The distance between A and E is 60 m.

Point of disturbance A		Point of disturbance E	
Distance from A (m)	Time of first arrival (ms)	Distance from E (m)	Time of first arrival (ms)
5	12.1	5	11.5
10	25.2	10	22.8
15	35.3	15	34.5
20	48.0	20	44.8
30	60.2	30	69.1
40	68.5	40	78.1
50	76.8	50	82.8
60	85.1	60	87.7

Determine
a. v_{p_1} and v_{p_2},
b. z' and z'', and
c. β.

Solution

The time-distance records have been plotted in Figure 4.44.
a. From branch Oa,

$$v_{p_1} = \frac{10}{25 \times 10^{-3}} = 400 \text{ m/s}$$

From branch $O'a'$,

$$v_{p_1} = \frac{10}{22 \times 10^{-3}} = 454 \text{ m/s}$$

The average value of v_{p_1} is $\underline{427 \text{ m/s}}$.

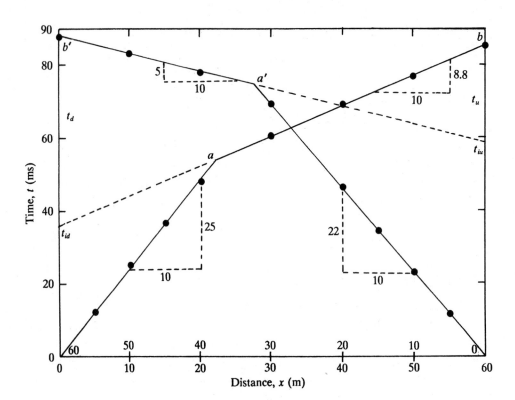

Figure 4.44

From the slope of branch ab,

$$m_d = \frac{8.8 \times 10^{-3}}{10} = 0.88 \times 10^{-3}$$

Again, from the slope of branch $a'b'$,

$$m_u = \frac{5 \times 10^{-3}}{10} = 0.5 \times 10^{-3}$$

From Eq. (4.64),

$$i_c = \tfrac{1}{2}[\sin^{-1}(v_{p_1} m_d) + \sin^{-1}(v_{p_1} m_u)]$$

$$\sin^{-1}(v_{p_1} m_d) = \sin^{-1}[(427)(0.88 \times 10^{-3})] = 22.07°$$

$$\sin^{-1}(v_{p_1} m_u) = \sin^{-1}[(427)(0.5 \times 10^{-3})] = 12.33°$$

Hence

$$i_c = \tfrac{1}{2}(22.07° + 12.33°) = \underline{17.2°}$$

Using Eq. (4.66),

$$v_{p_2} = \frac{v_{p_1}}{\sin i_c} = \frac{427}{\sin(17.2)} = \underline{1444 \text{ m/s}}$$

b. From Eq. (4.67)

$$z' = \frac{(t_{id})(v_{p_1})}{2\cos i_c}$$

$t_{id} = 35.9 \times 10^{-3}$ s (from Figure 4.44). So

$$z' = \frac{(35.9 \times 10^{-3})(427)}{2\cos(17.2)} = \underline{8.03 \text{ m}}$$

Again, from Eq. (4.68),

$$z'' = \frac{(t_{iu})(v_{p_1})}{2\cos i_c}$$

From Figure 4.44, $t_{iu} = 59.8 \times 10^{-3}$ s.

$$z'' = \frac{(59.8 \times 10^{-3})(427)}{2\cos(17.2)} = \underline{13.37 \text{ m}}$$

c. From Eq. (4.65),

$$\beta = \tfrac{1}{2}[\sin^{-1}(v_{p_1}m_d) - \sin^{-1}(v_{p_1}m_u)]$$
$$= \tfrac{1}{2}(22.07° - 12.33°) = \underline{4.87°}$$ ∎

4.13   Reflection Survey in Soil (Horizontal Layering)

Reflection surveys can also be conducted to obtain information about the soil layers. Figure 4.45a shows a two-layered soil system. A is the point of disturbance. If a recorder is placed at C at a distance x away from A, the travel time for the reflected P-wave can be given as

$$t = \frac{AB + BC}{v_{p_1}} = \frac{2}{v_{p_1}}\sqrt{z^2 + \left(\frac{x}{2}\right)^2} \tag{4.69}$$

where t = total travel time for the ray path ABC.

From Eq. (4.69), the thickness of layer 1 can be obtained as

$$\boxed{z = \tfrac{1}{2}\sqrt{(v_{p_1}t)^2 - x^2}} \tag{4.70}$$

If the travel times t for the reflected P-waves at various distances x are obtained, they can be plotted in a graphical form, as shown in Figure 4.45b. Note that the time-distance curve obtained from Eq. (4.69) will be a hyperbola. The line Oa shown in Figure 4.45b is the time-distance plot for the direct P-waves traveling through layer 1 (compare line Oa in Figure 4.45b to the line Oa in Figure 4.39b). The slope of this line will give $1/v_{p_1}$.

If the time-distance curve obtained from the reflection data is extended back, it will intersect the time axis at t_0. From Eq. (4.69) it can be seen that at $x = 0$,

$$t_0 = \frac{2z}{v_{p_1}}$$

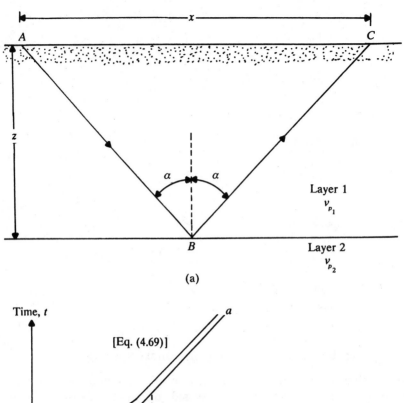

Figure 4.45   (Continued)

or

$$z = \frac{t_0 v_{p_1}}{2} \tag{4.71}$$

With v_{p_1} and t_0 known, the thickness of the top layer z can be calculated.

Another convenient way to interpret the reflection survey record is to plot a graph of t^2 versus x^2. From Eq. (4.69),

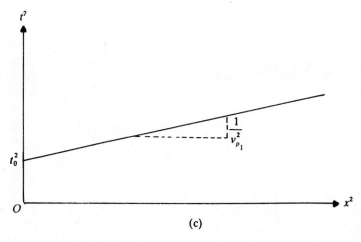

Figure 4.45   Reflection survey in soil—horizontal layering

$$t^2 = \frac{4}{v_{p_1}^2}\left[z^2 + \left(\frac{x}{2}\right)^2\right] = \frac{1}{v_{p_1}^2}(4z^2 + x^2)$$

(4.72)

This relation indicates that the plot of t^2 versus x^2 will be a straight line, as shown in Figure 4.45c. The slope of this line gives $1/v_{p_1}^2$ and the intercept on the t^2 axis will be equal to t_0^2. Substituting $t = t_0$ and $x = 0$ into Eq. (4.72),

$$t_0^2 = \frac{4z^2}{v_{p_1}^2}$$

or

$$z^2 = \frac{t_0^2 v_{p_1}^2}{4}$$

(4.73)

With t_0^2 and $v_{p_1}^2$ known, the thickness of the top layer can now be calculated.

Example 4.3

The results of a reflection survey on a relatively flat area (shale underlain by granite) are given here. Determine the velocity of P-waves in the shale.

Distance from point of disturbance (ft)	Time for first reflection (s)
100	1.0
300	1.002
500	1.003
700	1.007
900	1.011
1100	1.017
1300	1.023

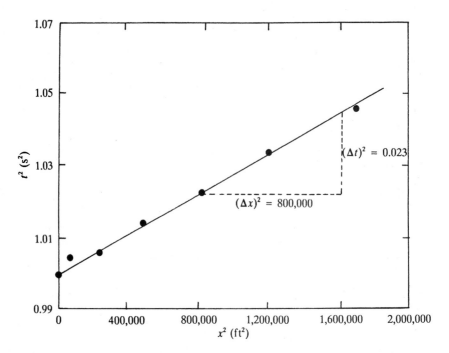

Figure 4.46

Solution

Using the time-distance records, the following table can be prepared.

x (ft)	x^2 (ft^2)	t (s)	t^2 (s^2)
100	10,000	1.0	1.0
300	90,000	1.002	1.004
500	250,000	1.003	1.006
700	490,000	1.007	1.014
900	810,000	1.011	1.022
1,100	1,210,000	1.017	1.034
1,300	1,690,000	1.023	1.046

A plot of t^2 versus x^2 is shown in Figure 4.46. From the plot,

$$v_{p_1} = \sqrt{\frac{(\Delta x)^2}{(\Delta t)^2}} = \sqrt{\frac{800,000}{0.023}} = \underline{5898 \text{ ft/s}}$$ ∎

4.14   Reflection Survey in Soil (Inclined Layering)

Figure 4.47 considers the case of a reflection survey where the reflecting boundary is inclined at an angle β with respect to the horizontal. A is the point for the source of disturbance. The reflected P-ray reaching point C will take the path

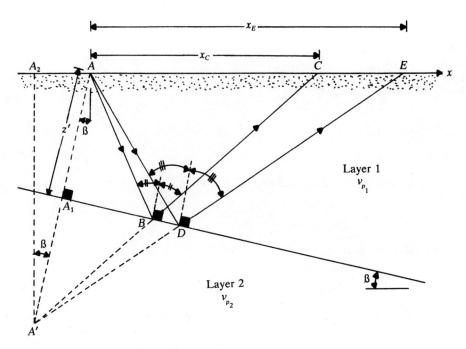

Figure 4.47   Reflection survey in soil—inclined layering

ABC. Referring to Figure 4.47,

$$AB + BC = A'B + BC = A'C$$

But

$$(A'C)^2 = (A'A_2)^2 + (A_2C)^2 \tag{4.74}$$

$$A'A_2 = AA' \cos \beta = 2z' \cos \beta \tag{4.75}$$

$$A_2C = A_2A + AC = 2z' \sin \beta + x_C \tag{4.76}$$

Substituting Eqs. (4.75) and (4.76) into Eq. (4.74),

$$A'C = \sqrt{(2z' \cos \beta)^2 + (2z' \sin \beta + x_C)^2}$$
$$= \sqrt{4z'^2 + x_C^2 + 4z'x_C \sin \beta}$$

Thus, the travel time for the reflected *P*-wave along the path *ABC* will be

$$t_C = \frac{A'C}{v_{p_1}}$$

So

$$t_C = \frac{1}{v_{p_1}} \sqrt{4z'^2 + x_C^2 + 4z'x_C \sin \beta} \tag{4.77}$$

In a similar manner, the time of arrival for the reflected *P*-waves received at point *E* can be given as

$$t_E = \frac{1}{v_{p_1}}\sqrt{4z'^2 + x_E^2 + 4z'x_E \sin \beta} \qquad (4.78)$$

Combining Eqs. (4.77) and (4.78),

$$\sin \beta = \frac{v_{p_1}^2(t_E^2 - t_C^2)}{4z'(x_E - x_C)} - \frac{x_E + x_C}{4z'} \qquad (4.79)$$

Now, let $\bar{t} = (t_E + t_C)/2$ and

$$\Delta t = t_E - t_C$$

Substitution of the preceding relations in Eq. (4.79) gives

$$\sin \beta = \frac{v_{p_1}^2 \bar{t}(\Delta t)}{2z'(x_E - x_C)} - \frac{x_E + x_C}{4z'} \qquad (4.80)$$

If x_C is equal to zero, Eq. (4.80) will transform to

$$\sin \beta = \frac{v_{p_1}^2 \bar{t}(\Delta t)}{2z'x_E} - \frac{x_E}{4z'} \qquad (4.81)$$

If $x_C = 0$ and $\beta = 0$ (that is, the reflecting layer is horizontal) then, from Eq. (4.81),

$$\Delta t = \frac{x_E^2}{2v_{p_1}^2 \bar{t}} \qquad (4.82)$$

If $x_C = 0$ and $\Delta t > x_E^2/2v_{p_1}^2 \bar{t}$, the reflecting layer is sloping down in the direction of positive x, as shown in Figure 4.47. If $x_C = 0$ and $\Delta t < x_E^2/2v_{p_1}^2 \bar{t}$, the

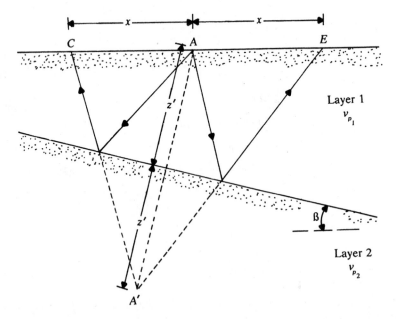

Figure 4.48

reflecting layer is sloping down in the direction of negative x (that is, opposite to that shown in Figure 4.47).

In actual practice, the point of disturbance A (Figure 4.48) is generally placed midway between the two detectors, so $x_E = -x_C = x$. So, from Eq. (4.80)

$$\sin \beta = \frac{v_{p_1}^2 \bar{t}(\Delta t)}{4z'x} \tag{4.83}$$

Referring to Figure 4.48, $AA' = 2z' = (A'C + A'E)/2$. So

$$\frac{2z'}{v_{p_1}} = \frac{1}{2}\left(\frac{A'C}{v_{p_1}} + \frac{A'E}{v_{p_1}}\right) = \frac{1}{2}(t_C + t_E) = \bar{t} \tag{4.84}$$

Combining Eqs. (4.83) and (4.84),

$$\sin \beta = \frac{v_{p_1}(\Delta t)}{2x} \tag{4.85}$$

Example 4.4

Refer to Figure 4.48. Given: $x = 280$ ft, $t_C = 0.026$ s, and $t_E = 0.038$ s. Determine β and z'. (The value of v_{p_1} has been previously determined to be 1350 ft/s.)

Solution

$$t = \frac{t_C + t_E}{2} = \frac{0.026 + 0.038}{2} = 0.032 \text{ s}$$

$$\Delta t = t_E - t_C = 0.038 - 0.026 = 0.012 \text{ s}$$

From Eq. (4.85)

$$\beta = \sin^{-1}\left[\frac{v_{p_1}(\Delta t)}{2x}\right] = \sin^{-1}\left[\frac{(1350)(0.012)}{(2)(280)}\right] = \underline{1.66°}$$

From Eq. (4.84)

$$\frac{2z'}{v_{p_1}} = \bar{t}$$

or

$$z' = \frac{\bar{t}v_{p_1}}{2} = \frac{(0.032)(1350)}{2} = \underline{21.6 \text{ ft}} \qquad ∎$$

4.15   Subsoil Exploration by Steady-State Vibration

In steady-state vibration, a circular plate placed on the ground surface is vibrated vertically by a sinusoidal loading (Figure 4.49a). This vibration will send out *Rayleigh waves* (Section 3.13) and the vertical motion of the ground surface will predominantly be due to these waves. This can be picked up by motion transducers. The velocity of the *Rayleigh waves* can be given as

$$v_r = fL \tag{4.86}$$

where f = frequency of vibration of the plate and L = wavelength.

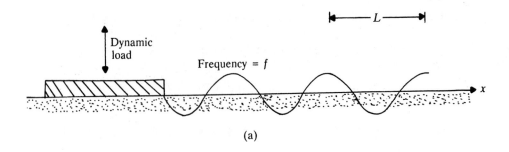

(a)

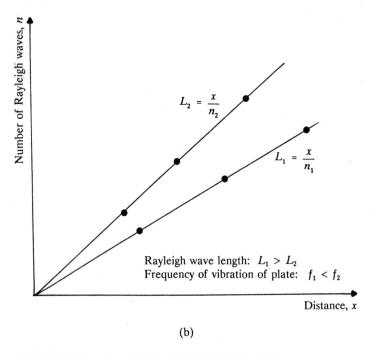

$$L_2 = \frac{x}{n_2}$$

$$L_1 = \frac{x}{n_1}$$

Rayleigh wave length: $L_1 > L_2$
Frequency of vibration of plate: $f_1 < f_2$

Distance, x

(b)

Figure 4.49   Subsoil exploration by steady-state vibration

If the wavelength L can be measured, the velocity of Rayleigh waves can be calculated. The wavelength is generally determined by the number of waves occurring at a given distance x. For a given frequency f_1 the wavelength can be given as

$$L_1 = \frac{x}{n_1} \tag{4.87}$$

where n_1 = number of waves at a distance x for frequency f_1 (as shown in Figure 4.49b).

It was shown in Chapter 3 that the Rayleigh wave velocity is approximately equal to the shear wave velocity. So

$$v_r \approx v_s \tag{4.88}$$

$v_s = v_r = fL$

It was also discussed in Chapter 3 that, for all practical purposes, the Rayleigh wave travels through the soil within a depth of one wavelength. Hence for a given frequency f, if the wavelength L is known, the value of v_s determined by the preceding technique will represent the soil conditions at an average depth of $L/2$. Thus for a *large value of f*, the value of v_s is representative of soil conditions at a *smaller depth*; and, for a *small value of f*, the value of v_s obtained is representative of the soil conditions at a *larger depth*. Figure 4.50 shows the results of wave propagation on a stratified pavement system obtained using this technique.

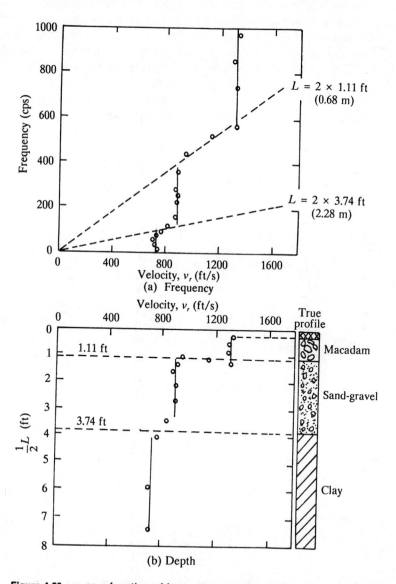

Figure 4.50 v_r as a function of frequency and depth determined by the steady-state vibration technique (after Heukelom and Foster, 1960)

4.16   Soil Exploration by "Shooting Up the Hole," "Shooting Down the Hole," and "Cross-Hole Shooting"

Shooting Up the Hole

In the technique of shooting up the hole, a hole is drilled into the ground and a detector is placed at the ground surface. Charges are exploded at various depths in the hole and the direct travel time of body waves (P or S) along the boundary of the hole is measured. Thus the values of v_p and v_s of various soil layers can be easily obtained. There is a definite advantage in this technique, since it determines the *shear wave velocities of various soil layers*. The refraction and reflection techniques give only the P-wave velocity. However, below the groundwater table the compression waves will travel through water. The first arrival for points below the water table will usually be for this type, and the wave velocity will generally be higher than the compression wave velocity in soils. On the other hand, shear waves cannot travel through water, and the shear wave velocity measured above or below the water table will be the same.

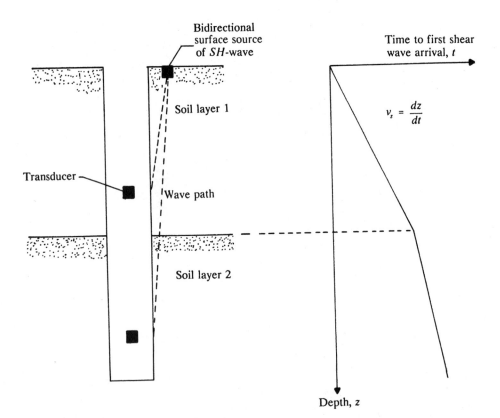

Figure 4.51 The down-hole method of seismic wave testing

Shooting Down the Hole

Shear wave velocity determination of various soil layers by shooting down the hole has been described by Schwarz and Musser (1972), Beeston and McEvilly (1977), and Larkin and Taylor (1979). Figure 4.51 shows a schematic diagram for the down-hole method of seismic wave testing as presented by Larkin and Taylor, which relies on measuring the time interval for *SH*-waves to travel between the ground surface and the subsurface points. A bidirectional impulsive source for the propagation of *SH*-waves is placed on the surface adjacent to a borehole. A horizontal sensitive transducer is located at a depth in the borehole. The depth of the transducer is varied throughout the length of the borehole. The shear wave velocity can then be obtained as

$$v_s = \frac{\Delta z}{\Delta t}$$

(4.89)

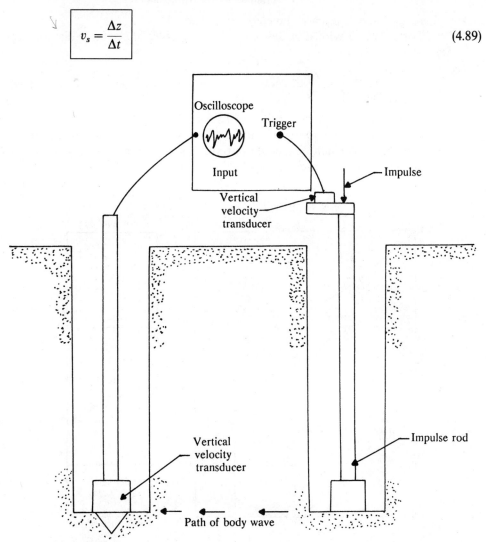

Figure 4.52 Schematic diagram of cross-hole seismic survey technique

where z = depth below the ground surface and t = time of travel of the shear wave from the surface impulsive source to the transducer.

During the process of field investigation, Larkin and Taylor (1979) determined that the shear strains at depths of 3 m and 50 m were about 1×10^{-6} and 0.3×10^{-6}, respectively. In order to compare the field and laboratory values of v_s, some undisturbed samples from various depths were collected. The shear wave velocity of various specimens at a shear strain level of 1×10^{-6} was determined. A comparison of the laboratory and field test results showed that, for similar soils, the value of $v_{s(lab)}$ is considerably lower than that obtained in the field. For *the range of soil tested,*

$$v_{s(lab)} \approx 0.25 v_{s(field)} + 83$$

where $v_{s(lab)}$ and $v_{s(field)}$ are in meters per second.

Larkin and Taylor also defined a quantity called the sample disturbance factor S_D:

$$S_D = \left[\frac{v_{s(field)}}{v_{s(lab)}} \right]^2 = \frac{G_{field}}{G_{lab}} \qquad (4.90)$$

The average value of S_D in Larkin and Taylor's investigation varied from about 1 for $v_{s(field)} = 140$ m/s to about 4 for $v_{s(field)} = 400$ m/s. This shows that small disturbances in the sampling could introduce large errors in the evaluation of representative shear moduli of soils.

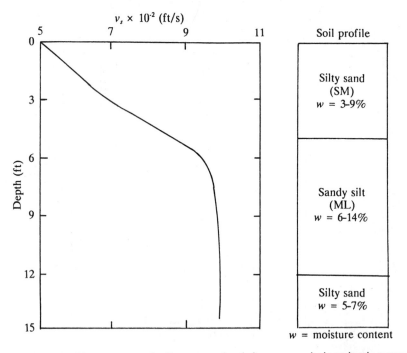

Figure 4.53 Shear wave velocity versus depth from cross-hole seismic survey (redrawn after Stokoe and Woods, 1972)

Cross-Hole Shooting

The technique of cross-hole shooting relies on the measurement of SV-wave velocity. In this procedure of seismic surveying, two vertical boreholes at a given distance apart are advanced into the ground (Figure 4.52). Shear waves are generated by a vertical impact at the bottom of one borehole. The arrival of the body wave is recorded by a vertically sensitive transducer placed at the bottom of another borehole at the same depth. Thus

$$v_s = \frac{L}{t}$$

(4.91)

where t = travel time for the shear wave and L = length between the two boreholes.

Figure 4.53 shows the plot of the shear wave velocity against depth for a test site obtained from the cross-hole shooting technique of seismic surveying.

4.17   Cyclic Plate Load Test

The cyclic field plate load test is similar to the plate bearing test conducted in the field for evaluation of the allowable bearing capacity of soil for foundation design purposes. The plates used for tests in the field are usually made of steel and are 1 in. ($\approx$ 25 mm) thick and 6 in. ($\approx$ 150 mm) to 30 in. ($\approx$ 762 mm) in diameter.

To conduct a test, a hole is excavated to the desired depth. The plate is placed at the center of the hole, and load is applied to the plate in steps—about one-fourth to one-fifth of the estimated ultimate load—by a jack. Each step load is kept constant until the settlement becomes negligible. The final settlement is recorded by dial gauges. Then the load is removed and the plate is allowed to rebound. At the end of the rebounding period, the settlement of the plate is recorded. Following that, the load on the plate is increased to reach a magnitude of the next proposed stage of loading. The process of settlement recording is then repeated.

Figure 4.54 shows the nature of the plot of q versus settlement (s) obtained from a cyclic plate load test. Note that

$$q = \frac{\text{load on the plate, } Q}{\text{area of the plate, } A}$$

Based on field test results, the magnitude of the spring constant k [see Chapter 2, Eq. (2.3)] and the shear modulus G of the soil can be calculated in the following manner.

Spring Constant k

1. Referring to Figure 4.54, calculate the elastic settlement $[s_{e(1)}, s_{e(2)}, \ldots]$ for each loading stage.

2. Plot a graph of q versus s_e, as shown in Figure 4.55.

3. Calculate the spring constant of the plate as

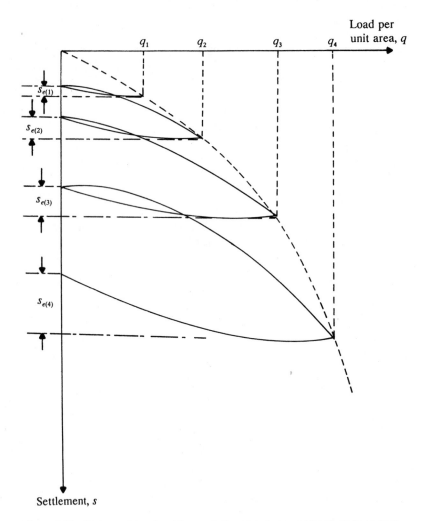

Figure 4.54 Nature of load-settlement diagram for a cyclic plate load test

$$k_{\text{plate}} = \frac{qA}{s_e} \qquad (4.92)$$

4. The spring constant for vertical loading for a proposed foundation can then be extrapolated as follows (Terzaghi, 1955).

Cohesive soil:

$$k_{\text{foundation}} = k_{\text{plate}} \left(\frac{\text{foundation width}}{\text{plate width}} \right) \qquad (4.93)$$

Cohesionless soil:

$$k_{\text{foundation}} = k_{\text{plate}} \left(\frac{\text{foundation width} + \text{plate width}}{2 \times \text{plate width}} \right)^2 \qquad (4.94)$$

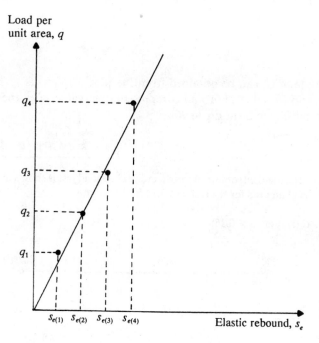

Figure 4.55

Shear Modulus, G

It can be shown theoretically (Barkan, 1962) that

$$C_z = \frac{q}{s_e} = 1.13 \frac{E}{1 - \mu^2} \frac{1}{\sqrt{A}}$$

(4.95)

where

C_z = subgrade modulus

E = modulus of elasticity

μ = Poisson's ratio

A = area of the plate

However,

$$G = \frac{E}{2(1 + \mu)}$$

So

$$C_z = \frac{2.26G(1 + \mu)}{1 - \mu^2} \frac{1}{\sqrt{A}}$$

or

$$G = \frac{(1 - \mu)C_z\sqrt{A}}{2.26} \qquad\qquad (4.96)$$

The magnitude of C_z can be obtained from the plot of q versus s_e (Figure 4.55). With the known value of A and a representative value of μ, the shear modulus can be calculated from Eq. (4.96).

Example 4.5

The plot of q versus s (settlement) obtained from a cyclic plate load test is shown in Figure 4.56. The area of the plate used for the test was 3.14 ft². Calculate

a. k_{plate}, and
b. shear modulus G (assume $\mu = 0.35$).

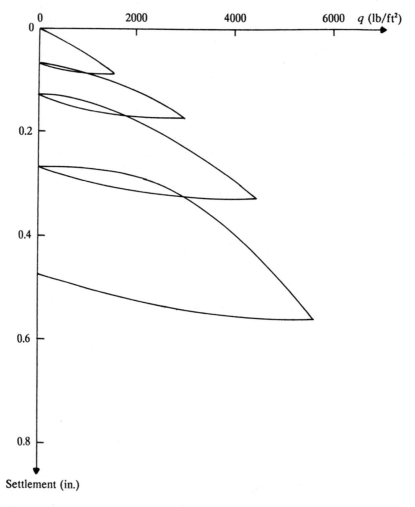

Settlement (in.)

Figure 4.56

Solution

a. From Figure 4.56, the following can be determined.

Load per unit area, q $(\text{lb/ft})^2$	Elastic settlement, s_e (in.)
1500	0.021
3000	0.043
4500	0.059
6000	0.082

Figure 4.57 shows a plot of q versus s_e. From the average plot,

$$C_z = \frac{q}{s_e} = \frac{(5000/144)}{0.069} = 503.2 \text{ lb/in.}^3$$

From Eq. (4.92)

$$k_{\text{plate}} = \frac{qA}{s_e} = (503.2)(3.14 \times 144) = \underline{\underline{22.75 \times 10^4 \text{ lb/in.}}}$$

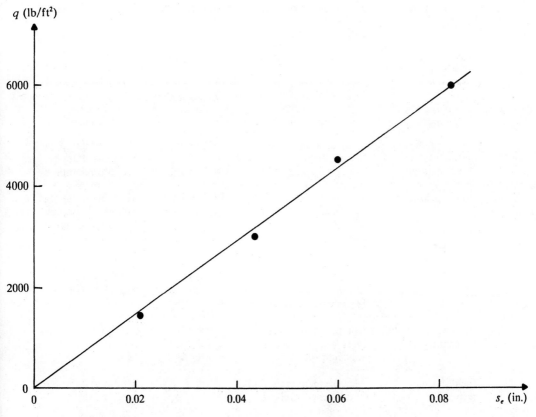

q (lb/ft^2)

s_e (in.)

Figure 4.57

b. From Eq. (4.96),

$$G = \frac{(1 - \mu)C_z\sqrt{A}}{2.26} = \frac{(1 - 0.35)(503.2)(\sqrt{3.14 \times 144})}{2.26}$$

$$\approx \underline{3080 \text{ lb/in.}^2}$$ ∎

CORRELATIONS FOR SHEAR MODULUS AND DAMPING RATIO

4.18   Test Procedures for Measurement of Moduli and Damping Characteristics

For design of machine foundations subjected to vibration, calculation of ground response during an earthquake, analysis of the stability of slopes during an earthquake, and other dynamic analysis of soil, it is required that the shear modulus and the damping ratio of the soil be known. The shear modulus G and the damping ratio D of soil are dependent on several factors, such as type of soil, confining pressure, level of dynamic strain, degree of saturation, frequency, and number of cycles of dynamic load application, magnitude of dynamic stress, and dynamic prestrain (Hardin and Black, 1968).

From the preceding discussions in this chapter, it is obvious that a wide

Table 4.4   Test Procedures for Measuring Moduli and Damping Characteristics (after Seed and Idriss, 1970)

General procedure	Test condition	Approximate strain range	Properties determined
Determination of hysteretic stress-strain relationships	Triaxial compression	10^{-2} to 5%	Modulus; damping
	Simple shear	10^{-2} to 5%	Modulus; damping
	Torsional shear	10^{-2} to 5%	Modulus; damping
Forced vibration	Longitudinal vibrations	10^{-4} to $10^{-2}\%$	Modulus; damping
	Torsional vibrations	10^{-4} to $10^{-2}\%$	Modulus; damping
	Shear vibrations—lab	10^{-4} to $10^{-2}\%$	Modulus; damping
	Shear vibrations—field	10^{-4} to $10^{-2}\%$	Modulus
Free vibration tests	Longitudinal vibrations	10^{-3} to 1%	Modulus; damping
	Torsional vibrations	10^{-3} to 1%	Modulus; damping
	Shear vibrations—lab	10^{-3} to 1%	Modulus; damping
	Shear vibrations—field	10^{-3} to 1%	Modulus
Field wave velocity measurements	Compression waves	$\approx 5 \times 10^{-4}\%$	Modulus
	Shear waves	$\approx 5 \times 10^{-4}\%$	Modulus
	Rayleigh waves	$\approx 5 \times 10^{-4}\%$	Modulus
Field seismic response	Measurement of motions at difference levels in deposit		Modulus; damping

variety of procedures, including laboratory and field tests, can be used to obtain the shear moduli and damping characteristics of soils. A summary of those test conditions, range of applicability, and the parameters obtained are given in Table 4.4. Based on these studies several correlations for estimation of G and D have evolved during the last 25 to 30 years. Some of these correlations are summarized in the following sections.

In general, the shear-stress-versus-shear-strain relationship for soils will be of the nature as shown in Figure 4.58. The following can be seen from this figure:

1. The shear modulus G decreases with the increased level of shear strain.

2. At a very low strain level, the magnitude of the shear modulus is maximum (that is, $G = G_{max}$).

3. The shear-stress-versus-shear-strain relationship shown in Figure 4.58 can be approximated as (Hardin and Drnevich, 1972)

$$\tau = \frac{\gamma'}{1/G_{max} + \gamma'/\tau_{max}} \qquad (4.97)$$

where τ = shear stress and γ' = shear strain.

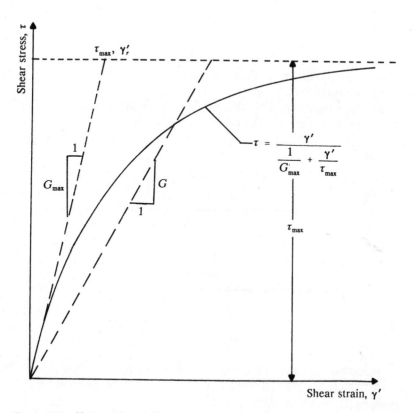

Figure 4.58   Nature of variation of shear modulus with strain

4.19   Shear Modulus and Damping Ratio in Sand

Hardin and Richart (1963) reported the results of several resonant column tests conducted in dry Ottawa sand. The shear wave velocities v_s determined from some of these tests are shown in Figure 4.15. The peak-to-peak shear strain amplitude for these tests was 10^{-3} rad. From Figure 4.15 it may be seen that the values of v_s are independent of the gradation, grain-size distribution, and also the relative density of compaction. However, v_s is dependent on the void ratio e and the effective confining pressure $\bar{\sigma}_0$ and can be expressed by the following empirical relations:

$$v_s = (170 - 78.2e)\bar{\sigma}_0^{1/4} \qquad \text{for } \bar{\sigma}_0 \geq 2000 \text{ lb/ft}^2 \tag{4.98}$$

and

$$v_s = (119 - 56e)\bar{\sigma}_0^{0.3} \qquad \text{for } \bar{\sigma}_0 < 2000 \text{ lb/ft}^2 \tag{4.99}$$

where e = void ratio and v_s and $\bar{\sigma}_0$ are in feet per second and pounds per square foot, respectively.

In SI units, the relations may be expressed as

$$v_s = (19.7 - 9.06e)\bar{\sigma}_0^{1/4} \qquad \text{for } \bar{\sigma}_0 \geq 95.8 \text{ kN/m}^2 \tag{4.100}$$

and

$$v_s = (11.36 - 5.35e)\bar{\sigma}_0^{0.3} \quad \text{for } \bar{\sigma}_0 < 95.8 \text{ kN/m}^2 \tag{4.101}$$

In Equations (4.100) and (4.101), the units of v_s and $\bar{\sigma}_0$ are meters per second and newtons per square meter, respectively.

Several experimental results for shear wave velocity in extremely *angular crushed quartz sands* were also reported by Hardin and Richart (1963). Based on these results, the value of v_s for *angular sands* can be expressed by the empirical relation

$$\underset{\substack{\uparrow \\ \text{(ft/s)}}}{v_s} = (159 - 53.5e) \ \underset{\substack{\uparrow \\ \text{(lb/ft}^2)}}{\bar{\sigma}_0^{1/4}} \tag{4.102}$$

In SI units,

$$\underset{\substack{\uparrow \\ \text{(m/s)}}}{v_s} = (18.43 - 6.2e) \ \underset{\substack{\uparrow \\ \text{(N/m}^2)}}{\bar{\sigma}_0^{1/4}} \tag{4.103}$$

Based on the shear wave velocity relations presented here, the shear modulus of sands for *low amplitudes of vibration* can be given by the following relations (Hardin and Black, 1968):

$$G_{max} = \frac{2630(2.17 - e)^2}{1 + e} \bar{\sigma}_0^{1/2} \quad \text{(for round-grained sands)} \qquad (4.104)$$

and

$$G_{max} = \frac{1230(2.97 - e)^2}{1 + e} \bar{\sigma}_0^{1/2} \quad \text{(for angular-grained sands)} \qquad (4.105)$$

The units of G_{max} and $\bar{\sigma}_0$ in Equations (4.104) and (4.105) are pounds per square inch. In SI units (kilonewtons per square meter) these relations can be expressed as

$$G_{max} = \frac{6908(2.17 - e)^2}{1 + e} \bar{\sigma}_0^{1/2} \quad \text{(round-grained)} \qquad (4.106)$$

and

$$G_{max} = \frac{3230(2.97 - e)^2}{1 + e} \bar{\sigma}_0^{1/2} \quad \text{(angular-grained)} \qquad (4.107)$$

For a soil specimen subjected to a stress condition such that $\bar{\sigma}_1 \neq \bar{\sigma}_2 \neq \bar{\sigma}_3$ (where $\bar{\sigma}_1$, $\bar{\sigma}_2$, and $\bar{\sigma}_3$ are the major, intermediate, and minor effective principal stresses, respectively), note that the average effective confining pressure is

$$\bar{\sigma}_0 = \tfrac{1}{3}(\bar{\sigma}_1 + \bar{\sigma}_2 + \bar{\sigma}_3) = \text{effective octahedral stress}$$

This value of $\bar{\sigma}_0$ can be used in Eqs. (4.98)–(4.107).

For field conditions at any given depth,

$$\bar{\sigma}_1 = \text{effective vertical stress} = \bar{\sigma}_v$$

$$\bar{\sigma}_2 = \bar{\sigma}_3 = K_0 \bar{\sigma}_v$$

where $K_0 = $ at-rest earth pressure coefficient $\approx 1 - \sin\phi$ (where $\phi = $ drained friction angle). So

$$\bar{\sigma}_0 = \tfrac{1}{3}[\bar{\sigma}_v + 2\bar{\sigma}_v(1 - \sin\phi)]$$

$$= \frac{\bar{\sigma}_v}{3}(3 - 2\sin\phi) \qquad (4.108)$$

Several investigators (e.g., Weissman and Hart, 1961; Richart, Hall, and Lysmer (1962); Drnevich, Hall, and Richart, 1966; Silver and Seed, 1969; Hardin and Drnevich, 1972; Seed and Idriss, 1970; Shibata and Soelarno, 1975; and Iwasaki, Tatsuoka, and Takagi, 1976) have reported the results of shear modulus and damping ratio measurements using various types of test techniques. From these test results it appears that the shear modulus at a given strain level can be expressed as (Seed and Idriss, 1970)

$$G = 1000K_2(\bar{\sigma}_0)^{0.5}$$

(4.109)

where G and $\bar{\sigma}_0$ are in pounds per square foot.

For low strain amplitudes ($\gamma' \leq 10^{-4}\%$), the preceding equation will be

$$G_{max} = 1000K_{2(max)}(\bar{\sigma}_0)^{0.5}$$

(4.110)

The magnitudes of $K_{2(max)}$ vary from about 30 for loose sands to about 75 for dense sands. Seed and Idriss (1970) recommended the following values of $K_{2(max)}$.

Relative density, R_D (%)	$K_{2(max)}$
30	34
40	40
45	43
60	52
75	61
90	70

Hence,

$$\frac{G}{G_{max}} = \frac{K_2}{K_{2(max)}} = F'$$

(4.111)

Figure 4.59 shows the variation of F' with shear strain γ' (%) obtained from several studies. These values fall in a rather narrow band and, for all practical purposes, the average plot can be used for design and estimation purposes. Thus Eqs. (4.104), (4.105), (4.109), (4.110), and (4.111) can be combined to estimate the shear modulus at any required shear strain level.

Studies by Hardin and Drnevich (1972) and Seed and Idriss (1970) show that the damping ratios for sands are affected by factors such as (a) grain-size characteristics, (b) degree of saturation, (c) void ratio, (d) earth pressure coefficient at rest (K_0), (e) angle of internal friction (ϕ), (f) number of stress cycles (N), (g) level of strain, and (h) effective confining pressure. The last two factors, however, have the major effect on the magnitude of the damping ratio. Figure 4.60 shows a compilation of past studies (Seed et al. 1986) to determine D. For most practical cases the average plot of the variation of D versus γ' can be used for most calculation purposes.

Based on tests on dry sands using a torsional simple shear device, Sherif, Ishibashi, and Gaddah (1977) proposed the following relationship for damping ratio.

$$D = \frac{50 - 0.6\bar{\sigma}_0}{38}(73.3F - 53.3)(\gamma')^{0.3}(1.01 - 0.046\log N)$$

(4.112)

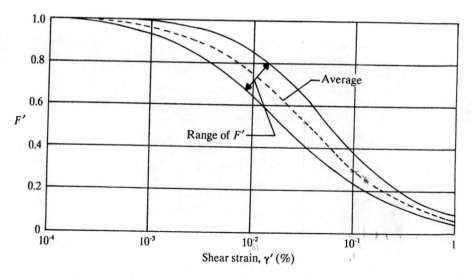

Figure 4.59   Variation of F' with shear strain for sands (after Seed et al., 1986)

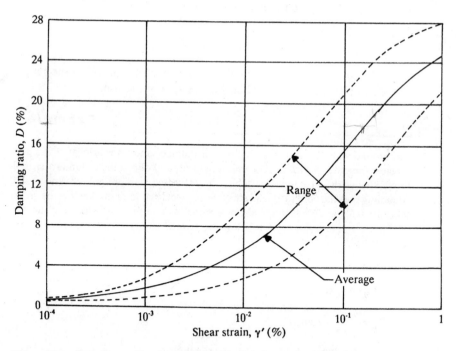

Figure 4.60   Damping ratios for sands (after Seed et al., 1986)

where

D = damping ratio (%)

$\bar{\sigma}_0$ = effective confining pressure (lb/in.^{2})

γ' = shear strain (%)

F = sphericity factor of the soil grains

N = number of cycles of strain application

The sphericity factor is defined as

$$F = \frac{1}{\psi^2 C_g} \qquad (4.113)$$

where

$$C_g = \frac{D_{30}^2}{(D_{10})(D_{60})} \qquad (4.114)$$

D_{10}, D_{30}, D_{60} = diameters, respectively, through which 10%, 30%, and 60% of the soil will pass

$$\psi = \frac{S'}{S}$$

where S' and S are, respectively, the surface area of a sphere of the same volume as the soil particle and the actual surface area of the soil.

Example 4.6

The ground water table in a normally consolidated sand layer is located at a depth of 10 ft below the ground surface. The unit weight of sand above the groundwater table is 100 lb/ft^3. Below the groundwater table, the saturated unit weight of sand is 120 lb/ft^3. Assuming that the void ratio and effective angle of friction of sand below the groundwater table are 0.6 and 36°, respectively, determine the damping ratio and the shear modulus of this sand at a depth of 25 ft below the ground surface if the strain is expected to be about 0.12%.

Solution

From Eq. (4.108)

$$\bar{\sigma}_0 = \frac{\bar{\sigma}_v}{3}(3 - 2\sin\phi)$$

$$\bar{\sigma}_v = 10(100) + 15(120 - 62.4) = 1864 \text{ lb/ft}^2$$

$$\bar{\sigma}_0 = \frac{1864}{3}[3 - (2)(\sin 36)] = 1133.6 \text{ lb/ft}^2$$

When ϕ is equal to 36°, R_D is about 40 to 50%. Assuming $R_D \approx 45\%$, $K_{2(\max)} \approx 43$. So, from Eq. (4.110)

$$G_{\max} = 1000 K_{2(\max)}(\bar{\sigma}_0)^{0.5}$$

or

$$G_{max} = (1000)(43)(1133.6)^{0.5} = 1,447,767 \text{ lb/ft}^2$$

$$\approx 10,054 \text{ lb/in.}^2$$

Referring to Figure 4.59, for $\gamma' = 0.12\%$, the value of F' is about 0.28. So

$$G = F'G_{max} = (0.28)(10,054) \approx \underline{2815 \text{ lb/in.}^2}$$

Referring to the average curve in Figure 4.60, for $\gamma' = 0.12\%$,

$$D \approx \underline{17\%}$$ ■

4.20   Correlation of G_{max} of Sand with Standard Penetration Resistance

The standard penetration test is used in soil-exploration programs in the United States and other countries. In granular soils the standard penetration numbers (N in blows/foot) are widely used for the design of foundations. The standard penetration number can be correlated (Seed et al., 1986) in the following form to predict the maximum shear modulus:

$$G_{max} \approx 35 \times 1000 N_{60}^{0.34}(\bar{\sigma}_0)^{0.4}$$ (4.115)

$$\text{(lb/ft}^2)\qquad\qquad\qquad\text{(lb/ft}^2)$$

where

$\bar{\sigma}_v$ = effective vertical stress (lb/ft^2)

N_{60} = N-value measured in SPT test delivering 60% of the theoretical free-fall energy to the drill rod

Equation (4.115) is very useful in predicting the variation of the maximum shear modulus with depth for a granular soil deposit.

4.21   Shear Modulus and Damping Ratio for Gravels

Seed et al. (1986) provided the experimental results of several well-graded gravels. An example of such a study on well-graded Oroville material is shown in Figure 4.61. Based on several studies of this type, Seed et al. concluded that Eqs. (4.110) and (4.111) can also be used to predict the variation of shear modulus with shear strain. However, the magnitude of $K_{2(max)}$ for gravels ranges between 80 to 180 (as compared to a range of 30 to 75 for sand). Thus,

$$G = G_{max}F' = 1000F'K_{2(max)}(\bar{\sigma}_0)^{0.5}$$ (4.116)

$$\text{(lb/ft}^2)\qquad\qquad\qquad\text{(lb/ft}^2)$$

The variation of F' with the level of shear strain is shown in Figure 4.62.

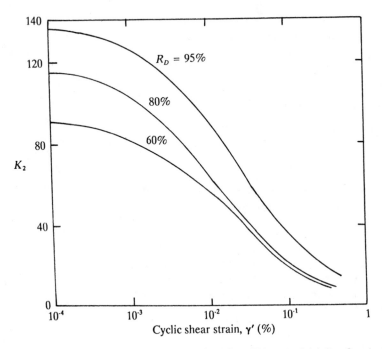

Figure 4.61   Shear moduli of well-graded Oroville material (after Seed et al., 1986)

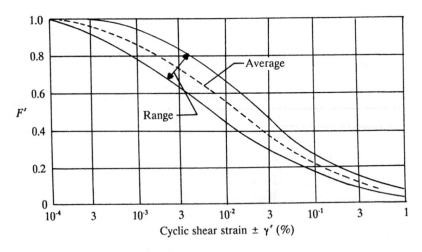

Figure 4.62   Variation of F' with shear strain for gravelly soils (after Seed et al., 1986)

The equivalent damping ratio of gravelly soils determined in the laboratory from the hysteresis loops at the fifth cycle of each strain amplitude is shown in Figure 4.63. It can be seen that, for a given value of γ', the equivalent damping ratio increases with the increase of the relative density R_D of the gravel. Seed et al. (1986) also observed that

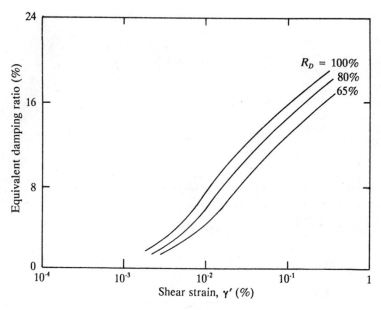

Figure 4.63 Effect of relative density on damping ratio of gravelly soils (after Seed et al., 1986)

a. there is not significant effect of gradation on the equivalent damping ratios of gravelly soil, and

b. the damping ratio is not significantly affected by the number of cycles at very small strain amplitudes. However it decreases to approximately three-fourths of its original value after 60 cycles at any axial strain amplitude of $\pm 0.2\%$.

Seed et al. showed that the range and the average plot of the damping ratio D with strain amplitude γ' for gravelly soils is approximately the same as that for sands (Figure 4.60).

4.22   Shear Modulus and Damping Ratio for Clays

For low amplitudes of strain, the shear modulus $G = G_{max}$ for clays of moderate sensitivity can be expressed in a modified form of Eq. (4.105) (Hardin and Drnevich, 1972):

$$G_{max} = \frac{1230(2.97 - e)^2}{1 + e}(OCR)^K \bar{\sigma}_0^{1/2} \qquad (4.117)$$

LOW PLASTIC, STIFF

PSI

where OCR = overconsolidation ratio and K = a constant = f(plasticity index, PI).

Following are the recommended values of K for use in the preceding equation.

Plasticity index, PI (%)	K
0	0
20	0.18
40	0.30
60	0.41
80	0.48
≥ 100	0.5

In SI units, Equation (4.117) can be written as

$$G_{max} = \frac{3230(2.97 - e)^2}{1 + e}(OCR)^K \bar{\sigma}_0^{1/2} \qquad (4.118)$$

$$\underset{(kN/m^2)}{\uparrow} \qquad\qquad \underset{(kN/m^2)}{\uparrow}$$

For field conditions

$$\bar{\sigma}_0 = \tfrac{1}{3}(\bar{\sigma}_v + 2K_0\bar{\sigma}_v) \qquad (4.119)$$

where $\bar{\sigma}_v$ = effective vertical stress
K_0 = at-rest earth pressure coefficient
For normally consolidated clays (Booker and Ireland, 1965)

$$K_0 = 0.4 + 0.007(PI) \qquad \text{(for } 0 \le PI \le 40\%) \qquad (4.120)$$

and

$$K_0 = 0.68 + 0.001(PI - 40) \qquad \text{(for } 40\% \le PI \le 80\%) \qquad (4.121)$$

In order to estimate the shear modulus at larger shear strain levels, Hardin and Drnevich (1972) suggested the following procedure. Referring to Figure 4.58,

$$G = \frac{\tau}{\gamma'}$$

and

$$G_{max} = \frac{\tau_{max}}{\gamma'_r}$$

where γ'_r = reference strain.
Substituting the preceding relationship into Eq. (4.97), one obtains

$$G = \frac{G_{max}}{1 + \gamma'/\gamma'_r} \qquad (4.122)$$

For real soils, the stress-strain relationship deviates somewhat from Eq. (4.122), and it can be modified as

$$G = \frac{G_{max}}{1 + \gamma'_h} \qquad (4.123)$$

where γ'_h = hyperbolic strain

$$= \left(\frac{\gamma'}{\gamma'_r}\right)[1 + ae^{-b(\gamma'/\gamma'_r)}] \qquad (4.124)$$

where, for saturated cohesive soils,

$$a = 1 + 0.25 \log N \qquad (4.125)$$

$$b = 1.3 \qquad (4.126)$$

N = number of cycles of loading

Figure 4.64 gives the variation of $\gamma'_r/(\bar{\sigma}_v)^{1/2}$ with the plasticity index for saturated cohesive soils. Once the magnitudes of γ'_h and G_{max} are calculated from Eqs. (4.124) and (4.117) [or (4.118)], they can be substituted in Eq. (4.123) to obtain G (at a strain level γ').

Hardin and Drnevich (1972) presented the relationship between the damping ratio and the shear modulus as

$$D = D_{max}\left(1 - \frac{G}{G_{max}}\right) \qquad (4.127)$$

where D_{max} is the maximum damping ratio, which occurs when $G = 0$. For saturated cohesive soils,

$$D_{max}(\%) = 31 - (3 + 0.03f)\bar{\sigma}_0^{1/2} + 1.5f^{1/2} - 1.5(\log N) \qquad (4.128)$$

where f = frequency (in cps)

$\bar{\sigma}_0$ = effective confining pressure (in kg/cm^2)

N = number of cycles of loading

For real soils, Eq. (4.127) can be rewritten as

$$\frac{D}{D_{max}} = \frac{\gamma''_h}{1 + \gamma''_h} \qquad (4.129)$$

where γ''_h = the hyperbolic strain, or

$$\gamma''_h = \left(\frac{\gamma'}{\gamma'_r}\right)[1 + a_1 e^{-b_1(\gamma'/\gamma'_r)}] \qquad (4.130)$$

$$a_1 = 1 + 0.2f^{1/2} \qquad (4.131)$$

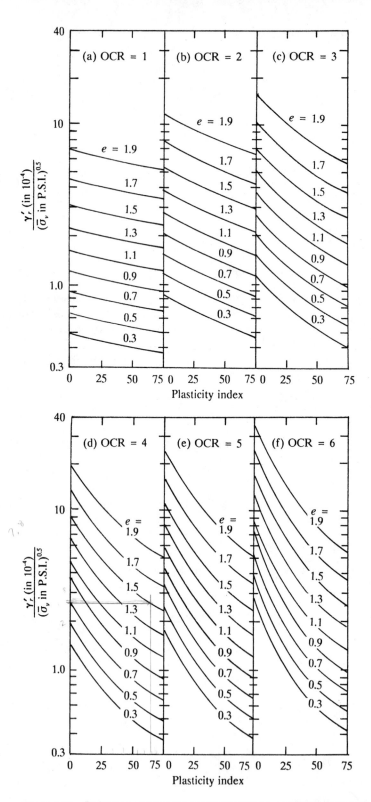

Figure 4.64 Reference strain for geostatic stress condition (after Hardin and Drnevich, 1972)

and

$$b_1 = 0.2f(e^{-\bar{\sigma}_0}) + 2.25\bar{\sigma}_0 + 0.3\log N \tag{4.132}$$

In the preceding two equations,

  f = frequency (in cps)

  $\bar{\sigma}_0$ = effective confining pressure (in kg/cm^2)

  N = number of cycles of loading

Hence, in order to calculate the damping ratio D at a strain level γ', the following procedure may be used.

  1. Calculate D_{max} using Eq. (4.128).
  2. Calculate γ_h'' using Eqs. (4.130), (4.131), and (4.132).
  3. Calculate D using Eq. (4.129).

Correlation of Seed and Idriss

Seed and Idriss (1970) collected the experimental results for shear modulus and damping ratio from various sources for saturated cohesive soils. Based on these results the variation of G/c_u (where c_u = undrained cohesion) with shear strain is shown in Figure 4.65. Also, Figure 4.66 shows the upper limit, average, and lower limit for the damping ratio at various strain levels.

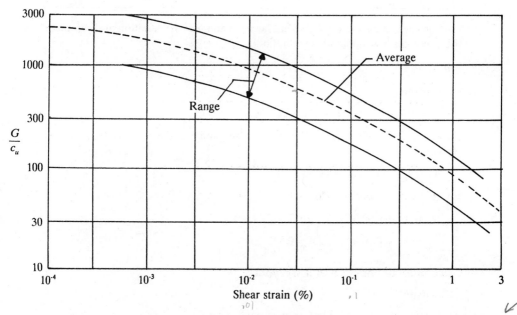

Figure 4.65   *In situ* shear modulus for saturated clays (after Seed and Idriss, 1970)

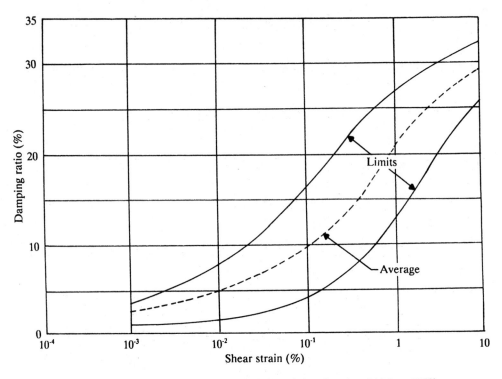

Figure 4.66   Damping ratio for saturated clays (after Seed and Idriss, 1970)

Example 4.7
_____

A soil profile is shown in Figure 4.67a. Calculate and plot the variation of shear modulus with depth (for low amplitude of vibration).

Solution

At any depth z

$$\bar{\sigma}_0 = \tfrac{1}{3}(\bar{\sigma}_1 + \bar{\sigma}_2 + \bar{\sigma}_3) \qquad \bar{\sigma}_2 = \bar{\sigma}_3 = K_0\bar{\sigma}_1$$

where K_0 is the coefficient of earth pressure at rest and $\bar{\sigma}_1$ is the vertical effective pressure. For sands,

$$K_0 = 1 - \sin\phi = 1 - \sin 30° = 0.5$$

For normally consolidated clays,

$$K_0 = 0.4 + 0.007(\text{PI}) \qquad \text{for } 0 \le \text{PI} \le 40\%$$

$$= 0.4 + 0.007(48 - 23) = 0.575$$

Calculation of Effective Unit Weights

$z = 0$–10 ft:

$$\gamma_{\text{dry}} = \gamma_{\text{eff}} = \frac{G_s\gamma_w}{1 + e} = \frac{(2.65)(62.4)}{1.7} = 97.27 \text{ lb/ft}^3$$

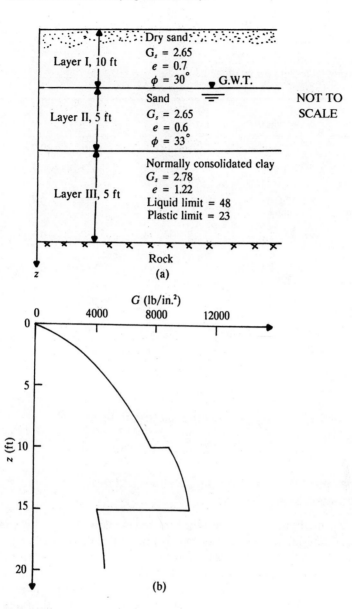

Figure 4.67

$z = 10$–15 ft:

$$\gamma_{\text{eff}} = \gamma_{\text{sat}} - \gamma_w = \frac{(G_s + e)\gamma_w}{1 + e} - \gamma_w = \left(\frac{G_s - 1}{1 + e}\right)\gamma_w$$

$$= \frac{(2.65 - 1)(62.4)}{1.6} = 64.35 \text{ lb/ft}^3$$

$z = 15$–20 ft:

$$\gamma_{eff} = \frac{(G_s - 1)\gamma_w}{1 + e} = \frac{(2.78 - 1)(62.4)}{2.22} = 50.03 \text{ lb/ft}^3$$

The following table can now be prepared.

Depth z (ft)	$\bar{\sigma}_1$ (lb/ft^2)	$\bar{\sigma}_2 = \bar{\sigma}_3 = K_0\bar{\sigma}_1$ (lb/ft^2)	$\bar{\sigma}_0$ (lb/ft^2)	e	$G = G_{max}$ (lb/ft^2)
0	0	0	0	0.7	0
5	(97.27)(5) = 486.35	243.18	324.24	0.7	5595[a]
10 (in layer I)	(97.27)(10) = 972.7	486.35	648.47	0.7	7912[a]
10 (in layer II)	972.7	442.58	619.29	0.6	8955[a]
15 (in layer II)	972.7 + (64.35)(5) = 1294.45	588.97	824.04	0.6	10329[a]
15 (in layer III)	1294.45	744.30	927.69	1.22	4306[b]
20	1294.45 + (50.03)(5) = 1544.6	888.15	1106.96	1.22	4704[b]

[a] Eq. (4.105)
[b] Eq. (4.117)   ■

The variation of $G = G_{max}$ with depth is plotted in Figure (4.67b).

4.23   Shear Modulus and Damping Ratio for Lightly Cemented Sand

Lightly cemented sand deposits are encountered in many parts of the world. The cementing material in the sand deposits is primarily calcium carbonate. More recently, the results of several research projects relating to the properties of lightly cemented sands have been published. From these studies it appears that the behavior of lightly cemented sands can be duplicated in the laboratory by mixing sand and Portland cement in required proportions. The maximum shear modulus can be expressed as (Saxena, Avramidis, and Reddy, 1988)

$$G_{max(CS)} = G_{max(S)} + \Delta G_{max(C)} \tag{4.133}$$

where

$G_{max(CS)}$ = maximum shear modulus of lightly cemented sand

$G_{max(S)}$ = maximum shear modulus of sand alone

$\Delta G_{max(C)}$ = increase of maximum shear modulus due to cementation effect

According to Saxena, Avramidis, and Reddy, the magnitudes of $G_{max(S)}$ and $\Delta G_{max(C)}$ can be obtained from the following empirical relationships.

$$G_{max(S)} = \frac{428.2}{0.3 + 0.7e^2}(P_a)^{0.426}(\bar{\sigma}_0)^{0.574}$$

(kN/m^2) $\qquad\qquad$ $(kN/m^2)(kN/m^2)$

(4.134)

where P_a = atmospheric pressure in the same units as $G_{max(S)}$.

$$\frac{\Delta G_{max(C)}}{P_a} = \frac{172}{(e - 0.5168)}(CC)^{0.88}\left(\frac{\bar{\sigma}_0}{P_a}\right)^{0.515e-0.13CC+0.285}$$
(for CC < 2%)

(4.135)

$$\frac{\Delta G_{max(C)}}{P_a} = \frac{773}{e}(CC)^{1.2}\left(\frac{\bar{\sigma}_0}{P_a}\right)^{0.698e-0.04CC-0.2}$$
(for 2% ≤ CC ≤ 8%)

(4.136)

where CC = cement content (in percent) and e = void ratio. When using Eqs. (4.135) and (4.136), the units of $G_{max(S)}$, P_a, and $\bar{\sigma}_0$ need to be consistent.

The damping ratio at low strain amplitudes ($\gamma' \leq 10^{-3}\%$) can be expressed as (Saxena, Avramidis, and Reddy, 1988)

$$D_{CS} = D_S + \Delta D_C \tag{4.137}$$

where

D_{CS} = damping ratio of cemented sand (%)

D_S = damping ratio of sand alone (%)

ΔD_C = increase in the damping ratio due to cementation effect

$$D_S = 0.94\left(\frac{\bar{\sigma}_0}{P_a}\right)^{-0.38}$$

(4.138)

$$\Delta D_C = 0.49(CC)^{1.07}\left(\frac{\bar{\sigma}_0}{P_a}\right)^{-0.36}$$

(4.139)

where CC = cement content (in percent). The units of P_a and $\bar{\sigma}_0$ need to be consistent.

Example 4.8

If a lightly cemented sand specimen is subjected to an effective confining pressure of 98 kN/m^2, estimate the value of $G_{max(CS)}$ given $e = 0.7$ and CC = 3%.

Solution

From Eq. (4.134),

$$G_{max(S)} = \frac{428.2}{0.3 + 0.7e^2}(P_a)^{0.426}(\bar{\sigma}_0)^{0.574}$$

Given $e = 0.7$, $P_a = 101.43$ kN/m^2, and $\bar{\sigma}_0 = 98$ kN/m^2,

$$G_{max(S)} = \frac{428.2}{0.3 + (0.7)(0.7)^2}(101.43)^{0.426}(98)^{0.574}$$

$$= 66{,}267 \text{ kN/m}^2 = 0.066 \text{ GN/m}^2$$

From Eq. (4.136),

$$\frac{\Delta G_{max(C)}}{P_a} = \frac{773}{e}(CC)^{1.2}\left(\frac{\bar{\sigma}_0}{P_a}\right)^{0.698e-0.04CC-0.2}$$

or

$$\frac{\Delta G_{max(C)}}{101.43} = \frac{773}{0.7}(3)^{1.2}\left(\frac{98}{101.43}\right)^{[0.698(0.7)-0.04(3)-0.2]}$$

$$= (1104.3)(3.737)(0.994)$$

$$= 416{,}067 \text{ kN/m}^2 = 0.416 \text{ GN/m}^2$$

So

$$G_{max(CS)} = G_{max(S)} + \Delta G_{max(C)}$$

$$= 0.066 + 0.416 = \underline{0.482 \text{ GN/m}^2} \qquad \blacksquare$$

PROBLEMS

4.1   A uniformly graded dry sand specimen was tested in a resonant column device. The shear wave velocity v_s determined by torsional vibration of the specimen was 760 ft/s. The longitudinal wave velocity determined by using a similar specimen was 1271 ft/s. Determine each of the following.
a. Poisson's ratio
b. Modulus of elasticity (E) and shear modulus (G) if the void ratio and the specific gravity of soil solids of the specimen were 0.5 and 2.65, respectively

4.2   A clayey soil specimen was tested in a resonant column device (torsional vibration; free–free end condition) for determination of shear modulus. Given: length of specimen = 90 mm, diameter of specimen = 35.6 mm, mass of specimen = 170 g, frequency at normal mode of vibration ($n = 1$) = 790 cycles/s. Determine the shear modulus of the specimen in kilonewtons per square meter.

4.3   The Poisson's ratio for the clay specimen described in Problem 4.2 is 0.52. If a similar specimen is vibrated longitudinally in a resonant column device (free–free end condition), what would be its frequency at normal mode of vibration ($n = 1$)?

4.4   Following are the results of a refraction survey. Assuming that the soil layers are horizontal, determine the P-wave velocities in the soil layers and the thickness of the top layer.

Distance (ft)	Time of first arrival (ms)
25	49.08
50	81.96
75	122.8
100	148.2
150	174.2
200	202.8
250	228.6
300	256.7

4.5 Repeat Problem 4.4 for the following:

Distance (m)	Time of first arrival (ms)	Distance (m)	Time of first arrival (ms)
10	19.23	100	125.82
20	38.40	150	138.72
30	57.71	200	152.61
40	76.90	250	166.81
60	115.40	300	178.31
80	120.71		

Comment regarding the material encountered in the second layer.

4.6 Repeat Problem 4.4 with the following results. Also determine the thickness of the second layer of soil encountered.

Distance (m)	Time of first arrival (ms)	Distance (m)	Time of first arrival (ms)
10	41.66	60	119.21
15	62.51	70	128.11
20	83.37	80	136.22
30	91.82	90	141.00
40	101.22	100	143.81
50	110.16	120	152.00

4.7 Refer to Figure 4.43 for the results of the following refraction survey:

Distance from point of disturbance, A (ft)	Time of first arrival (ms)	Distance from point of disturbance, E (ft)	Time of first arrival (ms)
0	0	0	0
20	20	20	20
40	40	40	40.1
60	60	60	59.8
80	78.2	80	79.7
120	92.8	120	121.0
200	122.2	200	167.2
280	149.8	280	175.1
Point E 360	177.9	Point A 360	180.2

Determine:
a. the P-wave velocities in the two layers,
b. z' and z'', and
c. the angle β.

4.8 The results of a reflection survey are given here. Determine the velocity of P-waves in the top layer and its thickness.

Distance from shot point (m)	Time for first arrival of reflected wave (ms)
10	32.5
20	39.05
30	48.02
40	58.3
60	80.78
100	128.55

4.9 For a reflection survey refer to Figure 4.48, in which A is the shot point. Distance $AC = AE = 180$ m. The times for arrival of the first reflected wave at points C and E are 45.0 ms and 64.1 ms, respectively. If the P-wave velocity in layer 1 is 280 m/s, determine β and z'.

4.10 The results of a subsoil exploration by steady-state vibration technique are given here (Section 4.15).

Distance from the plate vibrated x (m)	Number of waves per second	Frequency of vibration of the plate (cps)
10	41	900
10	18	400
10	9	200
10	4.55	100
10	2.65	90
10	2.3	75
10	1.77	60
10	1.47	50

Make necessary calculations and plot the variation of the wave velocity with depth.

4.11 An angular-grained sand has maximum and minimum void ratios of 1.1 and 0.55, respectively. Using Eq. (4.107), determine and plot the variation of maximum shear modulus G_{max} versus relative density ($R_D = 0–100\%$) for mean confining pressures of 50, 100, 150, 200 and 300 kN/m^2.

4.12 A 20-m-thick sand layer in the field is underlain by rock. The groundwater table is located at a depth of 5 m measured from the ground surface. Determine the maximum shear modulus of this sand at a depth of 10 m below the ground surface. Given: void ratio = 0.6, specific gravity of soil solids = 2.68, angle of friction of sand = 36°. Assume the sand to be round-grained.

4.13 For a deposit of sand, at a certain depth in the field the effective vertical pressure is 120 kN/m^2. The void ratio and the relative density are 0.72 and 30° respectively. Determine the shear modulus and damping ratio for a shear strain level of $5 \times 10^{-2}\%$.

4.14 A remolded clay specimen was consolidated by a hydrostatic pressure of 30 lb/in^2. The specimen was then allowed to swell under a hydrostatic pressure of 15 lb/in^2. The void ratio at the end of swelling was 0.8. If this clay is subjected to a torsional vibration in a resonant column test, what would be its maximum shear modulus (G_{max})? The liquid and plastic limits of the clay are 58 and 28, respectively.

4.15   Refer to Figure P4.15. Given:

$$H_1 = 2\,m \qquad\qquad G_{s(1)} = 2.68$$

$$H_2 = 8\,m \qquad\qquad G_{s(2)} = 2.65$$

$$H_3 = 3\,m \qquad\qquad \phi_1 = 35°$$

$$e_1 = 0.6 \qquad\qquad \phi_2 = 30°$$

$$e_2 = 0.7 \qquad \text{PI of clay} = 32$$

Estimate and plot the variation of the maximum shear modulus (G_{max}) with depth for the soil profile.

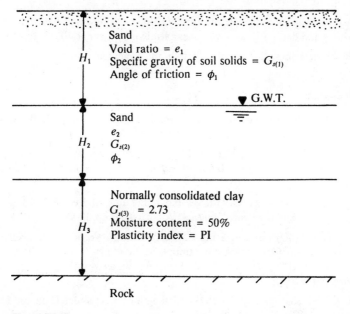

Figure P4.15

4.16   Repeat Problem 4.15 given

$$H_1 = H_2 = H_3 = 20\,ft \qquad\qquad G_{s(1)} = G_{s(2)} = 2.66$$

$$e_1 = 0.8 \qquad\qquad\qquad \phi_1 = 28°$$

$$e_2 = 0.68 \qquad\qquad\qquad \phi_2 = 32°$$

$$\text{PI of clay} = 20$$

4.17   A layer of clay deposit extends to a depth of 50 ft below the ground surface. The groundwater table coincides with the ground surface. Given, for the clay: void ratio = 1.0, specific gravity of soil solids = 2.78, plasticity index = 25%, overconsolidation ratio = 2. Determine the shear modulus and damping ratio of this clay at a depth of 25 ft for the fifth cycle at a strain level of 0.1%, assuming that the frequency (f) is about 1 cps. (*Note*: Use Figure 4.64.) Given:

$$K_{0(\text{overconsol.})} \approx K_{0(\text{norm consol.})}(\sqrt{OCR})$$

4.18   The unit weight of a sand deposit is 108 lb/ft^3 at a relative density of 60%. Assume that, for this sand

$$\phi = 30 + 0.15R_D$$

where ϕ is the drained friction angle and R_D is the relative density (in percent). At a depth of 20 ft below the ground surface, estimate its shear modulus and damping ratio at a shear strain level of 0.01%. Use the equation proposed by Seed and Idriss (1970).

4.19   The results of a standard unconsolidated undrained triaxial test on an undistributed saturated clay specimen are as follows:

$$\text{Confining pressure} = 70 \, \text{kN/m}^2$$

$$\text{Total axial stress at failure} = 166.6 \, \text{kN/m}^2$$

Using the method proposed by Seed and Idriss (1970), determine and plot the variation of shear modulus and damping ratio with shear strain (strain range $10^{-3}\%$ to 1%).

4.20   For Example 4.8, determine the damping ratio of the cemented sand.

REFERENCES

Barkan, D. D. (1962). *Dynamics of Bases and Foundations*, McGraw-Hill Book Company, New York.

Beeston, H. E., and McEvilly, T. V. (1977). "Shear Wave Velocities from Down Hole Measurements," *Journal of the International Association for Earthquake Engineering*, Vol. 5, No. 2, pp. 181–190.

Booker, E. W., and Ireland, H. O. (1965). "Earth Pressure at Rest Related to Stress History," *Canadian Geotechnical Journal*, Vol. 2, No. 1, pp. 1–15.

Carroll, W. F. (1963). "Dynamic Bearing Capacity of Soils. Vertical Displacements of Spread Footing on Clay: Static and Impulsive Loadings," *Technical Report No. 3-599*, Report 5, U.S. Army Corps of Engineers, Waterways Experiment Station, Vicksburg, Mississippi.

Casagrande, A., and Shannon, W. L. (1949). "Strength of Soils under Dynamic Loads," *Transactions*, ASCE, Vol. 114, pp. 755–772.

Das, B. M. (1990). *Principles of Geotechnical Engineering*, 2nd ed., PWS-KENT, Boston.

Drnevich, V. P. (1972). "Undrained Cyclic Shear of Saturated Sand," *Journal of the Soil Mechanics and Foundations Division*, ASCE, Vol. 98, No. SM8, pp. 807–825.

Drnevich, V. P., Hall, J. R., Jr., and Richart F. E., Jr. (1966). "Large Amplitude Vibration Effects on the Shear Modulus of Sand," University of Michigan Report to Waterways Experiment Station, Corps of Engineers, U.S. Army Contract DA-22-079-Eng-340, October 1966.

Drnevich, V. P., Hall, J. R., Jr., and Richart, F. E., Jr. (1967). "Effects of the Amplitude of Vibration on the Shear Modulus of Sand," *Proceedings of the International Symposium on Wave Propagation and Dynamic Properties of Earth Materials*, Ed. G. E. Triandafilidis, University of New Mexico Press, pp. 189–199.

Hall, J. R., Jr., and Richart, F. E., Jr. (1963). "Dissipation of Elastic Wave Energy in Granular Soils," *Journal of the Soil Mechanics and Foundations Division*, ASCE, Vol. 89, No. SM6, pp. 27–56.

Hardin, B. O. (1965). "The Nature of Damping in Sands," *Journal of the Soil Mechanics and Foundations Division*, ASCE, Vol. 91, No. SM1, pp. 63–97.

Hardin, B. O., and Black, W. L. (1968). "Vibration Modulus of Normally Consolidated Clays," *Journal of the Soil Mechanics and Foundations Division*, ASCE, Vol. 94, No. SM2, pp. 353–369.

Hardin, B. O., and Drnevich, V. P. (1972). "Shear Modulus and Damping in Soils: Design Equations and Curves," *Journal of the Soil Mechanics and Foundations Division*, ASCE, Vol. 98, No. SM7, pp. 667–692.

Hardin, B. O., and Richart, F. E., Jr. (1963). "Elastic Wave Velocities in Granular Soils," *Journal of the Soil Mechanics and Foundations Division*, ASCE, Vol. 89, No. SM1, pp. 33–65.

Heukelom, W., and Foster, C. R. (1960). "Dynamic Testing of Pavements," *Journal of the Soil Mechanics and Foundations Division*, ASCE, Vol. 86, No. SM1, Part 1, pp. 1–28.

Iida, K. (1938). "The Velocity of Elastic Waves in Sand," *Bulletin of the Earthquake Research Institute*, Tokyo Imperial University, Vol. 16, pp. 131–144.

Iida, K. (1940). "On the Elastic Properties of Soil Particularly in Relation to Its Water Contents," *Bulletin of the Earthquake Research Institute*, Tokyo Imperial University, Vol. 18, pp. 675–690.

Ishibashi, I., and Sherif, M. A. (1974). "Soil Liquefaction by Torsional Simple Shear Device," *Journal of the Geotechnical Engineering Division*, ASCE, Vol. 100, No. GT8, pp. 871–888.

Ishimato, M., and Iida, K. (1937). "Determination of Elastic Constants of Soils by Means of Vibration Methods," *Bulletin of the Earthquake Research Institute*, Tokyo Imperial University, Vol. 15, p. 67.

Iwasaki, T., Tatsuoka, F., and Takagi, Y. (1976). "Dynamic Shear Deformation Properties of Sand for Wide Strain Range," *Report of the Civil Engineering Institute, No. 1085*, Ministry of Construction, Tokyo, Japan.

Kolsky, H. (1963). *Stress Waves in Solids*, Dover Publications, Inc., New York.

Larkin, T. J., and Taylor, P. W. (1979). "Comparison of Down Hole and Laboratory Shear Wave Velocities," *Canadian Geotechnical Journal*, Vol. 16, No. 1, pp. 152–162.

Matsui, T., O-Hara, H., and Ito, T. (1980). "Cyclic Stress-Strain History and Shear Characteristics of Clay," *Journal of the Geotechnical Engineering Division*, ASCE, Vol. 106, No. GT10, pp. 1101–1120.

Richart, F. E., Jr., Hall, J. R. Jr., and Lysmer, J. (1962). "Study of the Propagation and Dissipation of 'Elastic' Wave Energy in Granular Soils," University of Florida Report to Waterways Experiment Station, Corps of Engineers, U.S. Army Contract DA-22-070-Eng-314.

Saxena, S. K., Avramidis, A. S., and Reddy, K. R. (1988). "Dynamic Moduli and Damping Ratios for Cemented Sands at Low Strains," *Canadian Geotechnical Journal*, Vol. 25, No. 2, pp. 353–368.

Schwarz, S. D., and Musser, J. (1972). "Various Techniques for Making *In Situ* Shear Wave Velocity Measurements: A Description and Evaluation," *Proceedings*, Microzonation Conference, Seattle, Washington, Vol. 2, p. 593.

Seed, H. B., and Chan, C. K. (1966). "Clay Strength under Earthquake Loading Conditions," *Journal of the Soil Mechanics and Foundations Division*, ASCE, Vol. 92, No. SM2, pp. 53–78.

Seed, H. B., and Idriss, I. M. (1970). "Soil Moduli and Damping Factors for Dynamic Response Analysis," *Report No. EERC 75-29*, Earthquake Engineering Research Center, University of California, Berkeley, California.

Seed, H. B., Wong, R. T., Idriss, I. M., and Tokimatsu, K. (1986). "Moduli and Damping Factors for Dynamic Analyses of Cohesive Soils," *Journal of Geotechnical Engineering*, ASCE, Vol. 112, No. GT11, pp. 1016–1032.

Sherif, M. A., Ishibashi, I., and Gaddah, A. H. (1972). "Damping Ratio for Dry Sands," *Journal of the Geotechnical Engineering Division*, ASCE, Vol. 103, No. GT7, pp. 743–756.

Shibata, T., and Soelarno, D. S. (1975). "Stress-Strain Characteristics of Sands under Cyclic Loading," *Proceedings of the Japanese Society of Civil Engineers*, No. 239.

Silver, M. L. (1981). "Load Deformation and Strength Behavior of Soils Under Loading," State-of-the-Art Paper, *Proceedings of the International Conference on Recent Advances in Geotechnical Earthquake Engineering and Soil Dynamics*, Vol. 3, pp. 873–896.

Silver, M. L., and Seed, H. B. (1969). "The Behavior of Sands under Seismic Loading Conditions," *Report No. EERC 69-16*, Earthquake Engineering Research Center, University of California, Berkeley, California.

Silver, M. L., and Seed, H. B. (1971). "Deformation Characteristics of Sands under Cyclic Loading," *Journal of the Soil Mechanics and Foundations Division*, ASCE, Vol. 94, No. SM8, pp. 1081–1098.

Stokoe, K. H., and Woods, R. D. (1972). "*In Situ* Shear Wave Velocity by Cross-Hole Method," *Journal of Soil Mechanics and Foundations Division*, ASCE, Vol. 98, No. SM5, pp. 443–460.

Terzaghi, K. (1955). "Evaluation of Coefficient of Subgrade Reaction," *Geotechnique*, No. 5, pp. 297–326.

Thiers, G. R., and Seed, H. B. (1968). "Cyclic Stress-Strain Characteristics of Clay," *Journal of the Soil Mechanics and Foundations Division*, ASCE, Vol. 94, No. SM6, pp. 555–569.

Vesic, A. S. (1973). "Analysis of Ultimate Loads of Shallow Foundations," *Journal of the Soil Mechanics and Foundations Division*, ASCE, Vol. 99, No. SM1, pp. 45–73.

Weissman, G. F., and Hart, R. R. (1961). "The Damping Capacity of Some Granular Soils," *ASTM Special Technical Publication No. 305*, Symposium of Soil Dynamics, pp. 45–54.

Whitman, R. V. and Healy, K. A. (1963). "Shear Strength of Sands During Rapid Loadings," *Transactions*, ASCE, Vol. 128, Part 1, pp. 1553–1594.

Whitman, R. V. and Lawrence, F. V. (1963). "Discussion on Elastic Wave Velocities in Granular Soils," *Journal of the Soil Mechanics and Foundations Division*, ASCE, Vol. 89, No. SM5, pp. 112–118.

Wilson, S. D. and Dietrich, R. J. (1960). "Effect of Consolidation Pressure on Elastic and Strength Properties of Clay," *Proceedings of the Research Conference on Shear Strength of Cohesive Soils*, ASCE, pp. 419–435.

Chapter 5

FOUNDATION VIBRATION

5.1   Introduction

In Chapter 2 (Figure 2.1), it was briefly mentioned that foundations supporting vibrating equipment do experience rigid body displacements. The cyclic displacement of a foundation can have six possible modes. They are

1. translation in the vertical direction,
2. translation in the longitudinal direction,
3. translation in the lateral direction,
4. rotation about the vertical axis (that is, yawing),
5. rotation about the longitudinal axis (that is, rocking), and
6. rotation about the lateral axis (that is, pitching).

In this chapter, the fundamentals of the vibration of foundations, in various modes, supported on an elastic medium will be developed. The elastic medium that supports the foundation will be considered to be homogeneous and isotropic. In general, the behavior of soils departs considerably from that of an elastic material; only at low strain levels may it be considered as a reasonable approximation to an elastic material. Hence, the theories developed here should be considered as applicable only to the cases where foundations undergo low amplitudes of vibration.

5.2   Vertical Vibration of Circular Foundations Resting on Elastic Half-Space—Historical Development

In 1904, Lamb studied the problem of vibration of a single vibrating force acting at a point on the surface of an elastic half-space. This study included cases in which the oscillating force R acts in the vertical direction and in the horizontal direction, as shown in Figure 5.1a and b. This is generally referred to as the *dynamic Boussinesq problem*.

In 1936, Reissner analyzed the problem of vibration of a *uniformly loaded flexible circular area* resting on an elastic half-space. The solution was obtained by integration of Lamb's solution for a point load. Based on Reissner's work, the vertical displacement at the *center* of the flexible loaded area (Figure 5.2a)

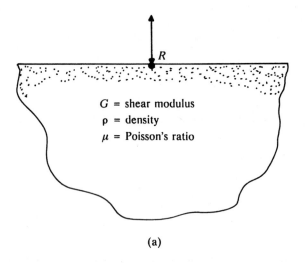

(a)

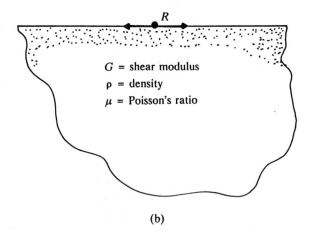

(b)

Figure 5.1   Vibrating force on the surface of an elastic half-space

can be given by

$$z = \frac{Q_0 e^{i\omega t}}{G r_0}(f_1 + if_2)$$

(5.1)

where

Q_0 = amplitude of the exciting force acting on the foundation

z = periodic displacement at the center of the loaded area

ω = circular frequency of the applied load

r_0 = radius of the loaded area

G = shear modulus of the soil

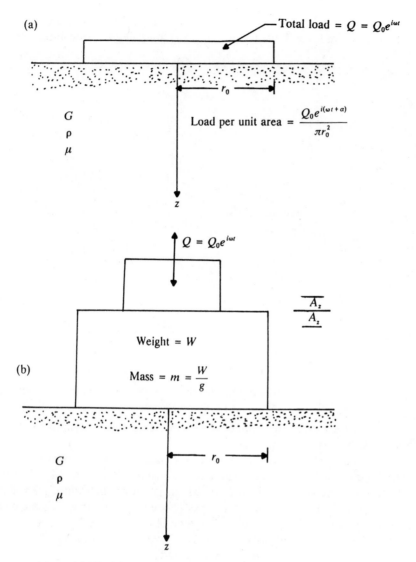

Figure 5.2 (a) Vibration of a uniformly loaded circular flexible area; (b) flexible circular foundation subjected to forced vibration

Q = exciting force, which has an amplitude of Q_0

f_1, f_2 = Reissner's displacement functions

The displacement functions f_1 and f_2 are related to the Poisson's ratio of the medium and the frequency of the exciting force.

Now, consider a flexible circular foundation of weight W (mass $= m = W/g$) resting on an elastic half-space and subjected to an exciting force of magnitude of $Q_0 e^{i(\omega t + \alpha)}$, as shown in Figure 5.2b. (*Note*: α = phase difference between the exciting force and the displacement of the foundation.)

Using the displacement relation given in Eq. (5.1) and solving the equation of equilibrium of force, Reissner obtained the following relationships:

$$A_z = \frac{Q_0}{Gr_0} Z \tag{5.2}$$

where

A_z = the amplitude of the vibration

Z = dimensionless amplitude

$$= \sqrt{\frac{f_1^2 + f_2^2}{(1 - ba_0^2 f_1)^2 + (ba_0^2 f_2)^2}} \tag{5.3}$$

b = dimensionless mass ratio

$$= \frac{m}{\rho r_0^3} = \left(\frac{W}{g}\right)\left[\frac{1}{(\gamma/g)r_0^3}\right] = \frac{W}{\gamma r_0^3} \tag{5.4}$$

ρ = density of the elastic material

γ = unit weight of the elastic material (for this problem, it is soil)

a_0 = dimensionless frequency

$$= \omega r_0 \sqrt{\frac{\rho}{G}} = \frac{\omega r_0}{v_s} \tag{5.5}$$

v_s = velocity of shear waves in the elastic material on which the foundation is resting

The classical work of Reissner was further extended by Quinlan (1953) and Sung (1953). As mentioned before, Reissner's work related only to the case of flexible circular foundations where the soil reaction is uniform over the entire area (Figure 5.3a). Both Quinlan and Sung considered the cases of rigid circular foundations, the contact pressure of which is shown in Figure 5.3b, flexible foundations (Figure 5.3a), and the types of foundations for which the contact pressure distribution is parabolic, as shown in Figure 5.3c. The distribution of contact pressure q for all three cases may be expressed as follows.

For flexible circular foundations (Figure 5.3a):

$$q = \frac{Q_0 e^{i(\omega t + \alpha)}}{\pi r_0^2} \qquad (\text{for } r \le r_0) \tag{5.6}$$

For rigid circular foundations (Figure 5.3b):

$$q = \frac{Q_0 e^{i(\omega t + \alpha)}}{2\pi r_0 \sqrt{r_0^2 - r^2}} \qquad (\text{for } r \le r_0) \tag{5.7}$$

For foundations with parabolic contact pressure distribution (Figure 5.3c):

$$q = \frac{2(r_0^2 - r^2)Q_0 e^{i(\omega t + \alpha)}}{\pi r_0^4} \qquad (\text{for } r \le r_0) \tag{5.8}$$

where

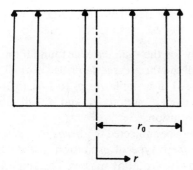

(a) Uniform pressure distribution

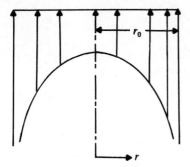

(b) Pressure distribution under
     rigid foundation

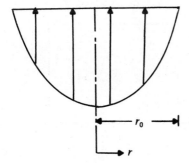

(c) Parabolic pressure distribution

Figure 5.3   Contact pressure distribution under a circular foundation of radius r_0

q = contact pressure at a distance r measured from the center
of the foundation

Quinlan derived the equations only for the *rigid circular* foundation; however, Sung presented the solutions for all the three cases described. For all cases, the amplitude of motion can be expressed in a similar form to Eqs. (5.2), (5.3), (5.4), and (5.5). However, the displacement functions f_1 and f_2 will change, depending on the contact pressure distribution.

Foundations, on some occasions, may be subjected to a *frequency-dependent excitation*, in contrast to the *constant-force* type of excitation just discussed. Figure 5.4 shows a foundation excited by two rotating masses. The amplitude of the exciting force can be given as

$$Q = 2m_e e\omega^2 = m_1 e\omega^2 \tag{5.9}$$

where

m_1 = total of the rotating masses

ω = circular frequency of the rotating masses

For this condition, the amplitude of vibration A_z may be given by the relation

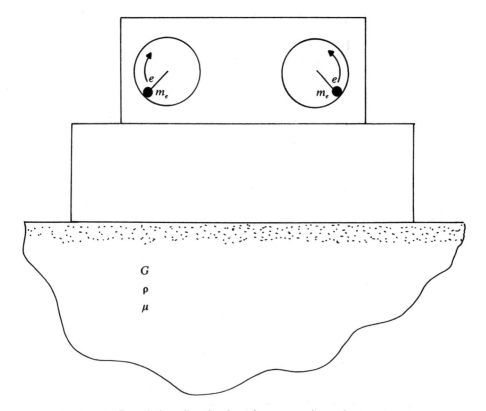

Figure 5.4   Foundation vibration by a frequency-dependent exciting force

$$A_z = \frac{m_1 e \omega^2}{G r_0} \sqrt{\frac{f_1^2 + f_2^2}{(1 - b a_0^2 f_1)^2 + (b a_0^2 f_2)^2}} \tag{5.10}$$

From Eq. (5.5)

$$a_0 = \omega r_0 \sqrt{\frac{\rho}{G}}$$

or

$$\omega^2 = \frac{a_0^2 G}{\rho r_0^2} \tag{5.11}$$

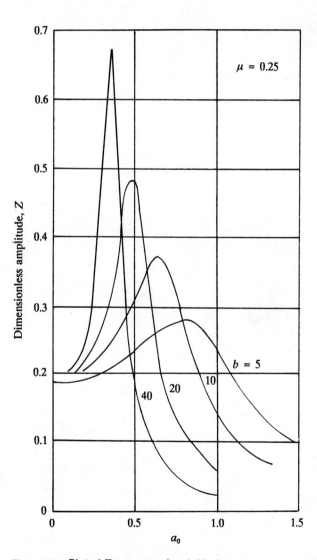

Figure 5.5   Plot of Z versus a_0 for rigid circular foundation (after Richart, 1962)

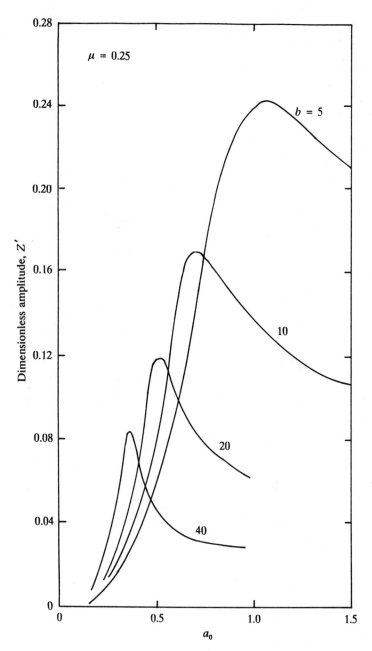

Figure 5.6   Variation of Z' with a_0 for rigid circular foundation (redrawn after Richart, 1962)

Substituting Eq. (5.11) into (5.10), one obtains

$$A_z = \frac{m_1 e a_0^2}{\rho r_0^3} \sqrt{\frac{f_1^2 + f_2^2}{(1 - b a_0^2 f_1)^2 + (b a_0^2 f_2)^2}} = \frac{m_1 e}{\rho r_0^3} Z'$$ (5.12)

where

Z' = dimensionless amplitude

$$= a_0^2 \sqrt{\frac{f_1^2 + f_2^2}{(1 - b a_0^2 f_1)^2 + (b a_0^2 f_2)^2}}$$ (5.13)

Figures 5.5 and 5.6 show the plots of the variation of the dimensionless amplitude with a_0 (Richart, 1962) for *rigid circular* foundations (for $\mu =$ Poisson's ratio = 0.25 and $b = 5, 10, 20$, and 40).

Effect of Contact Pressure Distribution and Poisson's Ratio

The effect of the contact pressure distribution on the nature of variation of the nondimensional amplitude Z' with a_0 is shown in Figure 5.7 (for $b = 5$ and $\mu = 0.25$). As can be seen, for a given value of a_0, the magnitude of the amplitude is highest for the case of parabolic pressure distribution and lowest for rigid bases.

For a given type of pressure distribution and mass ratio (b), the magnitude

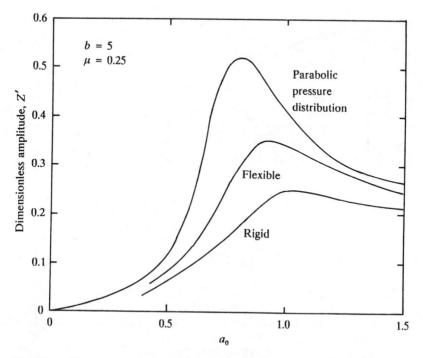

Figure 5.7 Effect of contact pressure distribution on the variation of Z' with a_0 (redrawn after Richart and Whitman, 1967)

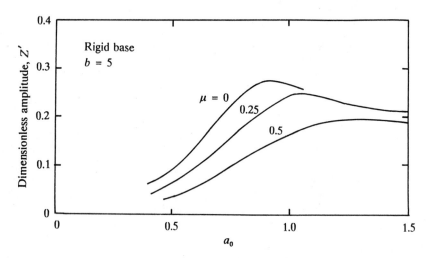

Figure 5.8 Effect of Poisson's ratio on the variation of Z' with a_0 (redrawn after Richart and Whitman, 1967)

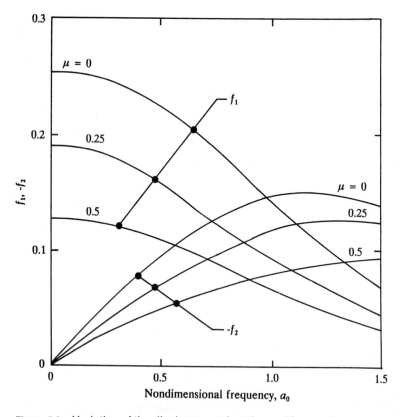

Figure 5.9 Variation of the displacement functions with a_0 and μ

of Z' also greatly depends on the assumption of the Poisson's ratio μ. This is shown in Figure 5.8.

Variation of Displacement Functions f_2 and f_2

As mentioned before, the displacement functions are related to the dimensionless frequency a_0 and Poisson's ratio μ. In Sung's original study, it was assumed that the contact pressure distribution remains the same throughout the range of frequency considered; however, for dynamic loading conditions, the rigid-base pressure distribution does not produce uniform displacement under the foundation. For that reason, Bycroft (1956) determined the weighted average of the displacements under a foundation. The variation of the displacement functions determined by that study is shown in Figure 5.9.

5.3   Analog Solution for Vertical Vibration of Foundations

Hsieh's Analog

Hsieh (1962) attempted to modify the original solution of Reissner in order to develop an equation similar to that for damped vibrations of single-degree free system [Eq. (2.72)]. Hsieh's analog can be explained with reference to Figure 5.10. Consider a rigid circular weightless disc on the surface of an elastic half-space. The disc is subjected to a vertical vibration by a force

$$P = P_0 e^{i\omega t} \tag{5.14}$$

The vertical displacement of the disk can be given by Eq. (5.1) as

$$z = \frac{P_0 e^{i\omega t}}{G r_0}(f_1 + if_2)$$

Now,

$$\frac{dz}{dt} = \frac{P_0 \omega e^{i\omega t}}{G r_0}(if_1 - f_2) \tag{5.15}$$

or

$$f_1 \omega z - f_2 \frac{dz}{dt} = \frac{P_0 \omega}{G r_0}(f_1^2 + f_2^2)e^{i\omega t}$$

Since $P = P_0 e^{i\omega t}$, the preceding relationship can be written as

$$f_1 \omega z - f_2 \frac{dz}{dt} = \frac{P \omega}{G r_0}(f_1^2 + f_2^2)$$

or

$$P = \underbrace{\left[(G r_0)\left(\frac{f_1}{f_1^2 + f_2^2} \right) \right]}_{k_z} z + \underbrace{\left[\left(\frac{G r_0}{\omega} \right)\left(\frac{-f_2}{f_1^2 + f_2^2} \right) \right]}_{c_z} \frac{dz}{dt}$$

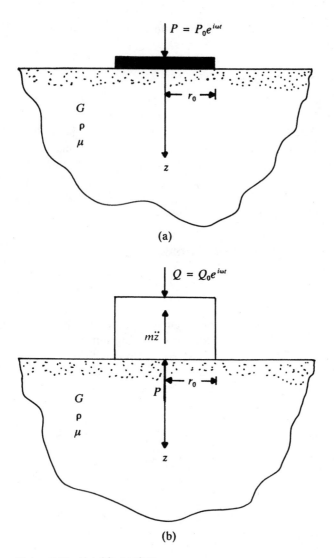

Figure 5.10   Hsieh's analog

So

$$P = k_z z + c_z \dot{z} \tag{5.16}$$

Now consider a rigid circular foundation having a mass m and radius r_0 placed on the surface of the elastic half-space (Figure 5.10b). The foundation undergoes vibration by a periodic force

$$Q = Q_0 e^{i\omega t} \tag{5.17}$$

For dynamic equilibrium

$$m\ddot{z} = Q - P \tag{5.18}$$

Combining Equations (5.16), (5.17), and (5.18)

$$m\ddot{z} + c_z\dot{z} + k_z z = Q_0 e^{i\omega t}$$

(5.19)

The preceding relationship is an equivalent mass-spring-dashpot model similar to Eq. (2.72). However, the spring constant k_z and the dashpot coefficient c_z are frequency dependent.

Lysmer's Analog

A simplified model was also proposed by Lysmer and Richart (1966), in which the expressions for k_z and c_z were frequency independent. Lysmer and Richart (1966) redefined the displacement functions in the form

$$F = \frac{f}{\left(\dfrac{1-\mu}{4}\right)} = \frac{f_1 + if_2}{\left(\dfrac{1-\mu}{4}\right)} = F_1 + iF_2$$

(5.20)

The functions F_1 and F_2 are practically independent of Poisson's ratio, as shown

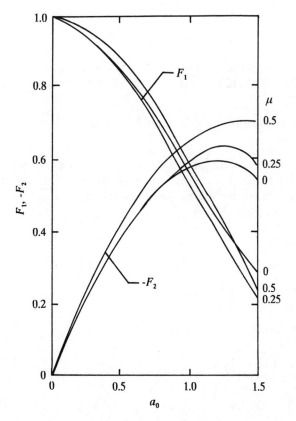

Figure 5.11   Plot of F_1 and $-F_2$ against a_0 for rigid circular foundation subjected to vertical vibration (after Lysmer and Richart, 1966)

in Figure 5.11. The term *mass ratio* [Eq. (5.4)] was also modified to the form

$$B_z = \left(\frac{1-\mu}{4}\right)b = \left(\frac{1-\mu}{4}\right)\left(\frac{m}{Gr_0^3}\right) \tag{5.21}$$

where B_z = modified mass ratio.

In this analysis, it was proposed that satisfactory results can be obtained within the range of practical interest by expressing the rigid circular foundation vibration in the form

$$\boxed{m\ddot{z} + c_z\dot{z} + k_z z = Q_0 e^{i\omega t}} \tag{5.22}$$

where

$$\boxed{\begin{aligned} k_z &= \text{static spring constant for rigid circular foundation} \\ &= \frac{4Gr_0}{1-\mu} \end{aligned}} \tag{5.23}$$

$$\boxed{c_z = \frac{3.4r_0^2}{1-\mu}\sqrt{G\rho}} \tag{5.24}$$

In Eqs. (5.23) and (5.24) the relationships for k_z and c_z are frequency independent. Equations (5.22), (5.23), and (5.24) are referred to as *Lysmer's analog*.

5.4   Calculation Procedure for Foundation Response— Vertical Vibration

Once the equation of motion of a rigid circular foundation is expressed in the form given in Equation (5.22), it is easy to obtain the resonant frequency and amplitude of vibration based on the mathematical expressions presented in Chapter 2. The general procedure is outlined next.

A. Resonant Frequency

1. *Calculation of natural frequency.* From Eqs. (2.6) and (2.18),

$$f_n = \frac{1}{2\pi}\sqrt{\frac{k_z}{m}} = \frac{1}{2\pi}\sqrt{\left(\frac{4Gr_0}{1-\mu}\right)\frac{1}{m}} \tag{5.25}$$

2. *Calculation of damping ratio D_z.* From Eq. (2.47a),

$$\begin{aligned} c_{cz} &= 2\sqrt{k_z m} = 2\sqrt{\left(\frac{4Gr_0}{1-\mu}\right)(m)} \\ &= 8\sqrt{\left(\frac{Gr_0}{1-\mu}\right)\left(\frac{B_z\rho r_0^3}{1-\mu}\right)} = \frac{8r_0^2}{1-\mu}\sqrt{GB_z\rho} \end{aligned} \tag{5.26}$$

From Eq. (2.47b)

$$D_z = \frac{c}{c_{cz}} = \frac{\dfrac{3.4 r_0^2}{1 - \mu} \sqrt{G\rho}}{\dfrac{8 r_0^2}{1 - \mu} \sqrt{G B_z \rho}} = \frac{0.425}{\sqrt{B_z}}$$ (5.27)

3. *Calculation of the resonance frequency* (*that is, frequency at maximum displacement*). From Eq. (2.86), for *constant force–type excitation*,

$$f_m = f_n \sqrt{1 - 2D_z^2} = \left[\frac{1}{2\pi} \sqrt{\left(\frac{4 G r_0}{1 - \mu} \right) \frac{1}{m}} \right] \left[\sqrt{1 - 2 \left(\frac{0.425}{\sqrt{B_z}} \right)^2} \right]$$ (5.28)

It has also been shown by Lysmer that, for $B_z \geq 0.3$, the following approximate relationship can be established:

$$f_m = \left(\frac{1}{2\pi} \right) \left(\sqrt{\frac{G}{\rho}} \right) \left(\frac{1}{r_0} \right) \sqrt{\frac{B_z - 0.36}{B_z}}$$ (5.29)

For *rotating mass–type excitation* [Eq. (2.98)]

$$f_m = \frac{f_n}{\sqrt{1 - 2D_z^2}} = \frac{\dfrac{1}{2\pi} \sqrt{\left(\dfrac{4 G r_0}{1 - \mu} \right) \left(\dfrac{1}{m} \right)}}{\sqrt{1 - 2 \left(\dfrac{0.425}{\sqrt{B_z}} \right)^2}}$$ (5.30)

Lysmer's corresponding approximate relationship for f_m is as follows:

$$f_m = \left(\frac{1}{2\pi} \right) \left(\sqrt{\frac{G}{\rho}} \right) \left(\frac{1}{r_0} \right) \sqrt{\frac{0.9}{B_z - 0.45}}$$ (5.31)

B. Amplitude of Vibration at Resonance

The amplitude of vibration A_z at resonance for *constant force–type excitation* can be determined from Eq. (2.87) as

$$A_{z(\text{resonance})} = \left(\frac{Q_0}{k_z} \right) \left(\frac{1}{2 D_z \sqrt{1 - D_z^2}} \right)$$ (5.32)

where

$$k_z = \frac{4 G r_0}{1 - \mu}$$

$$D_z = \frac{0.425}{\sqrt{B_z}}$$

Substitution of the relationships for k_z and D_z in Eq. (5.32) yields

$$A_{z(\text{resonance})} = \frac{Q_0(1-\mu)}{4Gr_0} \frac{B_z}{0.85\sqrt{B_z - 0.18}} \tag{5.33}$$

The amplitude of vibration for *rotating mass–type vertical excitation* can be given as [see Eq. (2.99)]

$$A_{z(\text{resonance})} = \frac{U}{m} \frac{1}{2D_z\sqrt{1 - D_z^2}}$$

where $U = m_1 e$ (m_1 = total rotating mass causing excitation), or

$$A_{z(\text{resonance})} = \frac{m_1 e}{m} \frac{B_z}{0.85\sqrt{B_z - 0.18}} \tag{5.34}$$

C. Amplitude of Vibration at Frequencies Other Than Resonance

For *constant force–type excitation*, Eq. (2.82) can be used for estimation of the amplitude of vibration, or

$$A_z = \frac{\dfrac{Q_0}{k_z}}{\sqrt{[1 - (\omega^2/\omega_n^2)]^2 + 4D_z^2(\omega^2/\omega_n^2)}} \tag{5.35}$$

The relationships for k_z and D_z are given by Eqs. (5.23) and (5.27) and

$$\omega_n = \sqrt{\frac{k_z}{m}} \tag{5.36}$$

Figure 5.12 shows the plot of $A_z/(Q_0/k_z)$ versus ω/ω_n. So, with known values of D_z and ω/ω_n, one can determine the value of $A_z/(Q_0/k_z)$ and, from that, A_z can be obtained.

In a similar manner, for *rotating mass-type excitation*, Eq. (2.95) can be used to determine the amplitude of vibration, or

$$A_z = \frac{(m_1 e/m)(\omega/\omega_n)^2}{\sqrt{[1 - (\omega^2/\omega_n^2)]^2 + 4D_z^2(\omega^2/\omega_n^2)}} \tag{5.37}$$

Figure 5.13 shows a plot of $A_z/(m_1 e/m)$ versus ω/ω_n, from which the magnitude of A_z can also be determined.

The procedure just described relates to a rigid circular foundation having a radius of r_0. If a foundation is rectangular in shape with length L and width B, it is conventional to obtain an equivalent radius, which can then be used in the preceding relationships. This can be done by equating the area of the given foundation to the area of an equivalent circle. Thus

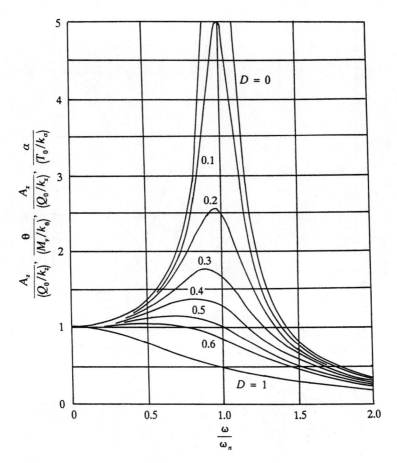

Figure 5.12   Plot of $A_z/(Q_0/k_z)$, $\theta/(M_y/k_\theta)$, $A_x/(Q_0/k_x)$, and $\alpha/(T_0/k_\alpha)$ against ω/ω_n for constant force–type vibrator (*Note:* $D = D_z$ for vertical vibration, $D = D_\theta$ for rocking, $D = D_x$ for sliding; $D = D_\alpha$ for torsional vibration.)

$$\pi r_0^2 = BL$$

or

$$r_0 = \sqrt{\frac{BL}{\pi}} \qquad (5.38)$$

where r_0 = radius of the equivalent circle.

The procedure for transforming areas of any shape to an equivalent circle of the same area gives good results in the evaluation of foundation response for $L/B \leq 2$. Dobry and Gazetas (1986) showed that this technique has limitations, and the L/B ratio has significant influence on dynamic stiffness and damping values, particularly for long foundations.

It is obviously impossible to eliminate vibration near a foundation. However, an attempt can be made to reduce the vibration problem as much as possible. Richart (1962) compiled guidelines for allowable vertical vibration

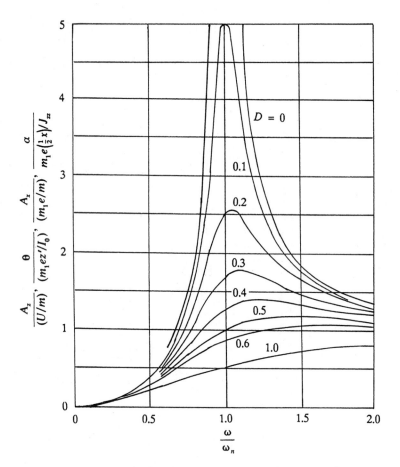

Figure 5.13 Plot of $A_z/(U/m)$, $\theta/(m_1 ez'/I_0)$, $A_x/(m_1 e/m)$, $\alpha/[m_1 e(\frac{1}{2}x)/J_{zz}]$ against ω/ω_n for rotating mass–type excitation (*Note*: $D = D_z$ for vertical vibration, $D = D_\theta$ for rocking, $D = D_x$ for sliding; $D = D_\alpha$ for torsional vibration.)

amplitude for a particular frequency of vibration, and this is given in Figure 5.14. The data presented in Figure 5.14 refer to the maximum allowable amplitudes of vibration. These can be converted to maximum allowable accelerations by

$$\text{Maximum acceleration} = (\text{maximum displacement})\omega^2$$

For example, in Figure 5.14, the limiting amplitude of displacement at an operating frequency of 2000 cpm is about 0.005 in. (0.127 mm). So the maximum operating acceleration for a frequency of 2000 cpm is

$$(0.005 \text{ in.})\left[\frac{(2\pi)(2000)}{60}\right]^2 = 219.3 \text{ in./s}^2 \ (5570 \text{ mm/s}^2)$$

In the design of machine foundations, the following general rules may be kept in mind to avoid possible resonance conditions:

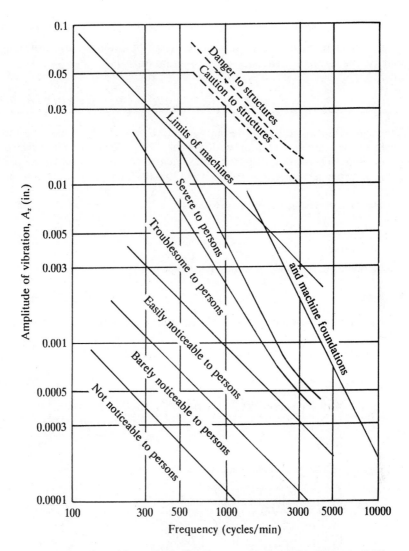

Figure 5.14   Allowable vertical vibration amplitudes (after Richart, 1962)

1. The resonant frequency of the foundation-soil system should be less than half the operating frequency for high-speed machines (that is, operating frequency $\geq$ 1000 cpm). For this case, during starting or stopping the machine will briefly vibrate at resonant frequency.

2. For low-speed machineries (speed less than about 350–400 cpm), the resonant frequency of the foundation-soil system should be at least two times the operating frequency.

3. In all types of foundations, the increase of weight will decrease the resonant frequency.

4. An increase of r_0 will increase the resonant frequency of the foundation.

5. An increase of shear modulus of soil (for example, by grouting) will increase the resonant frequency of the foundation.

Example 5.1

A foundation is subjected to a constant force–type vertical vibration. Given the total weight of the machinery and foundation block, $W = 150,000$ lb; unit weight of soil, $\gamma = 115$ lb/ft^3; $\mu = 0.4$; $G = 3000$ lb/in.^{2}; the amplitude of the vibrating force, $Q_0 = 1500$ lb; the operating frequency, $f = 180$ cpm; and that the foundation is 20 ft long and 6 ft wide:

a. Determine the resonant frequency. Check if

$$\frac{f_{resonance}}{f_{operating}} > 2$$

b. Determine the amplitude of vibration at resonance.

Solution

a. This is a rectangular foundation, so the equivalent radius [Eq. (5.38)] is

$$r_0 = \sqrt{\frac{BL}{\pi}} = \sqrt{\frac{(6)(20)}{\pi}} = 6.18 \text{ ft}$$

The mass ratio [Eq. (5.21)]

$$B_z = \left(\frac{1-\mu}{4}\right)\left(\frac{m}{Gr_0^3}\right) = \left(\frac{1-\mu}{4}\right)\left(\frac{W}{\gamma r_0^3}\right) = \left(\frac{1-0.4}{4}\right)\left[\frac{150,000}{(115)(6.18)^3}\right]$$

$$= 0.829$$

From Eq. (5.29), the resonant frequency is

$$f_m = \left(\frac{1}{2\pi}\right)\left(\sqrt{\frac{G}{\rho}}\right)\left(\frac{1}{r_0}\right)\sqrt{\frac{B_z - 0.36}{B_z}}$$

$$= \left(\frac{1}{2\pi}\right)\left[\sqrt{\frac{(3000)(144)}{(115/32.2)}}\right]\left(\frac{1}{6.18}\right)\sqrt{\frac{0.829 - 0.36}{0.829}}$$

$$= 6.737 \text{ cps} \approx \underline{404 \text{ cpm}}$$

Hence

$$\frac{f_{resonance}}{f_{operating}} = \frac{404}{180} = \underline{2.24 > 2}$$

b. From Eq. (5.33)

$$A_{z(resonance)} = \frac{Q_0(1-\mu)}{4Gr_0}\frac{B_z}{0.85\sqrt{B_z - 0.18}}$$

$$= \left[\frac{(1500)(1-0.4)}{(4)(3000 \times 144)(6.18)}\right]\left[\frac{0.829}{0.85\sqrt{0.829 - 0.18}}\right]$$

$$= 0.000102 \text{ ft} = \underline{0.00122 \text{ in.}}$$ ∎

Example 5.2

Figure 5.15a shows a single-cylinder reciprocating engine. The data for the engine are as follows: operating speed = 1500 cpm; connecting rod (r_2) = 0.3 m; crank (r_1) = 75 mm; total reciprocating weight = 54 N; total engine weight = 14 kN. Figure 5.15b shows the dimensions of the concrete foundation for the engine. The properties of the soil are as

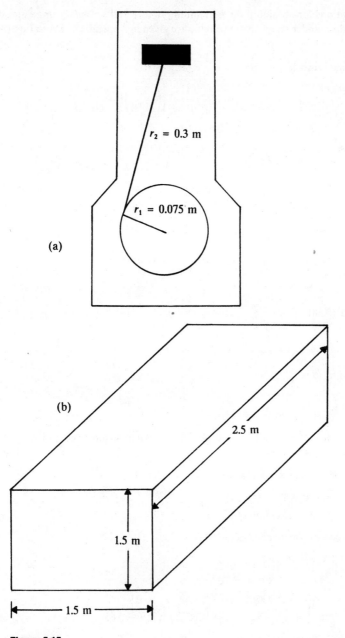

(a)

$r_2 = 0.3$ m

$r_1 = 0.075$ m

(b)

2.5 m

1.5 m

1.5 m

Figure 5.15

follows: $\gamma = 18.5$ kN/m^3; $G = 18,000$ kN/m^2; and $\mu = 0.5$. Calculate:

a. primary and secondary unbalanced forces at operating frequency (refer to Appendix A),
b. the resonance frequency, and
c. the vertical vibration amplitude at resonance.

Solution

a. The equations for obtaining the maximum *primary* and *secondary* unbalanced forces for a single cylinder reciprocating engine are given in Appendix A. From Eqs. (A.9) and (A.10)

$$\text{Primary unbalanced force} = m_{rec}r_1\omega^2$$

$$= \frac{54}{(1000)(9.81)}\left(\frac{75}{1000}\right)\left(\frac{2\pi 1500}{60}\right)^2$$

$$= \underline{10.19 \text{ kN}}$$

$$\text{Secondary unbalanced force} = \frac{m_{rec}r_1^2\omega^2}{r_2}$$

$$\frac{r_1}{r_2} = \frac{0.075}{0.3} = 0.25$$

So

$$\text{Secondary force} = (\text{primary force})\left(\frac{r_1}{r_2}\right) = (10.19)(0.25)$$

$$= \underline{2.55 \text{ kN}}$$

b. From Eq. (5.38),

$$r_0 = \sqrt{\frac{BL}{\pi}} = \sqrt{\frac{(1.5)(2.5)}{\pi}} = 1.093 \text{ m}$$

The mass ratio is

$$B_z = \left(\frac{1-\mu}{4}\right)\left(\frac{W}{\gamma r_0^3}\right)$$

Total weight is $W =$ weight of foundation + engine. Assume the unit weight of concrete is 23.58 kN/m^3. So

$$W = (1.5 \times 2.5 \times 1.5)(23.58) + 14 = 146.64 \text{ kN}$$

$$B_z = \left(\frac{1-0.5}{4}\right)\left[\frac{146.64}{(18.5)(1.093)^3}\right] = 0.759$$

The resonant frequency [Eq. (5.31)] is

$$f_m = \left(\frac{1}{2\pi}\right)\left(\sqrt{\frac{G}{\rho}}\right)\left(\frac{1}{r_0}\right)\sqrt{\frac{0.9}{B_z - 0.45}}$$

$$= \left(\frac{1}{2\pi}\right)\left[\sqrt{\frac{(18,000)(9.81)}{18.5}}\right]\left(\frac{1}{1.093}\right)\sqrt{\frac{0.9}{0.759 - 0.45}}$$

$$= 24.28 \text{ cps} \approx \underline{1457 \text{ cpm}}$$

c. From Eq. (5.34),

$$A_{z(\text{resonance})} = \frac{m_1 e}{m} \frac{B_z}{0.85\sqrt{B_z - 0.18}}$$

At 1500 cpm, the total unbalanced force = primary force + secondary force = 10.19 + 2.55 = 12.74 kN.

$$Q_{0(1457\,\text{cpm})} = Q_{0(1500\,\text{cpm})}\left(\frac{1457}{1500}\right)^2 = (12.74)\left(\frac{1457}{1500}\right)^2$$

$$= 12.02 \text{ kN}$$

$$Q_{0(1457\,\text{cpm})} = m_1 e\omega^2 = 12.02\text{kN}$$

Therefore,

$$m_1 e = \frac{12.02}{\omega^2}; \qquad \omega = \frac{2\pi(1457)}{60} = 152.58 \text{ rad/s}; \qquad m_1 e = \frac{12.02}{(152.58)^2}$$

Hence

$$A_{z(\text{resonance})} = \left[\frac{12.02/(152.58)^2}{146.64/9.81}\right]\left(\frac{0.759}{0.85\sqrt{0.759 - 0.18}}\right)$$

$$= 0.0000405 \text{ m} = \underline{0.0405 \text{ mm}} \qquad\blacksquare$$

5.5   Rocking Vibration of Foundations

Theoretical solutions for foundations subjected to rocking vibration have been presented by Arnold, Bycroft, and Wartburton (1955) and Bycroft (1956). For *rigid circular foundations* (Figure 5.16), the contact pressure can be described by the equation

$$q = \frac{3M_y r \cos\alpha}{2\pi r_0^3 \sqrt{r_0^2 - r^2}} e^{i\omega t}$$

where

q = pressure at any point defined by point a on the plan

M_y = the exciting moment about the y axis = $M_y e^{i\omega t}$

Hall (1967) developed a mass-spring-dashpot model for rigid circular foundations in the same manner as Lysmer and Richart (1966) did for vertical vibration. According to Hall, the equation of motion for a rocking vibration can be given as

$$\boxed{I_0\ddot{\theta} + c_\theta\dot{\theta} + k_\theta\theta = M_y e^{i\omega t}} \qquad (5.39)$$

where

θ = rotation of the vertical axis of the foundation at any time t

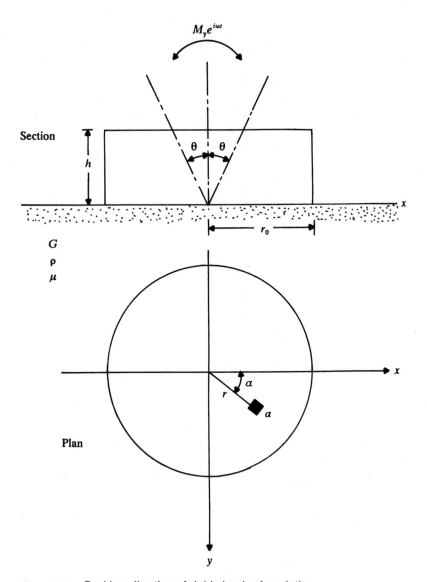

Figure 5.16 Rocking vibration of rigid circular foundation

I_0 = mass moment of inertia about the y axis (through its base)

$$= \frac{W_0}{g}\left(\frac{r_0^2}{4} + \frac{h^2}{3}\right) \tag{5.40}$$

where

W_0 = weight of the foundation

g = acceleration due to gravity

h = height of the foundation

$$k_\theta = \text{static spring constant} = \frac{8Gr_0^3}{3(1-\mu)} \tag{5.41}$$

$$c_\theta = \text{dashpot coefficient} = \frac{0.8r_0^4\sqrt{G}}{(1-\mu)(1+B_\theta)} \tag{5.42}$$

where

$$B_\theta = \text{inertia ratio} = \frac{3(1-\mu)}{8}\frac{I_0}{\rho r_0^5} \tag{5.43}$$

The calculation procedure for foundation response using Eq. (5.39) is as follows.

A. Resonant Frequency

1. Calculate the natural frequency:

$$f_n = \frac{1}{2\pi}\sqrt{\frac{k_\theta}{I_0}} \tag{5.44}$$

2. Calculate the damping ratio D_θ:

$$c_{c\theta} = 2\sqrt{k_\theta I_0}$$

$$D_\theta = \frac{c_\theta}{c_{c\theta}} = \frac{0.15}{\sqrt{B_\theta}(1+B_\theta)} \tag{5.45}$$

3. Calculate the resonant frequency:

$$f_m = f_n\sqrt{1-2D_\theta^2} \qquad \text{(for constant force excitation)}$$

$$f_m = \frac{f_n}{\sqrt{1-2D_\theta^2}} \qquad \text{(for rotating mass–type excitation)}$$

B. Amplitude of Vibration at Resonance

$$\theta_{\text{resonance}} = \frac{M_y}{k_\theta}\frac{1}{2D_\theta\sqrt{1-D_\theta^2}} \qquad \text{(for constant force excitation)} \tag{5.46}$$

$$\theta_{\text{resonance}} = \frac{m_1 ez'}{I_0}\frac{1}{2D_\theta\sqrt{1-D_\theta^2}} \qquad \begin{array}{l}\text{(for rotating mass–type excitation;}\\ \text{see Figure 5.17)}\end{array}$$

$$\tag{5.47}$$

where

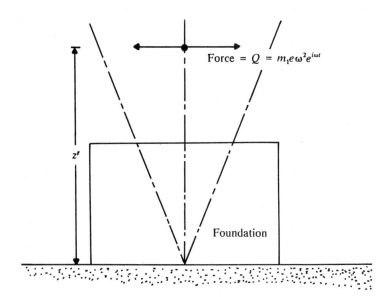

Force = $Q = m_1 e \omega^2 e^{i\omega t}$

Figure 5.17

m_1 = total rotating mass causing excitation

e = eccentricity of each mass

C. Amplitude of Vibration at Frequencies Other than Resonance

For constant force–type excitation [Eq. (2.82)]:

$$\theta = \frac{M_y/k_\theta}{\sqrt{[1 - (\omega^2/\omega_n^2)]^2 + 4D_\theta^2(\omega^2/\omega_n^2)}} \tag{5.48}$$

A plot of $\theta/(M_y/k_\theta)$ versus ω/ω_n is given in Figure 5.12.

For rotating mass–type excitation [Eq. (2.95)]:

$$\theta = \frac{(m_1 e z'/I_0)(\omega^2/\omega_n^2)}{\sqrt{[1 - (\omega^2/\omega_n^2)]^2 + 4D_\theta^2(\omega^2/\omega_n^2)}} \tag{5.49}$$

Figure 5.13 shows a plot of $\theta/(m_1 e z'/I_0)$ versus ω/ω_n.

In the case of rectangular foundations, the preceding relationships can be used by determining the equivalent radius as

$$r_0 = \sqrt[4]{\frac{BL^3}{3\pi}} \tag{5.50}$$

The definitions of B and L are shown in Figure 5.18.

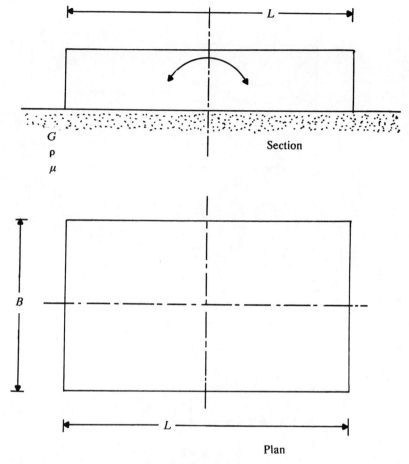

Figure 5.18   Equivalent radius of rectangular rigid foundation—rocking motion

Example 5.3

A horizontal piston-type compressor is shown in Figure 5.19. The operating frequency is 600 cpm. The amplitude of the horizontal unbalanced force of the compressor is 30 kN, and it creates a rocking motion of the foundation about point O (see Figure 5.19b). The mass moment of inertia of the compressor assembly about the axis $b'Ob'$ is 16×10^5 kg-m² (see Figure 5.19c). Determine

a. the resonant frequency, and
b. the amplitude of rocking vibration at resonance.

Solution

Moment of inertia of the foundation block and the compressor assembly about $b'Ob'$:

$$I_0 = \left(\frac{W_{\text{foundation block}}}{3g}\right)\left[\left(\frac{L}{2}\right)^2 + h^2\right] + 16 \times 10^5 \text{ kg-m}^2$$

Assume the unit weight of concrete is 23.58 kN/m³.

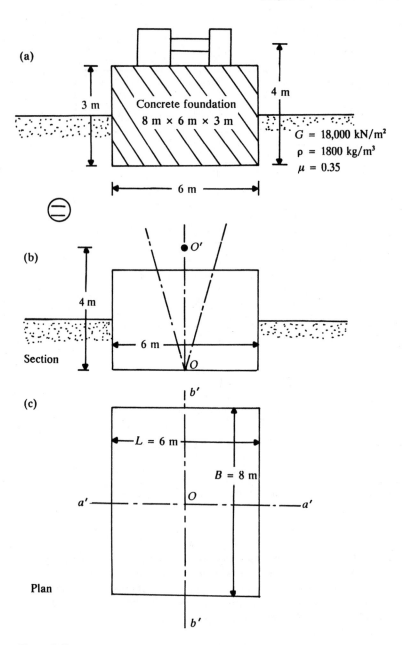

(a)

3 m

Concrete foundation
8 m × 6 m × 3 m

4 m

$G = 18,000 \text{ kN/m}^2$
$\rho = 1800 \text{ kg/m}^3$
$\mu = 0.35$

6 m

(b)

O'

4 m

6 m

O

Section

(c)

b'

$L = 6 \text{ m}$

$B = 8 \text{ m}$

a' O a'

Plan

b'

Figure 5.19

$$W_{\text{foundation block}} = (8 \times 6 \times 3)(23.58) = 3395.52 \text{ kN}$$

$$= 3395.52 \times 10^3 \text{ N}$$

$$I_0 = \frac{3395.52 \times 10^3}{(3)(9.81)}(3^2 + 3^2) + 16 \times 10^5$$

$$= 36.768 \times 10^5 \text{ kg-m}^2$$

Calculation of equivalent radius of the foundation: From Eq. (5.50), the equivalent radius is

$$r_0 = \sqrt[4]{\frac{BL^3}{3\pi}} = \sqrt[4]{\frac{8 \times 6^3}{3\pi}} = 3.67 \text{ m}$$

a. *Determination of resonant frequency*:

$$k_\theta = \frac{8Gr_0^3}{3(1 - \mu)} = \frac{(8)(18,000)(3.67)^3}{(3)(1 - 0.35)} = 3650279 \text{ kN-m/rad}$$

$$f_n = \frac{1}{2\pi}\sqrt{\frac{k_\theta}{I_0}} = \frac{1}{2\pi}\sqrt{\frac{3650279 \times 10^3 \text{ N-m/rad}}{36.768 \times 10^5}} = 5.01 \text{ cps}$$

$$= 300 \text{ cpm}$$

$$B_\theta = \frac{3(1 - \mu)}{8}\frac{I_0}{\rho r_0^5} = \frac{3(1 - 0.35)}{8}\frac{36.768 \times 10^5}{1800(3.67)^5} = 0.748$$

$$D_\theta = \frac{0.15}{\sqrt{B_\theta(1 + B_\theta)}} = \frac{0.15}{\sqrt{0.748(1 + 0.748)}} = 0.099$$

$$f_m = \frac{f_n}{\sqrt{1 - 2D_\theta^2}} = \frac{300}{\sqrt{1 - 2D_\theta^2}} = \frac{300}{\sqrt{1 - 2(0.099)^2}} = \underline{303 \text{ cpm}}$$

b. *Calculation of amplitude of vibration at resonance*:

$$M_{y(\text{operating frequency})} = \text{unbalanced force} \times 4$$

$$= 30 \times 4 = 120 \text{ kN-m}$$

$$M_{y(\text{at resonance})} = 120\left(\frac{f_m}{f_{\text{operating}}}\right)$$

$$= 120\left(\frac{303}{600}\right)^2 = 30.6 \text{ kN-m}$$

$$(m_1 e\omega^2)z' = M_y$$

$$\omega_{\text{resonance}} = \frac{(2\pi)(303)}{60} = 31.73 \text{ rad/s}$$

$$m_1 ez' = \frac{M_y}{\omega^2} = \frac{30.6 \times 10^3 \text{ N-m}}{(31.73)^3} = 0.0304 \times 10^3$$

From Eq. (5.47)

$$\theta_{\text{resonance}} = \frac{m_1 ez'}{I_0}\frac{1}{2D_\theta\sqrt{1 - D_\theta^2}}$$

$$= \left(\frac{0.0304 \times 10^3}{36.768 \times 10^5}\right)\left[\frac{1}{(2)(0.099)\sqrt{1 - (0.099)^2}}\right] = \underline{4.2 \times 10^{-5} \text{ rad}} \quad \blacksquare$$

5.6   Sliding Vibration of Foundations

Arnold, Bycroft, and Wartburton (1955) have provided theoretical solutions for sliding vibration of *rigid circular* foundations (Figure 5.20) acted on by a force $Q = Q_0 e^{i\omega t}$. Hall (1967) developed the mass-spring-dashpot analog for this type of vibration. According to this analog, the equation of motion of the foundation can be given in the form

$$m\ddot{x} + c_x \dot{x} + k_x x = Q_0 e^{i\omega t}$$ (5.51)

where

$m = $ mass of the foundation

$k_x = $ static spring constant for sliding

$$= \frac{32(1-\mu)Gr_0}{7-8\mu}$$ (5.52)

$c_x = $ dashpot coefficient for sliding

$$= \frac{18.4(1-\mu)}{7-8\mu} r_0^2 \sqrt{\rho G}$$ (5.53)

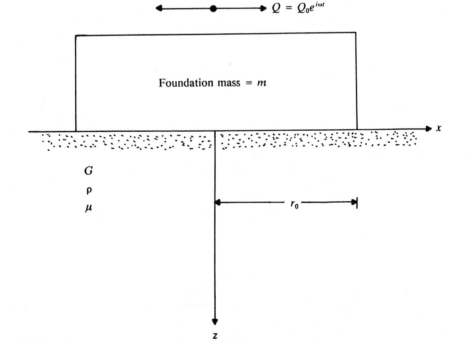

Figure 5.20   Sliding vibration of rigid circular foundation

Based on Eqs. (5.51), (5.52), and (5.53), the natural frequency of the foundation for sliding can be calculated as

$$f_n = \frac{1}{2\pi}\sqrt{\frac{k_x}{m}} = \frac{1}{2\pi}\sqrt{\frac{32(1-\mu)Gr_0}{(7-8\mu)m}} \tag{5.54}$$

The critical damping and damping ratio in sliding can be evaluated as

c_{cx} = critical damping in sliding

$$= 2\sqrt{k_x m} = 2\sqrt{\frac{32(1-\mu)Gr_0 m}{(7-8\mu)}} \tag{5.55}$$

D_x = damping ratio in sliding

$$= \frac{c_x}{c_{cx}} = \frac{0.288}{\sqrt{B_x}} \tag{5.56}$$

where the dimensionless mass ratio

$$\boxed{B_x = \frac{7-8\mu}{32(1-\mu)}\frac{m}{\rho r_0^3}} \tag{5.57}$$

For rectangular foundations, the preceding relationships can be used by obtaining the equivalent radius r_0, or

$$r_0 = \sqrt{\frac{BL}{\pi}}$$

where B and L are the length and width of the foundation, respectively.

Calculation Procedure for Foundation Response Using Eq. (5.51)

Resonant Frequency

1. Calculate the natural frequency f_n using Eq. (5.54).

2. Calculate the damping ratio D_x using Eq. (5.56). [*Note:* B_x can be obtained from Eq. (5.57)].

3. For constant force excitation (that is, Q_0 = constant), calculate

$$f_m = f_n\sqrt{1-2D_x^2}$$

4. For rotating mass type excitation, calculate

$$f_m = \frac{f_n}{\sqrt{1-2D_x^2}}$$

Amplitude of Vibration at Resonance

1. For constant force excitation, amplitude of vibration at resonance is

$$A_{x(\text{resonance})} = \frac{Q_0}{k_x} \frac{1}{2D_x\sqrt{1 - D_x^2}} \qquad (5.58)$$

where $A_{x(\text{resonance})}$ = amplitude of vibration at resonance.

2. For rotating mass–type excitation,

$$A_{x(\text{resonance})} = \frac{m_1 e}{m} \frac{1}{2D_x\sqrt{1 - D_x^2}} \qquad (5.59)$$

where

m_1 = total rotating mass causing excitation

e = eccentricity of each rotating mass

Amplitude of Vibration at Frequency Other than Resonance

1. For constant force–type excitation,

$$A_x = \frac{Q_0/k_x}{\sqrt{[1 - (\omega^2/\omega_n^2)]^2 + 4D_x^2(\omega^2/\omega_n^2)}} \qquad (5.60)$$

Figure 5.12 can also be used to determine $A_x/(Q_0/k_x)$ for given values of ω/ω_n and D_x.

2. For rotating mass–type excitation,

$$A_x = \frac{(m_1 e/m)(\omega/\omega_n)^2}{\sqrt{[1 - (\omega^2/\omega_n^2)]^2 + 4D_x^2(\omega^2/\omega_n^2)}} \qquad (5.61)$$

Figure 5.13 provides a plot of $A_x/(m_1 e/m)$ versus ω/ω_n for various values of D_x.

5.7   Torsional Vibration of Foundations

Figure 5.21a shows a circular foundation of radius r_0 subjected to a torque $T = T_0 o^{i\omega t}$ about an axis z–z. Reissner (1937) solved the vibration problem of this type considering a linear distribution of shear stress $\tau_{z\theta}$ (shear stress zero at center and maximum at the periphery of the foundation), as shown in Figure 5.21b. This represents the case of a *flexible* foundation. In 1944 Reissner and Sagoli solved the same problem for the case of a *rigid* foundation considering a *linear variation of displacement from the center to the periphery* of the foundation. For this case, the shear stress can be given by (Figure 5.21c)

$$\tau_{z\theta} = \frac{3}{4\pi} \frac{Tr}{r_0^3\sqrt{r_0^2 - r^2}} \qquad \text{for } 0 < r < r_0 \qquad (5.62)$$

Similar to the cases of vertical, rocking, and sliding modes of vibration, the equation for the torsional vibration of a *rigid circular* foundation can be written

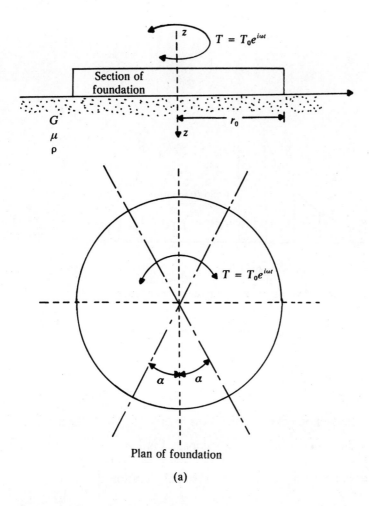

Plan of foundation

(a)

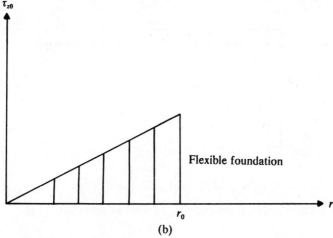

(b)

Figure 5.21   (Continued)

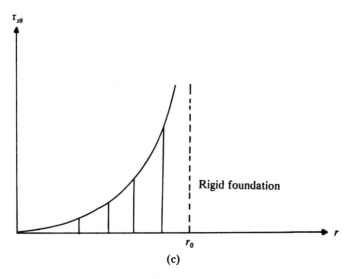

Figure 5.21   Torsional vibration of rigid circular foundation

as

$$J_{zz}\ddot{\alpha} + c_\alpha\dot{\alpha} + k_\alpha\alpha = T_0 e^{i\omega t} \tag{5.63}$$

where

J_{zz} = mass moment of inertia of the foundation about the axis z–z

c_α = dashpot coefficient for torsional vibration

k_α = *static* spring constant for torsional vibration

$$= \frac{16}{3} G r_0^3 \tag{5.64}$$

α = rotation of the foundation at any time due to the application of a torque $T = T_0 e^{i\omega t}$

The damping ratio D_α for this mode of vibration has been determined as (Richart, Hall, and Wood, 1970)

$$D_\alpha = \frac{0.5}{1 + 2B_\alpha} \tag{5.65}$$

where

B_α = the dimensionless mass ratio for torsion at vibration

$$= \frac{J_{zz}}{\rho r_0^5} \tag{5.66}$$

Calculation Procedure for Foundation Response Using Eq. (5.63)

Resonant Frequency

1. Calculate the natural frequency of the foundation as

$$f_n = \frac{1}{2\pi}\sqrt{\frac{k_\alpha}{J_{zz}}} \tag{5.67}$$

2. Calculate B_α using Eq. (6.66) and then D_α using Eq. (5.65).

3. For constant force excitation (that is, T_0 = constant)

$$f_m = f_n\sqrt{1 - 2D_\alpha^2}$$

For rotating mass–type excitation

$$f_m = \frac{f_n}{\sqrt{1 - 2D_\alpha^2}}$$

Amplitude of Vibration at Resonance: For constant force excitation, the amplitude of vibration at resonance is

$$\boxed{\alpha_{\text{resonance}} = \frac{T_0}{k_\alpha}\frac{1}{2D_\alpha\sqrt{1 - D_\alpha^2}}} \tag{5.68}$$

For rotating mass–type excitation

$$\boxed{\alpha_{\text{resonance}} = \frac{m_1 e\left(\dfrac{x}{2}\right)}{J_{zz}}\frac{1}{2D_\alpha\sqrt{1 - D_\alpha^2}}} \tag{5.69}$$

where

m_1 = total rotating mass causing the excitation

e = eccentricity of each rotating mass

For the definition of x in Eq. (5.69), see Figure 5.22.

Amplitude of Vibration at Frequency Other than Resonance: For constant force excitation, calculate ω/ω_n and then refer to Figure 5.12 to obtain $\alpha/(T_0/k_\alpha)$. For rotating mass–type excitation, calculate ω/ω_n and then refer to Figure 5.13 to obtain $\alpha/[m_1 e(x/2)/J_{zz}]$.

For a rectangular foundation with dimensions $B \times L$, the equivalent radius may be given by

$$r_0 = \sqrt[4]{\frac{BL(B^2 + L^2)}{6\pi}} \tag{5.70}$$

The torsional vibration of foundations is uncoupled motion and hence can be treated independently of any vertical motion. Also, Poisson's ratio does not influence the torsional vibration of foundations.

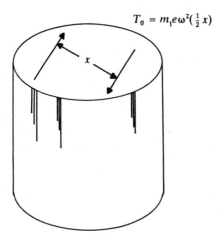

$$T_0 = m_1 e \omega^2 (\tfrac{1}{2} x)$$

Figure 5.22

Example 5.4

A radar antenna foundation is shown in Figure 5.23. For torsional vibration of the foundation, given

$$T_0 = 18 \times 10^4 \text{ ft-lb} \quad \text{(due to inertia)}$$

$$T_0 = 6 \times 10^4 \text{ ft-lb} \quad \text{(due to wind)}$$

mass moment of inertia of the tower about the axis z–z = 10×10^6 ft-lb-s^2, and the unit weight of concrete used in the foundation = 150 lb/ft^3. Calculate
a. the resonant frequency for torsional mode of vibration; and
b. angular deflection at resonance.

Solution

a. $J_{zz} = J_{zz\text{(tower)}} + J_{zz\text{(foundation)}}$

$$= 10 \times 10^6 + \frac{1}{2}\left[\pi r_0^2 h\left(\frac{150}{32.2}\right) \right] r_0^2$$

$$= 10 \times 10^6 + \frac{1}{2}\left[(\pi)(25)^2(8)\left(\frac{150}{32.2}\right) \right](25)^2$$

$$= 10 \times 10^6 + 22.87 \times 10^6 = 32.87 \times 10^6 \text{ ft-lb-s}^2$$

From Eq. (5.66)

$$B_\alpha = \frac{J_{zz}}{\rho r_0^5} = \frac{32.85 \times 10^6}{(110/32.2)(25)^5} = 0.985$$

Again from Eq. (5.65),

$$D_\alpha = \frac{0.5}{1 + 2B_\alpha} = \frac{0.5}{1 + (2)(0.985)} = 0.168$$

Also, k_α [Eq. (5.64)] is

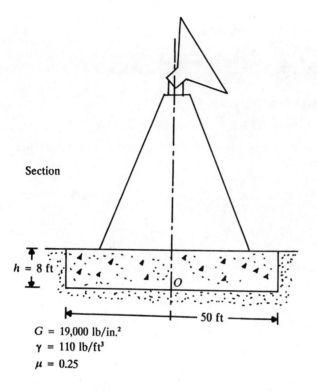

Section

$h = 8$ ft

|← —————————— 50 ft ——————————→|

$G = 19,000$ lb/in.^{2}
$\gamma = 110$ lb/ft^3
$\mu = 0.25$

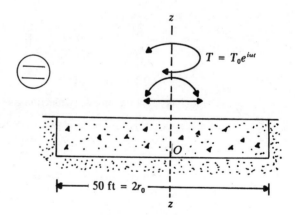

z

$T = T_0 e^{i\omega t}$

O

|← —————— 50 ft $= 2r_0$ —————→|

z

Figure 5.23

$$k_\alpha = \frac{16}{3} Gr_0^3 = \left(\frac{16}{3}\right)(19,000)(144)(25)^3$$

$$f_n = \frac{1}{2\pi}\sqrt{\frac{k_\alpha}{J_{zz}}} = \frac{1}{2\pi}\sqrt{\frac{(\frac{16}{3})(19,000)(144)(25)^3}{32.87 \times 10^6}} = 13.26 \text{ cps}$$

Thus, the damped natural frequency

$$f_m = f_n\sqrt{1 - 2D_\alpha^2} = (13.26)\sqrt{1 - (2)(0.168)^2} = 12.88 \text{ cps}$$

b. *Angular deflection at resonant frequency:* If the torque due to wind (T_0) is to be treated as a static torque, then

$$\frac{T_0}{\alpha_{\text{static}}} = k_\alpha$$

or

$$\alpha_{\text{static}} = \frac{T_0}{k_\alpha}$$

So

$$\alpha_{\text{static}} = \frac{3}{16Gr_0^3} T_{0(\text{static})} = \left[\frac{3}{(16)(19,000)(144)(25)^3}\right](6 \times 10^4)$$

$$= 0.0263 \times 10^{-5} \text{ rad}$$

Using Eq. (5.68), for the torque due to inertia

$$\alpha_{\text{resonance}} = \frac{T_0}{k_\alpha} \frac{1}{2D_\alpha\sqrt{1 - D_\alpha^2}}$$

$$= \left[\frac{18 \times 10^4}{(\frac{16}{3})(19,000)(144)(25)^3}\right]\left[\frac{1}{(2)(0.168)\sqrt{1 - (0.168)^2}}\right]$$

$$= 0.23 \times 10^{-5} \text{ rad}$$

At resonance, the total angular deflection is

$$\alpha = \alpha_{\text{inertia}} + \alpha_{\text{static}} = (0.23 + 0.0263) \times 10^{-5}$$

$$= 0.2563 \times 10^{-5} \text{ rad} \qquad\blacksquare$$

5.8   Comparison of Footing Vibration Tests with Theory

Richart and Whitman (1967) conducted a comprehensive study to evaluate the applicability of the preceding theoretical findings to actual field problems. Ninety-four large-scale field test results for large footings 5 ft to 16 ft (1.52 to 4.88 m) in diameter subjected to *vertical vibration* were reported by Fry (1963). Of these 94 test results, 55 were conducted at the U.S. Army Waterways Experiment Station, Vicksburg, Mississippi. The remaining 39 were conducted at Eglin Field, Florida. The classification of the soils for the Vicksburg site and Eglin site were CL and SP, respectively (unified soil classification system). For these tests, the *vertical dynamic force* on footings was generated by rotating mass vibrators.

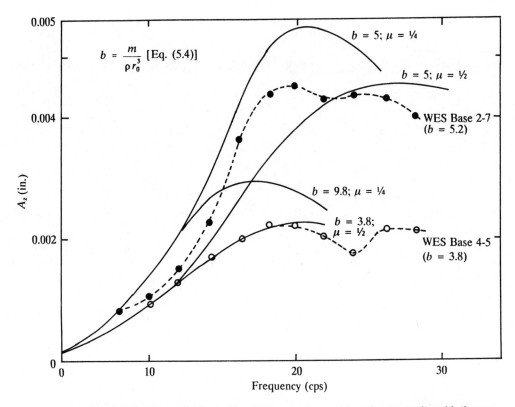

Figure 5.24   Vertical vibration of foundation—comparison of test results with theory (after Richart and Whitman, 1967)

Figure 5.24 shows a comparison of the theoretical amplitudes of vibration A_z as determined from elastic half-space theory with the experimental results obtained for two bases at the Vicksburg site. The nondimensional mass ratios b [Eq. (5.4)] of these two bases were 5.2 and 3.8. For the base with $b = 5.2$, the experimental results fall between the theoretical curves, with $\mu = 0.5$ and $\mu = 0.25$. However, for the base with $b = 3.8$, the experimental curve is nearly identical to the theoretical curve with $\mu = 0.5$. Figure 5.25 shows a comparison of the theory and experimental values reported by Fry in a nondimensional plot of $A_z m/m_1 e$ at resonance versus b. Similarly, a comparison of these test results with theory in a nondimensional plot of a_0 [Eq. (5.5)] at resonance versus b is shown in Figure 5.26.

From these two plots it may be seen that the results of the Vicksburg site follow the general trends indicated by the theoretical curve obtained from the elastic half-space theory for a *rigid base*. A considerable scatter, however, exists for the tests conducted at Eglin Field. This may be due to the clean fine sand found at that site, for which the shear modulus will change with depth. The fundamental assumption of the theoretical derivation of a homogeneous, elastic, isotropic body is very much different than the actual field conditions. Figure 5.27

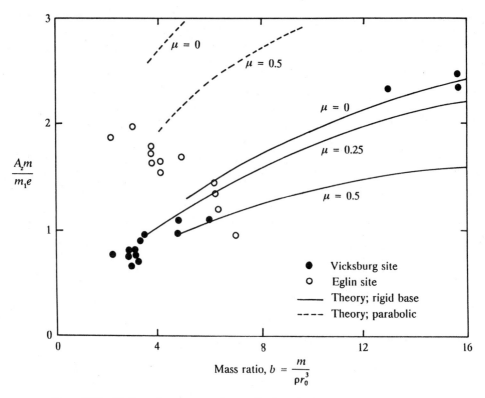

Figure 5.25 Motion at resonance for vertical excitation—comparison between theory and experiment (after Richart and Whitman, 1967)

shows a summary of all vertical vibration tests, which is a plot of

$$\frac{A_{z(\text{computed})}}{A_{z(\text{measured})}} \quad \text{versus} \quad \frac{A_z \omega^2}{g}$$

(that is, nondimensional acceleration, g, equals acceleration due to gravity). When the nondimensional acceleration reaches 1, the footing probably leaves the ground on the upswing and acts as a hammer. In any case, in actual design problems, a machine foundation is not subjected to an acceleration greater than $0.3g$. However, for dynamic problems of this nature, the general agreement between theory and experiment is fairly good.

Several large-scale field tests were conducted by the U.S. Army Waterways Experiment Station (Fry, 1963) in which footings were subjected to torsional vibration. Mechanical vibrators were set to produce pure torque on a horizontal plane. Figure 5.28 shows a plot of the dimensionless amplitude $\alpha J_{zz}/[m_1 e(x/2)]$ versus B_α (α = amplitude of torsional motion and m_1 = sum of the rotating masses; for definition of x, see the insert in Figure 5.28) for some of these tests that correspond to the lowest settings of the eccentric masses on the vibrator. The theoretical curve based on the elastic half-space theory is also plotted in this figure for comparison purposes. It can be seen that, for low

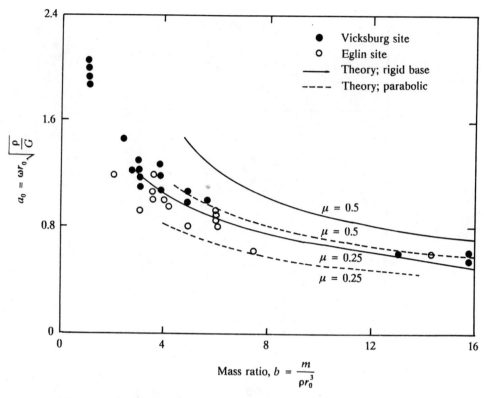

Figure 5.26 Plot of a_0 at resonance versus b—comparison of theory with field test results (after Richart and Whitman, 1967)

amplitudes of vibration, the agreement between theory and field test results is good. The limiting torsional motion in most practical cases is about 0.1 mil (1×10^{-4} in.). So the half-space theory generally serves well for most practical design considerations.

Comparisons between the elastic half-space theory and experimental results for footing vibration tests in rocking and sliding modes were also presented by Richart and Whitman (1967). The agreement seemed fairly good.

5.9   Comments on the Mass-Spring-Dashpot Analog Used for Solving Foundation Vibration Problems

The equations for the mass-spring-dashpot analog for various modes of vibration of *rigid circular* foundations developed in the preceding sections may be summarized as follows:

For vertical vibration,

$$m\ddot{z} + c_z\dot{z} + k_z z = Q_0 e^{i\omega t} \tag{5.22}$$

For rocking vibration,

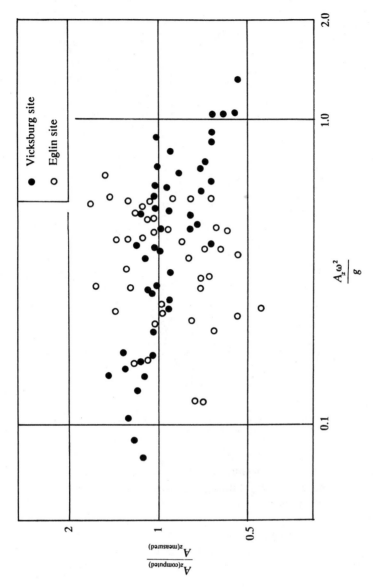

Figure 5.27   Summary of vertical vibration tests (after Richart and Whitman, 1967)

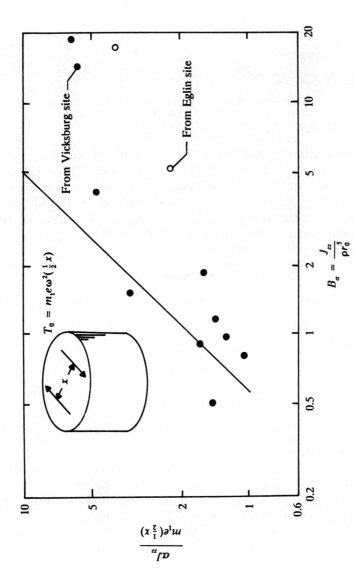

Figure 5.28   Comparison of amplitude for torsional vibration (redrawn after Richart and Whitman, 1967)

$$I_0\ddot{\theta} + c_\theta\dot{\theta} + k_\theta\theta = M_y e^{i\omega t} \tag{5.39}$$

For sliding vibration,

$$m\ddot{x} + c_x\dot{x} + k_x x = Q_0 e^{i\omega t} \tag{5.51}$$

For torsional vibration,

$$J_{zz}\ddot{\alpha} + c_\alpha\dot{\alpha} + k_\alpha\alpha = T_0 o^{i\omega t} \tag{5.63}$$

The mathematical approach for solution of the preceding equations is similar for determination of the natural frequency, resonant frequency, critical damping, damping ratio, and the amplitudes of vibration at various frequencies. The agreement of these solutions with the field conditions will depend on proper choice of the parameters (that is, m, I_0, J_{zz}, c_z, c_θ, c_x, c_α, k_z, k_θ, k_x, and k_α). In this section, we will make a critical evaluation of these parameters.

Choice of Mass and Mass Moment of Inertia

The mass terms m used in Eqs. (5.22) and (5.51) are actually the sum of

1. mass of the structural foundation block m_f, and
2. mass of all the machineries mounted on the block m_m.

During the vibration of foundations, there is a mass of soil under the foundation that vibrates along with the foundation. Thus, it would be reasonable to consider the term m in Eqs. (5.22) and (5.51) to be the sum of

$$m = m_f + m_m + m_s \tag{5.71}$$

where m_s = effective mass of soil vibrating with the foundation.

In a similar manner, the mass moment of inertia terms I_0 and J_{zz} included in Eqs. (5.39) and (5.63) include the contributions of the mass of the foundation and that of the machine mounted on the block. It appears reasonable also to add the contribution of the effective mass of the vibrating soil (m_s), that is, the effective soil mass moment of inertia. Thus

$$I_0 = I_{0(\text{foundation})} + I_{0(\text{machine})} + I_{0(\text{effective soil mass})} \tag{5.72}$$

and

$$J_{zz} = J_{zz(\text{foundation})} + J_{zz(\text{machine})} + J_{zz(\text{effective soil mass})} \tag{5.73}$$

Theoretically, calculated values of m_s, $I_{0(\text{effective soil mass})}$, and $J_{zz(\text{effective soil mass})}$ are given by Hsieh (1962). They are as follows:

1. Values of m_s for vertical vibration:

Poisson's ratio μ	m_s
0.0	$0.5\rho r_0^3$
0.25	$0.5\rho r_0^3$
0.5	$2.0\rho r_0^3$

2. Values of m_s for horizontal vibration:

Poisson's ratio μ	m_s
0.0	$0.2\rho r_0^3$
0.25	$0.2\rho r_0^3$
0.5	$0.1\rho r_0^3$

3. Values of $I_{0(\text{effective soil mass})}$ for rocking vibration: Poisson's ratio $\mu = 0$;
$I_{0(\text{effective soil mass})} = 0.4\rho r_0^5$

4. Values of $J_{zz(\text{effective soil mass})}$ for torsional vibration:

Poisson's ratio μ	$J_{zz(\text{effective soil mass})}$
0.0	$0.3\rho r_0^5$
0.25	$0.3\rho r_0^5$
0.5	$0.3\rho r_0^5$

In most cases, for design purposes the contribution of the effective soil mass is neglected. This will, in general, lead to answers that are *within 30% accuracy.*

Table 5.1   Values of Spring Constants for Rigid Foundations (after Whitman and Richart, 1967)

Motion	Spring constant	Reference
	Circular foundations	
Vertical	$k_z = \dfrac{4Gr_0}{1 - \mu}$	Timoshenko and Goodier (1951)
Horizontal (sliding)	$k_x = \dfrac{32(1 - \mu)Gr_0^3}{7 - 8\mu}$	Bycroft (1956)
Rocking	$k_\theta = \dfrac{8Gr_0^3}{3(1 - \mu)}$	Borowicka (1943)
Torsion	$k_\alpha = \dfrac{16}{3}Gr_0^3$	Reissner and Sagoci (1944)
	Rectangular foundation	
Vertical[a]	$k_z = \dfrac{G}{1 - \mu}F_z\sqrt{BL}$	Barkan (1962)
Horizontal[a] (sliding)	$k_x = 2(1 + \mu)GF_x\sqrt{BL}$	Barkan (1962)
Rocking[b]	$k_\theta = \dfrac{G}{1 - \mu}F_\theta BL^2$	Gorbunov-Possadov and Serebrajanyi (1961)

[a] B = width of foundation; L = length of foundation.
[b] For definition of B and L, refer to Figure 5.18. Refer to Figure 5.29 for values of F_z, F_x, and F_θ.

Choice of Spring Constants

In Equations (5.23), (5.41), (5.52), and (5.64), the spring constants defined were for the cases of *rigid circular* foundations. In examples where rigid rectangular foundations were encountered, the equivalent radii r_0 were first determined. These values of r_0 were then used to determine the value of the spring constants. However, more exact solutions for spring constants for rectangular foundations derived from the theory of elasticity can be used. These are given in Table 5.1 on p. 225 along with those for circular foundations. Dobry and Gazetas (1986) have more recently developed more realistic values of spring constants.

Another fact that needs to be kept in mind is that the foundation blocks are never placed at the surface. If the bottom of the foundation block is placed at a depth z measured from the ground surface, the spring constants will be higher than that calculated by theory. This fact is demonstrated in Figure 5.30 for the case of vertical motion of rigid foundations. The behavior of embedded foundations subjected to various types of vibration is presented in Sections 5.12 through 5.15.

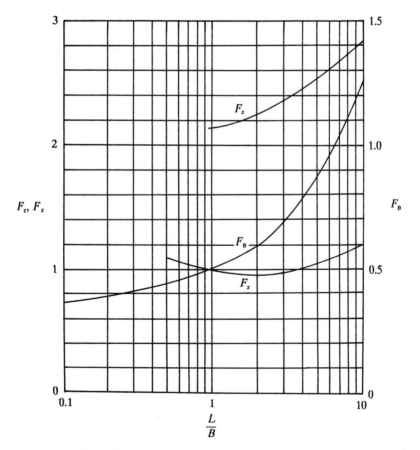

Figure 5.29   Plot of F_z, F_x, and F_θ against L/B (after Whitman and Richart, 1967)

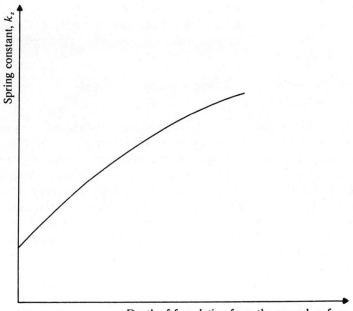

Figure 5.30   Nature of variation of k_z with the depth of the foundation

Choice of Poisson's Ratio

Whitman and Richart (1967) recommended the following values for Poisson's ratio:

Sand (dry, moist, partially saturated) $\mu = 0.35$ to 0.4

Clay (saturated) $\mu = 0.5$

A good value for most partially saturated soils is about 0.4.

Choice of Damping Ratio

In soils there are two types of damping, geometric damping (or radiation damping) and internal damping. The radiation damping depends on parameters such as Poisson's ratio, mass of the foundation, equivalent radius, and the density of the soil. The relations for the damping ratios given in Eqs. (5.27), (5.45), (5.56), and (5.65) are for radiation damping only.

The internal damping D_i varies over a wide range, depending on the type of soil. Generally the values of D_i are in the range of 0.01 to 0.1. Thus an average value of D_i would be a good estimate (Whitman and Richart, 1967). The damping ratio can then be approximated as

$$D = D_{\text{radiation}} + 0.05 \qquad\qquad (5.74)$$

For vertical and sliding motions, the contribution of internal damping can be somewhat neglected. However, for torsional and rocking modes of vibration, the contribution of the internal damping may be too large to be ignored.

5.10   Coupled Rocking and Sliding Vibration of Rigid Circular Foundations

In several cases of machine foundations, the rocking and sliding vibrations are coupled. This is because the center of gravity of the footing and vibrators are not coincident with the center of sliding resistance, as can be seen from Figure 5.31a. This is a case of vibration of a foundation with two degrees of freedom. The derivation given next for the coupled motion for rocking and sliding is based on the analysis of Richart and Whitman (1967). From Figure 5.31, it can be seen that the nature of foundation motion shown in Figure 5.31a is equal to the sum of the sliding motion shown in Figure 5.31b and the rocking motion shown in Figure 5.31c. Note that

$$x_b = x_g - h'\theta \tag{5.75}$$

For the sliding motion,

$$m\ddot{x}_g = P \tag{5.76}$$

where

P = horizontal resistance to sliding

$$= -c_x \frac{dx_b}{dt} - k_x x_b \tag{5.77}$$

Substitution of Eq. (5.75) into (5.77) yields

$$P = -c_x \frac{d}{dt}(x_g - h'\theta) - k_x(x_g - h'\theta)$$

$$= -c_x \dot{x}_g + c_x h'\dot{\theta} - k_x x_g + k_x h'\theta \tag{5.78}$$

Now, combining Eqs. (5.76) and (5.78)

$$\boxed{m\ddot{x}_g + c_x \dot{x}_g + k_x x_g - c_x h'\dot{\theta} - k_x h'\theta = 0} \tag{5.79}$$

For rocking motion about the center of gravity,

$$I_g \ddot{\theta} = M + M_r - h'P \tag{5.80}$$

where

I_g = mass moment of inertia about the horizontal axis passing through the center of gravity (at right angles to the cross section shown)

M_r = the soil resistance to rotational motion

But

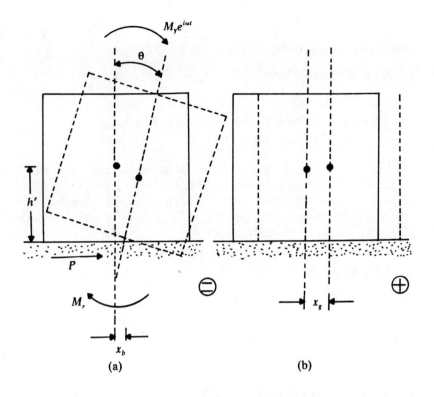

(a)   (b)

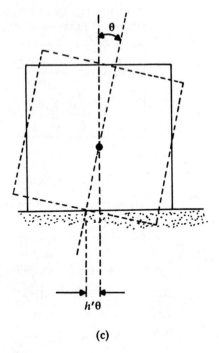

(c)

Figure 5.31   Coupled rocking and sliding vibration

$$M_r = -c_\theta \dot{\theta} - k_\theta \theta \tag{5.81}$$

Substitution of Eqs. (5.78) and (5.81) into Eq. (5.80) gives

$$I_g \ddot{\theta} = M - (c_\theta \dot{\theta} + k_\theta \theta) - h'(-c_x \dot{x}_g + c_x h' \dot{\theta} - k_x x_g + k_x h' \theta)$$

or

$$I_g \ddot{\theta} + (c_\theta + c_x h'^2)\dot{\theta} + (k_\theta + k_x h'^2)\theta - h'(c_x \dot{x}_g + k_x x_g)$$
$$= M = M_y e^{i\omega t} \tag{5.82}$$

For a foundation resting on an elastic half-space, the spring and dashpot coefficients are frequency dependent. They need to be calculated first for a given frequency before Eqs. (5.79) and (5.82) can be solved. However if they are assumed to be frequency independent, as in the case of analog solutions, Eqs. (5.79) and (5.82) can be easily solved. For that case, for determination of the damped natural frequency, one can make M in Eq. (5.82) equal zero. So

$$I_g \ddot{\theta} + (c_\theta + c_x h'^2)\dot{\theta} + (k_\theta + k_x h'^2)\theta - h'(c_x \dot{x}_g + k_x x_g) = 0 \tag{5.83}$$

For solving Eqs. (5.79) and (5.83), let

$$x_g = A_1 e^{i\omega_m t} \tag{5.84}$$

and

$$\theta = A_2 e^{i\omega_m t} \tag{5.85}$$

where ω_m = damped natural frequency.

Substituting Eqs. (5.84) and (5.85) into Eqs. (5.79) and (5.83) and rearranging, one obtains (Prakash and Puri, 1981, 1988)

$$\left[\omega_m^4 - \omega_m^2 \left(\frac{\omega_{n\theta}^2 + \omega_{nx}^2}{\delta} - \frac{4D_\theta D_x \omega_{n\theta} \omega_{nx}}{\delta} \right) + \frac{\omega_{n\theta}^2 \cdot \omega_{nx}^2}{\delta} \right]^2$$
$$+ 4 \left[\frac{D_x \omega_{nx} \omega_m}{\delta} (\omega_{n\theta}^2 - \omega_m^2) + \frac{D_\theta \omega_{n\theta} \omega_m}{\delta} (\omega_{nx}^2 - \omega_m^2) \right]^2 = 0 \tag{5.86}$$

where

D_x = damping ratio for sliding vibration [Eq. (5.56)]

D_θ = damping ratio for rocking vibration [Eq. (5.45)]

$$\delta = \frac{I_g}{I_0} \tag{5.87}$$

[*Note:* The term I_0 was defined in Eq. (5.40).]

$$\omega_{nx} = \sqrt{\frac{32(1 - \mu)Gr_0}{(7 - 8\mu)m}} \tag{5.88}$$

$\uparrow$

[from Eq. (5.54)]

$$\omega_{n\theta} = \sqrt{\frac{8Gr_0^3}{3(1-\mu)I_0}} \qquad (5.89)$$

↑

[from Eqs. (5.41) and (5.44)]

Equation (5.86) can then be solved to obtain two values of ω_m.

The damped amplitudes of rocking and sliding vibrations can be obtained as

$$A_x = \left(\frac{M_y}{I_g}\right) \frac{[(\omega_{nx}^2)^2 + (2D_x\omega_{nx})^2]^{1/2}}{\Delta(\omega^2)} \qquad (5.90)$$

and

$$A_\theta = \left(\frac{M_y}{I_g}\right) \frac{[(\omega_{nx}^2 - \omega^2)^2 + (2D_x\omega_{nx}\omega)^2]^{1/2}}{\Delta(\omega^2)} \qquad (5.91)$$

where

$$\Delta(\omega^2) = \left\{ \left[\omega^4 - \omega^2 \left(\frac{\omega_{n\theta}^2 + \omega_{nx}^2}{\delta} - \frac{4D_\theta D_x\omega_{n\theta}\omega_{nx}}{\delta} \right) + \frac{\omega_{n\theta}^2\omega_{nx}^2}{\delta} \right]^2 \right.$$
$$\left. + 4\left[\frac{D_x\omega_{nx}\omega}{\delta}(\omega_{n\theta}^2 - \omega^2) + \frac{D_\theta\omega_{n\theta}\omega}{\delta}(\omega_{nx}^2 - \omega^2) \right]^2 \right\}^{1/2} \qquad (5.92)$$

5.11   Vibration of Foundations for Impact Machines

There are several machines whose foundations are subjected to transient loads of short duration, often referred to as *impact*. Hammers are typical examples of this type of machine. Figure 5.32 shows a schematic diagram of a hammer foundation system. It consists of the following:

1. Foundation block
2. Anvil and a frame
3. Elastic padding between the anvil and the foundation block
4. Hammer, referred to as a *tup*

The hammer foundation system can be analyzed by assuming a simplified model as shown in Figure 5.33. The spring constant k_1 can be taken from Eq. (5.23) as

$$k_1 = k_z = \frac{4Gr_0}{1-\mu} \qquad (5.93)$$

The spring constant due to the elastic pad is

$$k_2 = \frac{E}{t}A \qquad (5.94)$$

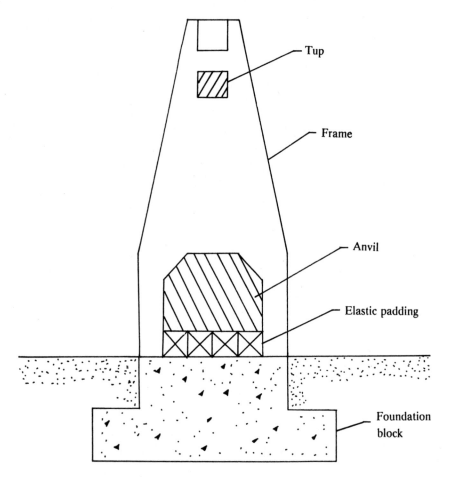

Figure 5.32 Schematic diagram of a hammer foundation

where

E = modulus of elasticity of the pad material

t = thickness of the pad

A = area of the anvil base in contact with the pad

When the tup drops on the anvil, due to the impact the following are the initial conditions:

$$z_1 = 0 \quad \text{and} \quad \dot{z}_1 = 0 \tag{5.95a}$$

$$z_2 = 0 \quad \text{and} \quad \dot{z}_2 = v_0 \tag{5.95b}$$

The equation of motion for the free vibrations may be given as [see Figure 2.13b and Eqs. (2.116) and (2.117)]

$$m_1 \ddot{z}_1 + k_1 z_1 + k_2(z_1 - z_2) = 0 \tag{5.96}$$

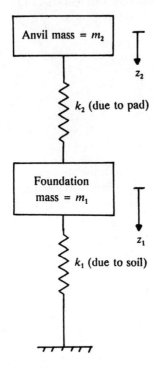

Figure 5.33   Simplified model for analysis of hammer foundation

$$m_2\ddot{z}_2 + k_2(z_2 - z_1) = 0 \tag{5.97}$$

where

m_1 = mass of the foundation + frame (if attached to the foundation block)

m_2 = mass of the anvil + frame (if attached to the anvil)

The solutions for the natural circular frequencies of the system have been given in Eq. (2.122) as

$$\boxed{\omega_n^4 - (1 + \eta)(\omega_{nl_1}^2 + \omega_{nl_2}^2)\omega_n^2 + (1 + \eta)(\omega_{nl_1}^2)(\omega_{nl_2}^2) = 0} \tag{5.98}$$

where

$$\omega_{nl_1} = \sqrt{\frac{k_1}{m_1 + m_2}} \tag{5.99}$$

$$\omega_{nl_2} = \sqrt{\frac{k_2}{m_2}} \tag{5.100}$$

$$\eta = \frac{m_2}{m_1} \tag{5.101}$$

The amplitudes of vibration due to impact can also be given by Eqs. (2.134a) and (2.134b) as

$$Z_1 = \frac{(\omega_{nl_2}^2 - \omega_{n_1}^2)(\omega_{nl_2}^2 - \omega_{n_2}^2)}{\omega_{nl_2}^2(\omega_{n_1}^2 - \omega_{n_2}^2)\omega_{n_2}} v_0 \tag{5.102}$$

$$Z_2 = \frac{(\omega_{nl_2}^2 - \omega_{n_1}^2)v_0}{(\omega_{n_1}^2 - \omega_{n_2}^2)\omega_{n_2}} \tag{5.103}$$

The preceding two equations can be solved to determine the amplitudes of vibration if v_0 is known. This value can be estimated in the following manner. Using the theory of conservation of momentum, the momentum of the tup and the anvil after impact is

$$m_h v_a + m_2 v_0$$

where

m_h = mass of the tup

v_a = velocity of the tup after impact

Thus

$$m_h v_b = m_h v_a + m_2 v_0 \tag{5.104}$$

where v_b = velocity of the tup before impact.

A second equation may be obtained from Newton's second law as

$$n = \frac{v_0 - v_a}{v_b} \tag{5.105}$$

where n is the coefficient of restitution. Combining Eqs. (5.104) and (5.105),

$$v_0 = \frac{1+n}{1+(m_2/m_h)} v_b \tag{5.106}$$

For a single-acting drop hammer, the magnitude of the coefficient of restitution may vary from 0.2 to about 0.5. Also,

$$v_b \text{ (m/s)} = E_f\sqrt{2gH} \tag{5.107}$$

where

H = height of fall of the tup (m)

g = acceleration due to gravity (9.81 m/s^2)

E_f = efficiency of drop (≈ 0.65 to about 1)

VIBRATION OF EMBEDDED FOUNDATIONS

In the theories for the vibration of foundations in various modes, as developed in Sections 5.2 through 5.9, it was assumed that the foundation rests on the ground surface. In reality, however, all foundations are constructed below the ground surface. For an embedded foundation, soil resistance is mobilized at its base and also along its sides. A limited number of theories has so far been developed for the dynamic response of embedded block foundations. The findings from these studies are summarized in the following four sections.

5.12   **Vertical Vibration of Rigid Cylindrical Foundations**

The dynamic response of vertically vibrating rigid cylindrical foundations (Figure 5.34) has been studied by Novak and Beredugo (1972). The foundation shown in Figure 5.34 has a radius of r_0. The shear modulus and the density of the side layer of soil are G_s and ρ_s, respectively. Similarly, the shear modulus and the density of the soil beneath the foundation are, respectively, G and ρ. If the

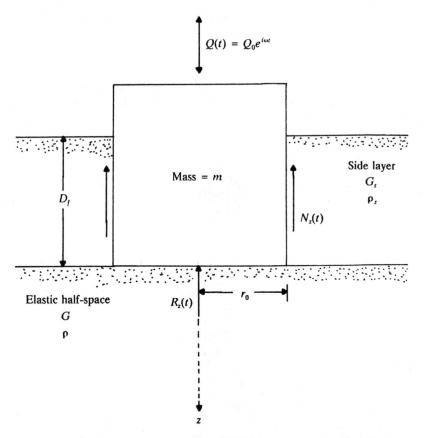

Figure 5.34 Embedded rigid cylindrical foundation—vertical vibration

foundation is subjected to a vertical exciting force, the equation of motion may be written in the form

$$m\ddot{z}(t) = Q(t) - R_z(t) - N_z(t) \tag{5.108}$$

The dynamic reaction $R_z(t)$ is considered to be independent of the depth of embedment. Using the elastic half-space solution, the dynamic reaction can be expressed as

$$R_z(t) = Gr_0(C_1 + iC_2)z(t) \tag{5.109}$$

where

$$C_1 = \frac{-f_1'}{f_1'^2 + f_2'^2} \tag{5.110}$$

and

$$C_2 = \frac{f_2'}{f_1'^2 + f_2'^2}$$

f_1', f_2' = functions of nondimensional frequency a_0 [Eq. (5.5)],   (5.111)
          Poisson's ratio, and stress distribution at the base

The dynamic soil reaction on the sides can be obtained as

$$N_z(t) = \int_0^{D_f} s(z,t)\,dz \tag{5.112}$$

where s is the dynamic reaction per unit depth of embedment.

If s is considered to be independent of depth (Baranov, 1967), then $s = s(t)$, or

$$s(t) = G_s(S_1 + iS_2)z(t) \tag{5.113}$$

where

$$S_1 = 2\pi a_0 \frac{J_1(a_0)J_0(a_0) + Y_1(a_0)Y_0(a_0)}{J_0^2(a_0) + Y_0^2(a_0)} \tag{5.114a}$$

and

$$S_2 = \frac{4}{J_0^2(a_0) + Y_0^2(a_0)} \tag{5.114b}$$

$J_0(a_0), J_1(a_0)$ = Bessel functions of the first kind of order
                    0 and 1, respectively

$Y_0(a_0), Y_1(a_0)$ = Bessel functions of the second kind of order
                    0 and 1, respectively

So

$$N_z(t) = \int_0^{D_f} G_s(S_1 + iS_2)z(t)\,dz = G_s D_f(S_1 + iS_2)z(t) \tag{5.115}$$

Now, combining Eqs. (5.108), (5.109), and (5.115)

$$m\ddot{z}(t) + Gr_0\left[C_1 + iC_2 + \frac{G_s}{G}\frac{D_f}{r_0}(S_1 + iS_2)\right]z(t)$$

$$= Q(t) = Q_0 e^{i\omega t} = Q_0(\cos\omega t + i\sin\omega t) \tag{5.116}$$

The steady-state response is

$$z(t) = ze^{i\omega t} \tag{5.117}$$

In the preceding two equations, Q_0 and z are, respectively, the real force amplitude and real response. The relationships for the spring constant and the damping coefficient can thus be derived as

$$k_z = Gr_0\left(C_1 + \frac{G_s}{G}\frac{D_f}{r_0}S_1\right) \tag{5.118}$$

$$c_z = \frac{Gr_0}{\omega}\left(C_2 + \frac{G_s}{G}\frac{D_f}{r_0}S_2\right) \tag{5.119}$$

Note that k_z and c_z, as expressed by the two preceding relationships, are frequency dependent. However, without losing much accuracy, one can assume that

$C_1 = \overline{C}_1 = $ constant

$S_1 = \overline{S}_1 = $ constant

$C_2 = a_0\overline{C}_2$ (where $\overline{C}_2$ is a constant)

$S_2 = a_0\overline{S}_2$ (where $\overline{S}_2$ is a constant)

When the preceding assumptions are substituted into Eqs. (5.118) and (5.119), one obtains the frequency-independent k_z and c_z as follows:

$$k_z = Gr_0\left(\overline{C}_1 + \frac{G_s}{G}\frac{D_f}{r_0}\overline{S}_1\right) \tag{5.120}$$

$$c_z = r_0^2\sqrt{\rho G}\left(\overline{C}_2 + \overline{S}_2\frac{D_f}{r_0}\sqrt{\frac{G_s\rho_s}{G\rho}}\right) \tag{5.121}$$

Hence, the damping ratio can be given as

$$D_z = \left(\frac{1}{2\sqrt{b}}\right)\frac{\left(\overline{C}_2 + \overline{S}_2\frac{D_f}{r_0}\sqrt{\frac{G_s\rho_s}{G\rho}}\right)}{\sqrt{\overline{C}_1 + \frac{G_s}{G}\frac{D_f}{r_0}\overline{S}_1}} \tag{5.122}$$

where

$$b = \text{mass ratio} = \frac{m}{\rho r_0^3} \quad \text{[Eq. (5.4)]}$$

Table 5.2   Values of $\bar{C}_1, \bar{C}_2, \bar{S}_1$, and $\bar{S}_2$

Poisson's ratio μ	$\bar{C}_1{}^a$	$\bar{C}_2{}^a$	$\bar{S}_1{}^b$	$\bar{S}_2{}^b$
0.0	3.9	3.5	2.7	6.7
0.25	5.2	5.0	2.7	6.7
0.5	7.5	6.8	2.7	6.7

[a] Validity range: $0 \le a_0 \le 1.5$
[b] Validity range: $0 \le a_0 \le 2$

The values of $\bar{C}_1, \bar{C}_2, \bar{S}_1$, and $\bar{S}_2$ (Novak and Beredugo, 1972) are given in Table 5.2.

Once the spring constant, dashpot coefficient, and the damping ratio are determined, the foundation response can be calculated as follows.

Undamped natural frequency:

$$\omega_n = \sqrt{\frac{k_z}{m}}$$

$$f_n = \frac{1}{2\pi}\sqrt{\frac{k_z}{m}}$$

Amplitude of vibration at resonance:

$$A_z = \frac{Q_0}{k_z}\frac{1}{2D_z\sqrt{1-D_z^2}} \qquad \text{(for constant force excitation)}$$

$$A_z = \frac{m_1 e}{m}\frac{1}{2D_z\sqrt{1-D_z^2}} \qquad \text{(for rotating mass excitation)}$$

Amplitude of vibration at frequency other than resonance:

$$A_z = \frac{Q_0/k_z}{\sqrt{[1-(\omega^2/\omega_n^2)]^2 + 4D_z^2(\omega^2/\omega_n^2)}} \qquad \text{(for constant force excitation)}$$

$$A_z = \frac{(m_1 e/m)(\omega/\omega_n)^2}{\sqrt{[1-(\omega^2/\omega_n^2)]^2 + 4D_z^2(\omega^2/\omega_n^2)}} \qquad \text{(for rotating mass excitation)}$$

5.13   Sliding Vibration of Rigid Cylindrical Foundations

Figure 5.35 shows an embedded rigid cylindrical foundation subjected to sliding vibration. The response of this type of system was analyzed by Beredugo and Novak (1972). The frequency-independent spring constant and dashpot coefficient suggested by them are as follows:

$$k_x = Gr_0\left(\bar{C}_{x1} + \frac{G_s}{G}\frac{D_f}{r_0}\bar{S}_{x1}\right) \tag{5.123}$$

$$c_x = r_0^2\sqrt{\rho G}\left(\bar{C}_{x2} + \bar{S}_{x2}\frac{D_f}{r_0}\sqrt{\frac{G_s\rho_s}{G\rho}}\right) \tag{5.124}$$

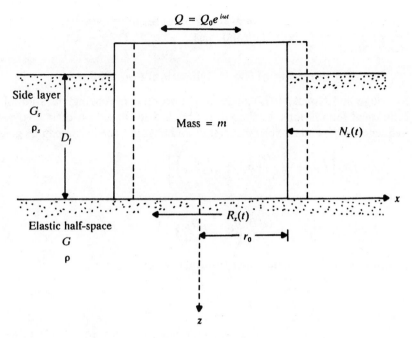

Figure 5.35   Embedded rigid cylindrical foundation—horizontal vibration

The variation of $\overline{C}_{x1}$, $\overline{C}_{x2}$, $\overline{S}_{x1}$, and $\overline{S}_{x2}$ as evaluated by Beredugo and Novak are as follows:

Poisson's ratio μ	Parameter
0	$\overline{C}_{x1} = 4.30;$ $\overline{C}_{x2} = 2.7$
0.5	$\overline{C}_{x1} = 5.10;$ $\overline{C}_{x2} = 3.15$
0	$\overline{S}_{x1} = 3.6;$ $\overline{S}_{x2} = 8.2$
0.25	$\overline{S}_{x1} = 4.0;$ $\overline{S}_{x2} = 9.1$
0.4	$\overline{S}_{x1} = 4.1;$ $\overline{S}_{x2} = 10.6$

The undamped frequency of vibration for this case can be given as

$$\omega_n = \sqrt{\frac{k_x}{m}}$$

and

$$f_n = \frac{1}{2\pi}\sqrt{\frac{k_x}{m}}$$

The damping ratio can be calculated as

$$D_x = \frac{c_x}{2\sqrt{k_x m}}$$

Once ω_n and D_x are calculated, the amplitudes of vibration can be estimated using Eqs. (5.58), (5.59), (5.60), and (5.61).

5.14   Rocking Vibration of Rigid Cylindrical Foundations

Beredugo and Novak (1972) analyzed the problem of rocking vibration of rigid cylindrical foundations, as shown in Figure 5.36. Based on their analysis, the frequency-independent spring constant and dashpot coefficient can be given as

$$k_\theta = Gr_0^3 \left[\overline{C}_{\theta 1} + \frac{G_s}{G} \frac{D_f}{r_0} \left(\overline{S}_{\theta 1} + \frac{D_f^2}{3r_0^2} \overline{S}_{x1} \right) \right] \tag{5.125}$$

and

$$c_\theta = r_0^4 \sqrt{\rho G} \left[\overline{C}_{\theta 2} + \frac{G_s}{G} \frac{D_f}{r_0} \left(\overline{S}_{\theta 2} + \frac{D_f^2}{3r_0^2} \overline{S}_{x2} \right) \right] \tag{5.126}$$

For this problem, the undamped natural frequency is

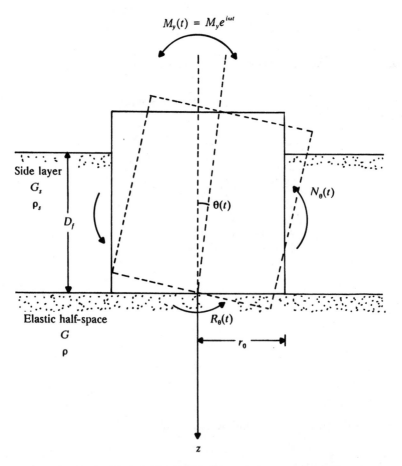

Figure 5.36 Embedded rigid cylindrical foundation—rocking vibration

$$\omega_n = \sqrt{\frac{k_\theta}{I_0}}$$

For the definition of I_0, see Eq. (5.40). The damping ratio is

$$D_\theta = \frac{c_\theta}{2\sqrt{k_\theta I_0}}$$

The amplitudes of vibration can be calculated using Eqs. (5.46), (5.47), (5.48), and (5.49).

The variation of $\bar{S}_{x1}$ and $\bar{S}_{x2}$ was given in the preceding section. For $\mu = 0$,

$$\bar{C}_{\theta 1} = 2.5 \qquad \bar{C}_{\theta 2} = 0.43$$

and, for any value of μ,

$$\bar{S}_{\theta 1} = 2.5 \qquad \bar{S}_{\theta 2} = 1.8$$

5.15   Torsional Vibration of Rigid Cylindrical Foundations

Figure 5.37 shows a rigid cylindrical foundation subjected to a torsional vibration. Novak and Sachs (1973) evaluated the frequency-independent spring constant and dashpot coefficient, and they are as follows:

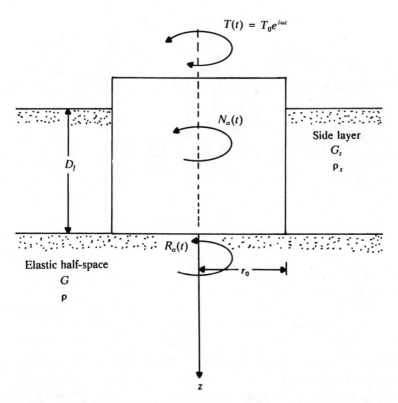

Figure 5.37 Embedded rigid cylindrical foundation—torsional vibration

$$k_\alpha = G r_0^3 \left(\overline{C}_{\alpha 1} + \frac{G_s}{G} \frac{D_f}{r_0} \overline{S}_{\alpha 1} \right) \tag{5.127a}$$

and

$$c_\alpha = r_0^4 \sqrt{\rho G} \left(\overline{C}_{\alpha 2} + \overline{S}_{\alpha 2} \frac{D_f}{r_0} \sqrt{\frac{G_s \rho_s}{G \rho}} \right) \tag{5.127b}$$

The values of the parameters $\overline{C}_{\alpha 1}$, $\overline{C}_{\alpha 2}$, $\overline{S}_{\alpha 1}$, and $\overline{S}_{\alpha 2}$ are

$$\left. \begin{array}{l} \overline{C}_{\alpha 1} = 4.3 \\ \overline{C}_{\alpha 2} = 0.7 \end{array} \right\} \quad \text{for } 0 \le a_0 \le 2.0$$

$$\left. \begin{array}{l} \overline{S}_{\alpha 1} = 12.4 \\ \overline{S}_{\alpha 2} = 2.0 \end{array} \right\} \quad \text{for } 0 \le a_0 \le 0.2$$

$$\left. \begin{array}{l} \overline{S}_{\alpha 1} = 10.2 \\ \overline{S}_{\alpha 2} = 5.4 \end{array} \right\} \quad \text{for } 0.2 \le a_0 \le 2.0$$

Once the magnitudes of k_α and c_α are calculated, the undamped natural circular frequency can be obtained as

$$\omega_n = \sqrt{\frac{k_\alpha}{J_{zz}}}$$

where

J_{zz} = mass moment of inertia of the foundation about the z axis
   [see Eq. (5.63)]

The damping ratio is

$$D_\alpha = \left(\frac{1}{2\sqrt{B_\alpha}} \right) \frac{\left(\overline{C}_{\alpha 2} + \overline{S}_{\alpha 2} \frac{D_f}{r_0} \sqrt{\frac{G_s \rho_s}{G \rho}} \right)}{\sqrt{\overline{C}_{\alpha 1} + \frac{G_s}{G} \frac{D_f}{r_0} \overline{S}_{\alpha 1}}} \tag{5.128}$$

where

$$B_\alpha = \text{mass ratio} = \frac{J_{zz}}{\rho r_0^5} \quad \text{[see Eq. (5.66)]}$$

The amplitudes of vibration can now be calculated using Eqs. (5.68) and (5.69) and Figures 5.12 and 5.13.

VIBRATION SCREENING

In Section 5.4, the allowable vertical vibration amplitudes for machine foundation were considered. It is sometimes possible that, for some rugged vibratory equipment, the intensity of vibration may not be objectionable for the equipment itself. However, the vibration may not be within a tolerable limit for sensitive equipment nearby. Under these circumstances it is desirable to control the vibration energy reaching the sensitive zone. This is referred to as *vibration screening*. It needs to be kept in mind that most of the vibratory energy affecting structures nearby is carried by *Rayleigh (surface) waves* traveling from the source of vibration. Effective screening of vibration may be achieved by proper *interception, scattering,* and *diffraction* of surface waves using barriers such as trenches, sheet pile walls, and piles.

5.16   Active and Passive Isolation: Definition

While studying the problem of vibration screening, it is convenient to group the screening problems into two major categories.

Active Isolation: Active isolation involves screening at the source of vibration, as shown in Figure 5.38, in which a circular trench of radius R and depth H surrounds the foundation that is the source of disturbance.

Passive Isolation: The passive isolation process involves providing a barrier at a point remote from the source of disturbance but near a site where vibration has to be reduced. An example of this is shown in Figure 5.39, in which an open trench of length L and depth H is used near a sensitive instrument foundation to protect it from damage.

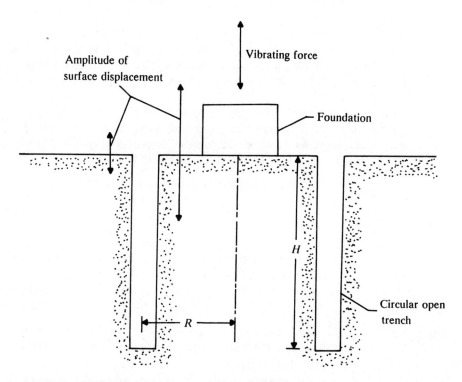

Figure 5.38 Schematic diagram of vibration isolation using a circular trench surrounding the source of vibration—active isolation

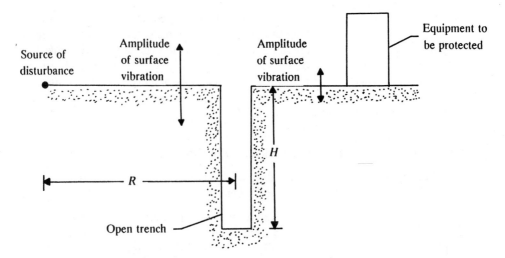

Figure 5.39 Passive isolation by using an open trench

5.17   Active Isolation by Use of Open Trenches

Woods (1968) reported the results of a field investigation for active isolation using open trenches. The field tests were conducted at a site with a deep stratum of silty sand. The experimental study consisted of applying vertical vibrations by a small vibrator [18 lb (80.1 N) maximum force] resting on a circular pad. Trenches were constructed around the circular pad to screen the surface displacement due to the surface waves. Vertical velocity transducers were used for measurement of surface displacement around the trench over a 25-ft- (7.62-m-) diameter area. Other conditions remaining the same, measurements for the surface displacement due to the vibration of the circular pad were also taken without the trenches surrounding the pad. Some results of this investigation are shown in Figure 5.40 in the form of *amplitude-reduction-factor* contour diagrams. The amplitude-reduction factor (ARF) is defined as

$$\text{ARF} = \frac{\text{vertical amplitude of vibration with trench}}{\text{vertical amplitude of vibration without trench}} \tag{5.129}$$

Also note that in Figure 5.40, θ is the angular length of the trench (in degrees) and λ_r is the length of Rayleigh waves. The value of λ_r for a given frequency of vibration at a given site can be determined in a manner similar to that described in Section 4.15. The tests of Woods (1968) were conducted for $R/\lambda_r = 0.222$ to 0.910 and $H/\lambda_r = 0.222$ to 1.82. For satisfactory isolation, Woods defined the ARF to be less than or equal to 0.25. The conclusions of this study can be summarized as follows:

   1. For $\theta = 360°$, a minimum value of $H/\lambda_r = 0.6$ is required to achieve ARFs less than or equal to 0.25.

   2. For $360° > \theta > 90°$, the screened zone may be defined as an area outside the trench bounded on the sides by radial lines from the center of the source

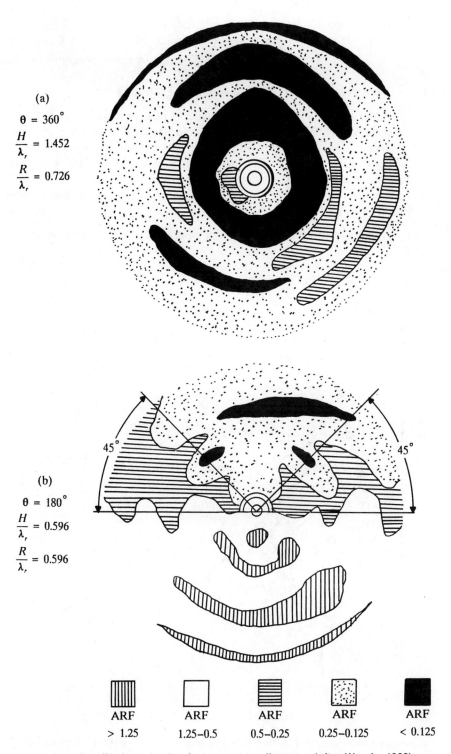

(a)

$\theta = 360°$

$\dfrac{H}{\lambda_r} = 1.452$

$\dfrac{R}{\lambda_r} = 0.726$

(b)

$\theta = 180°$

$\dfrac{H}{\lambda_r} = 0.596$

$\dfrac{R}{\lambda_r} = 0.596$

45° 45°

ARF	ARF	ARF	ARF	ARF
> 1.25	1.25–0.5	0.5–0.25	0.25–0.125	< 0.125

Figure 5.40 Amplitude-reduction factor contour diagrams (after Woods, 1968)

through points 45° from the ends of the trench. To obtain ARFs less than or equal to 0.25 in the screened zone, a minimum value of $H/\lambda_r = 0.6$ is required.

3. For $\theta \le 90°$, effective screen of vibration by trenches cannot be obtained.

5.18   Passive Isolation by Use of Open Trenches

Woods (1968) also investigated the case of passive isolation in the field using open trenches. The plan view of the field site layout used for screening at a distance is shown in Figure 5.41. The layout consisted of two vibrator exciter footings (used one at a time for the tests), a trench barrier, and 75 pickup benches. For these tests, it was assumed that the zone screened by the trench will be symmetrical about the 0° line. The variables used to study the passive isolation tests were

1. the distance from the source of vibration to the center of the open trench, R,

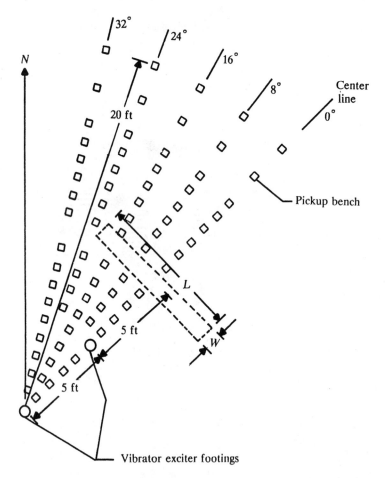

Figure 5.41 Plan view of the field site layout for passive isolation by use of open trench (after Woods, 1968)

2. the length of the trench, L,

3. the width of the trench, W, and

4. the depth of the trench, H.

In this investigation, the value of R/λ_r was varied from 2.22 to 9.10. For satisfactory isolation, it was defined that *ARF's* [*Eq.* (5.129)] *should be less than or equal to 0.25 in a semicircular zone of radius $L/2$ behind the trench.*

Figure 5.42 shows the ARF contour diagram for one of these tests. The conclusions of this study may be summarized as follows:

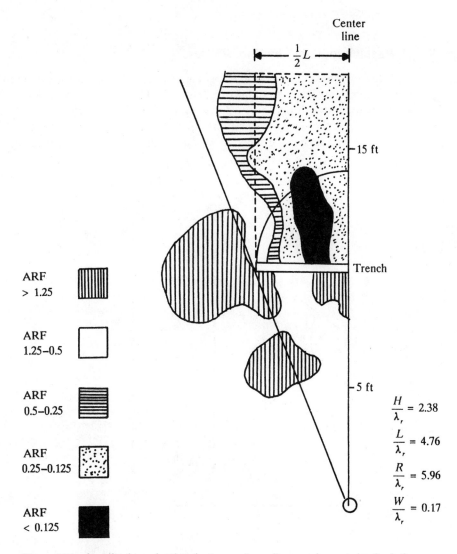

Figure 5.42   Amplitude-reduction-factor contour diagram for passive isolation (after Woods, 1968)

1. For a satisfactory passive isolation (for $R = 2\lambda_r$ to about $7\lambda_r$), the minimum trench depth H should be about $1.2\lambda_r$ to $1.5\lambda_r$. This means that, in general, H/λ_r should be about 1.33.

2. The trench width W has practically no influence on the effectiveness of screening.

3. To maintain the same degree of isolation, the least area of the trench in the vertical direction (that is, $LH = A_T$) should be as follows:

$$A_T = 2.5\lambda_r^2 \quad \text{at} \quad R = 2\lambda_r$$

and

$$A_T = 6.0\lambda_r^2 \quad \text{at} \quad R = 7\lambda_r$$

5.19   Passive Isolation by Use of Piles

There are several situations where Rayleigh waves that emanate from manufactured sources may be in the range of 120 to 150 ft ($\approx$ 40 to 50 m). For these types of problems, a trench depth of 1.33 times 120 to 150 ft (about 60 to 75 m) is needed for effective passive isolation. Open trenches or bentonite-slurry-filled trenches deep enough to be effective are not practical. At the same time, solidifi-

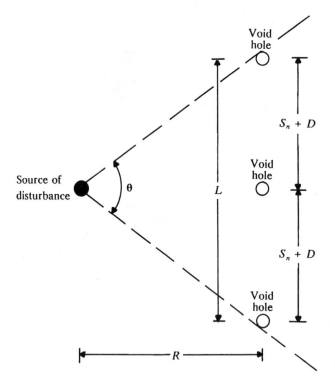

D = diameter of void holes

Figure 5.43   Void cylindrical obstacles for passive isolation

cation of the bentonite-slurry will also pose a problem. For this reason, possible use of rows of piles as an energy barrier was studied by Woods, Barnett, and Sagesser (1974) and Liao and Sangrey (1978). Woods, Barnett, and Sagesser used the principle of holography and observed vibrations in a model half-space in order to develop the criteria for *void cylindrical obstacles* for passive isolation (Figure 5.43). The model half-space was prepared in a fine sand medium in a box. In Figure 5.43, the diameter of the cylindrical obstacle is D, and the net space for the energy to penetrate between two consecutive void obstacles is equal to S_n. The numerical evaluation of the barrier effectiveness was made by obtaining the average ARFs from several lines beyond the barrier in a section $\pm 15°$ on both sides of an axis through the source of disturbance and perpendicular to the barrier. For all tests, H/λ_r and L/λ_r were kept at 1.4 and 2.5, respectively. These values of H/λ_r and L/λ_r are similar to those suggested in Section 5.18 for open trenches. A nondimensional plot of the *isolation effectiveness* developed from these tests is given in Figure 5.44. The isolation effectiveness is defined as

$$\text{Effectiveness} = 1 - \text{ARF} \tag{5.130}$$

Based on these test results, Woods, Barnett, and Sagesser (1974) suggested that a row of void cylindrical holes may act as an isolation barrier if

$$\frac{D}{\lambda_r} \geq \frac{1}{6} \tag{5.131}$$

and

$$\frac{S_n}{\lambda_r} < \frac{1}{4} \tag{5.132}$$

Liao and Sangrey (1978) used an acoustic model employing sound waves in a fluid medium to evaluate the possibility of the use of rows of piles as passive isolation barriers. Model piles for the tests were made from aluminum, steel, styrofoam, and polystyrene plastic. Based on their study, Liao and Sangrey determined that Eqs. (5.131) and (5.132) suggested by Woods, Barnett, and Sagesser are generally valid. They also determined that $S_n = 0.4\lambda_r$ may be the upper limit for a barrier to have some effectiveness. However, the degree of effectiveness of the barrier will depend on whether the piles are soft or hard compared to the soil in which they are embedded. The degree of softness or hardness may be determined by the term *impedance ratio* (IR), defined as

$$\text{IR} = \frac{\rho_P v_{r(P)}}{\rho_S v_{r(S)}} \tag{5.133}$$

where

ρ_P and ρ_S = the densities of the pile material and soil, respectively

$v_{r(P)}$ and $v_{r(S)}$ = the velocities of Rayleigh waves in the pile material and soil, respectively

The piles are considered *soft* if IR is less than 1 and *hard* if IR is greater than 1. Soft piles are more efficient as isolation barriers compared to hard piles. Figure

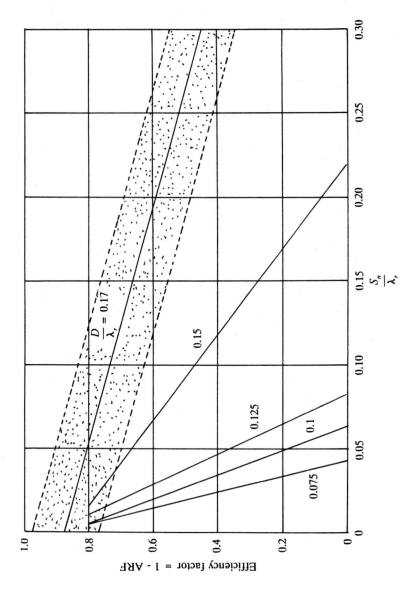

Figure 5.44 Isolation effectiveness as a function of hole diameter and spacing (redrawn after Woods, Barnett, and Sagesser, 1974)

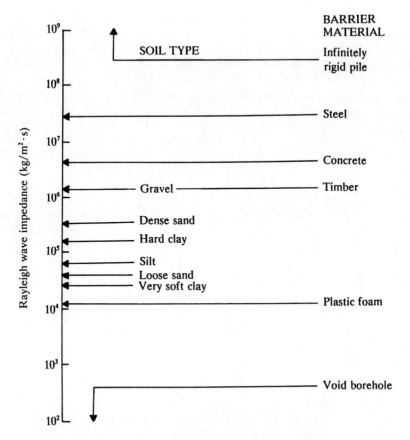

Figure 5.45 Estimated values of Rayleigh wave impedance for various soils and pile materials (after Liao and Sangrey, 1978)

5.45 gives a general range of the Rayleigh wave impedance ($= \rho v_r$) for various soils and pile materials. For a more detailed discussion, the reader is referred to the original paper of Liao and Sangrey.

PROBLEMS

5.1 A concrete foundation is 8 ft in diameter. The foundation is supporting a machine. The total weight of the machine and the foundation is 60,000 lb. The machine imparts a vertical vibrating force $Q = Q_0 \sin \omega t$. Given $Q_0 = 6000$ lb (not frequency dependent). The operating frequency is 150 cpm. For the soil supporting the foundation, unit weight = 120 lb/ft^3, shear modulus = 6500 lb/in.^{2}, and Poisson's ratio = 0.3. Determine:
a. the resonant frequency,
b. the amplitude of vertical vibration at resonant frequency, and
c. the amplitude of vertical vibration at the operating frequency.

5.2 Redo Problem 5.1 assuming the foundation is 8 ft × 6 ft in plan. Assume the total weight of the foundation and the machine is the same as in Problem 5.1.

5.3 A concrete foundation (unit weight = 23.5 kN/m^3) supporting a machine is 3.5 m × 2.5 m in plan and is subjected to a sinusoidal vibrating force (vertical) having an amplitude of

10 kN (not frequency dependent). The operating frequency is 2000 cpm. The weight of the machine and foundation is 400 kN. The soil properties are unit weight $= 18 \, kN/m^3$, shear modulus $= 38,000 \, kN/m^2$, and Poisson's ratio $= 0.25$. Determine
a. the resonant frequency of the foundation, and
b. the amplitude of vertical vibration at the operating frequency.

5.4   Consider the case of a single-cylinder reciprocating engine (Figure 5.15a). For the engine, operating speed $= 1000$ cpm, crank $(r_1) = 90$ mm, connecting rod $(r_2) = 350$ mm, weight of the engine $= 20$ kN, and reciprocating weight $= 65$ N. The engine is supported by a concrete foundation block of 3 m × 2 m × 1.5 m ($L \times B \times H$). The unit weight of concrete is $23.58 \, kN/m^3$. The properties of the soil supporting the foundation are unit weight $= 19 \, kN/m^3$, $G = 24,000 \, kN/m^2$, and $\mu = 0.25$. Calculate
a. the resonant frequency, and
b. the amplitude of vertical vibration at resonance.

5.5   Refer to Problem 5.4. What will be the amplitude of vibration at operating frequency?

5.6   Solve Example 5.2, parts (b) and (c) by assuming that the Poisson's ratio is $\mu = 0.25$. Also determine the amplitude of vertical vibration at operating frequency.

5.7   The concrete foundation (unit weight $= 23.5 \, kN/m^3$) of a machine has the following dimensions (refer to Figure 5.18): $L = 3$ m, $B = 4$ m, height of the foundation $= 1.5$ m. The foundation is subjected to a sinusoidal horizontal force from the machine having an amplitude of 10 kN at a height of 2 m measured from the base of the foundation. The soil supporting the foundation is sandy clay. Given $G = 30,000 \, kN/m^2$, $\mu = 0.2$, and $\rho = 1700$ kg/m^3. Determine
a. the resonant frequency for the rocking mode of vibration of the foundation, and
b. the amplitude of rocking vibration at resonance.
(*Note:* The amplitude of horizontal force is not frequency dependent. Neglect the moment of inertia of the machine.)

5.8   Solve Problem 5.7 assuming that the horizontal force is frequency dependent. The amplitude of the force at an operating speed of 800 cpm is 20 kN.

5.9   Refer to Problem 5.7. Determine
a. the resonant frequency for the sliding mode of vibration, and
b. amplitude for the sliding mode of vibration at resonance.
Assume the weight of the machinery on the foundation to be 100 kN.

5.10   Repeat Problem 5.9 assuming that the horizontal force is frequency dependent. The amplitude of the horizontal force at an operating frequency of 800 cpm is 40 kN. The weight of the machinery on the foundation is 100 kN.

5.11   A concrete foundation (unit weight $= 23.5 \, kN/m^3$) supporting a machine has the following dimensions: length $= 5$ m, width $= 4$ m, height $= 2$ m. The machine imparts a torque T on the foundation such that $T = T_0 e^{i\omega t}$. Given $T_0 = 3000 \, N \cdot m$. The mass moment of inertia of the machine about the vertical axis passing through the center of gravity of the foundation is 75×10^3 kg-m^2. The soil has the following properties: $\mu = 0.25$, unit weight $= 18 \, kN/m^3$, and $G = 28,000 \, kN/m^2$. Determine
a. the resonant frequency for the torsional mode of vibration, and
b. angular deflection at resonance.

5.12   Consider the case of a drop hammer foundation. For this system the frame is attached to the anvil. Given are the following:

weight of the anvil and frame $= 130,000$ lb
weight of foundation $= 200,000$ lb
spring constant for the elastic pad between the anvil and foundation $= 150 \times 10^6$ lb/ft
spring constant for the soil supporting the foundation $= 22 \times 10^6$ lb/ft

weight of tup $= 8000$ lb
velocity of tup before impact $= 10$ ft/s
coefficient of restitution, $n = 0.4$.

Determine the amplitude of vibration of the anvil and the foundation.

5.13 Refer to Figure 5.34 for the vertical vibration of a rigid cylindrical concrete foundation. Given the following:
Foundation

radius $= 4.5$ ft
height $= 5$ ft
depth of embedment, $D_f = 3.5$ ft
unit weight of concrete $= 150$ lb/ft^3

Vibratory machine

weight $= 25,000$ lb
amplitude of vibrating force $= 2500$ lb (not frequency dependent)
operating speed $= 600$ cpm

Soil

$G_s = 3200$ lb/in.^{2} $G = 3000$ lb/ft^2 $\mu = 0.25$
unit weight, $\gamma_s = 115$ lb/ft^3 (for side layer)
unit weight, $\gamma = 122$ lb/ft^3 (below the base)

Determine
a. the damped natural frequency,
b. the amplitude of vertical vibration at resonance, and
c. the amplitude of vibration at operating speed.

5.14 Solve Problem 5.13 with the following changes:
Concrete foundation

length $= 6$ ft
width $= 5$ ft
height $= 5$ ft
depth of embedment, $D_f = 4$ ft
unit weight of concrete $= 150$ lb/ft^3

Vibratory machine

weight $= 20,000$ lb
frequency-dependent amplitude of vibrating force
   $= 2000$ lb at an operating speed of 500 cpm

5.15 Refer to Figure 5.35 for the sliding vibration of a rigid cylindrical foundation. Given the following:
Concrete foundation

radius $= 3$ m
height $= 4$ m
depth of embedment, $D_f = 2.5$ m
unit weight of concrete $= 23$ kN/m^3

Vibrating machine

weight $= 100$ kN
frequency-dependent unbalanced force at an operating
   frequency of 600 cpm $= 40$ kN

Soil

$G_s = 16,000 \text{ lb/in.}^2$ $G = 18,000 \text{ lb/ft}^2$ $\mu = 0$
unit weight, $\gamma_s = 17.8 \text{ kN/m}^3$ (for side layer)
unit weight, $\gamma = 18.8 \text{ kN/m}^3$ (below the base)

Determine
a. the natural frequency,
b. the amplitude of horizontal vibration at resonance, and
c. the amplitude of horizontal vibration at operating speed.

5.16   A horizontal piston-type compressor is shown in Figure P5.16. The operating speed is
800 cpm. The amplitude of the horizontal unbalanced force of the compressor is 25 kN.

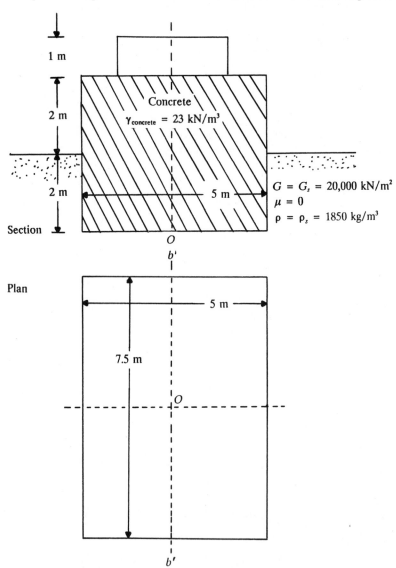

Figure P5.16

It creates a rocking motion of the foundation about O. The mass moment of inertia of the compressor assembly about the $b'Ob'$ axis is 20×10^5 kg·m². Determine
a. the natural frequency, and
b. the amplitude of rocking vibration at resonance. Use the theory developed in Section 5.14.

REFERENCES

Arnold, R. N., Bycroft, G. N., and Wartburton, G. B. (1955). "Forced Vibrations of a Body on an Infinite Elastic Solid," *Journal of Applied Mechanics*, ASME, Vol. 77, pp. 391–401.

Baranov, V. A. (1967). "On the Calculation of Excited Vibrations of an Embedded Foundation," (in Russian), *Vopr. Dyn. Prochn.*, Vol. 14, pp. 195–209.

Barkan, D. D. (1962). *Dynamic Bases and Foundations*, McGraw-Hill Book Company, New York.

Beredugo, Y. O., and Novak, M. (1972). "Coupled Horizontal and Rocking Vibration of Embedded Footings," *Canadian Geotechnical Journal*, Vol. 9, No. 4, pp. 477–497.

Borowicka, H. (1943). "Uber Ausmittig Belastere Starre Platten auf Elastischisotropem Undergrund," *Ingenieur-Archiv*, Berlin, Vol. 1, pp. 1–8.

Bycroft, G. N. (1956). "Forced Vibrations of a Rigid Circular Plate on a Semi-Infinite Elastic Space and on an Elastic Stratum," *Philosophical Transactions of the Royal Society*, London, Ser. A., Vol. 248, pp. 327–368.

Dobry, R., and Gazetas, G. (1986). "Dynamic Response of Arbitrarily Shaped Foundations," *Journal of the Geotechnical Engineering Division*, ASCE, Vol. 112, No. GT2, pp. 109–135.

Fry, Z. B. (1963). "Report 1: Development and Evaluation of Soil Bearing Capacity, Foundation of Structures, Field Vibratory Tests Data," *Technical Report No. 3-632*, U.S. Army Engineers Waterways Experiment Station, Vicksburg, Mississippi.

Gorbunov-Possadov, M. I., and Serebrajanyi, R. V. (1961). "Design of Structures upon Elastic Foundations," *Proceedings*, 5th International Conference on Soil Mechanics and Foundation Engineering, Vol. 1, pp. 643–648.

Hall, J. R., Jr. (1967). "Coupled Rocking and Sliding Oscillations of Rigid Circular Footings," *Proceedings*, International Symposium on Wave Propagation and Dynamic Properties of Earth Materials, Albuquerque, New Mexico, pp. 139–148.

Hsieh, T. K. (1962). "Foundation Vibrations," *Proceedings*, Institute of Civil Engineers, London, Vol. 22, pp. 211–226.

Lamb, H. (1904). "On the Propagation of Tremors over the Surface of an Elastic Solid," *Philosophical Transactions of the Royal Society*, London, Ser. A., Vol. 203, pp. 1–42.

Liao, S., and Sangrey, D. A. (1978). "Use of Piles as Isolation Barriers," *Journal of the Geotechnical Engineering Division*, ASCE, Vol. 104, No. GT9, pp. 1139–1152.

Lysmer, J., and Richart, F. E., Jr. (1966). "Dynamic Response of Footings to Vertical Loading," *Journal of the Soil Mechanics and Foundations Division*, ASCE, Vol. 92, No. SM1, pp. 65–91.

Novak, M., and Beredugo, Y. O. (1972). "Vertical Vibration of Embedded Footings," *Journal of the Soil Mechanics and Foundations Division*, ASCE, Vol. 98, No. SM12, pp. 1291–1310.

Novak, M., and Sachs, K. (1973). "Torsional and Coupled Vibrations of Embedded Footings," *International Journal of Earthquake Engineering and Structural Dynamics*, Vol. 2, No. 1, pp. 11–33.

Prakash, S., and Puri, V. K. (1981). "Observed and Predicted Response of a Machine Foundation," *Proceedings*, 10th International Conference on Soil Mechanics and Foundation Engineering, Stockholm, Vol. 3, pp. 269–272.

Prakash, S., and Puri, V. K. (1988). *Foundations for Machines: Analysis and Design*, John Wiley and Sons, New York.

Quinlan, P. M. (1953). "The Elastic Theory of Soil Dynamics," Symposium on Dynamic Testing of Soils, *Special Technical Publication 156*, ASTM, pp. 3–34.

Reissner, E. (1936). "Stationare, axialsymmetrische durch eine Schuttelnde Masseerregte Schwingungen eines homogenen elastischen halbraumes," *Ingenieur-Archiv.*, Vol. 7, No. 6, pp. 381–396.

Reissner, E. (1937). "Freie und erzwungene Torsionschwingungen des elastischen halbraumes," *Ingenieur-Archiv.*, Vol. 8, No. 4, pp. 229–245.

Reissner, E., and Sagochi, H. F. (1944). "Forced Torsional Oscillations of an Elastic Half Space," *Journal of Applied Physics*, Vol. 15, pp. 652–662.

Richart, F. E., Jr. (1962). "Foundation Vibrations," *Transactions*, ASCE, Vol. 127, Part 1, pp. 863–898.

Richart, F. E., Jr., Hall, J. R., and Woods, R. D. (1970). *Vibration of Soils and Foundations*, Prentice-Hall, Inc., Englewood Cliffs, New Jersey.

Richart, F. E., Jr. and Whitman, R. V. (1967). "Comparison of Footing Vibration Tests with Theory," *Journal of the Soil Mechanics and Foundations Division*, ASCE, Vol. 93, No. SM6, pp. 143–167.

Sung, T. Y. (1953). "Vibration in Semi-Infinite Solids Due to Periodic Surface Loadings," Symposium on Dynamic Testing of Soils, *Special Technical Publication No. 156*, ASTM, pp. 35–54.

Timoshenko, S. P., and Goodier, J. H. (1951). *Theory of Elasticity*. McGraw-Hill Book Company, New York.

Whitman, R. V., and Richart, F. E., Jr. (1967). "Design Procedures for Dynamically Loaded Foundations," *Journal of the Soil Mechanics and Foundations Division*, ASCE, Vol. 93, No. SM6, pp. 169–193.

Woods, R. D. (1968)."Screening of Surface Waves in Soils," *Journal of the Soil Mechanics and Foundations Division*, ASCE, Vol. 94, No. SM4, pp. 951–979.

Woods, R. D., Barnett, N. E., and Sagesser, R. (1974). "Holography—A New Tool for Soil Dynamics," *Journal of the Geotechnical Engineering Division*, ASCE, Vol. 100, No. GT11, pp. 1231–1247.

Chapter 6

DYNAMIC BEARING CAPACITY OF SHALLOW FOUNDATIONS

6.1   Introduction

The static bearing capacity of shallow foundations has been extensively studied and reported in the literature. However, foundations can be subjected to single pulse dynamic loads which may be in vertical or horizontal directions. The dynamic loads due to nuclear blasts are mainly vertical. Horizontal dynamic loads on foundations are due mostly to earthquakes. These types of loading may induce large permanent deformations in foundations. A fundamental definition of the *dynamic bearing capacity* has not yet been found. However, one must keep in mind that, during the analysis of the time-dependent motion of a foundation subjected to dynamic loading, several factors need to be considered. Most important of these factors are

a. nature of variation of the magnitude of the loading pulse,

b. duration of the pulse, and

c. strain-rate response of the soil during deformation

A rather limited amount of information on the dynamic bearing capacity of foundations is available in literature at this time. Most of the important works on this topic are summarized in this chapter.

ULTIMATE DYNAMIC BEARING CAPACITY

6.2   Bearing Capacity in Sand

The static ultimate bearing capacity of shallow foundations subjected to vertical loading (Figure 6.1) can be given by the equation

$$q_u = cN_c S_c d_c + qN_q S_q d_q + \tfrac{1}{2}\gamma B N_\gamma S_\gamma d_\gamma \qquad (6.1)$$

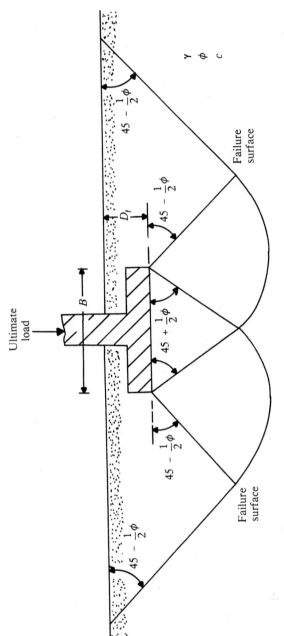

Figure 6.1   Static ultimate bearing capacity of continuous shallow foundations

where

q_u = ultimate load per unit area of the foundation

γ = effective unit weight of soil

$q = \gamma D_f$

D_f = depth of foundation

B = width of foundation

c = cohesion of soil

N_c, N_q, N_γ = bearing capacity factors which are only functions of the soil friction angle ϕ

S_c, S_q, S_γ = shape factors

d_c, d_q, d_γ = depth factors

In sands, with $c = 0$, Eq. (6.1) becomes

$$q_u = qN_qS_qd_q + \tfrac{1}{2}\gamma BN_\gamma S_\gamma d_\gamma \qquad\qquad (6.2)$$

The values of N_q (Reissner, 1924) and N_γ (Caquot and Kerisel, 1953; Vesic, 1973) can be represented by the following equations:

$$N_q = e^{\pi \tan \phi} \tan^2\left(45 + \frac{\phi}{2}\right) \qquad\qquad (6.3)$$

$$N_\gamma = 2(N_q + 1)\tan \phi \qquad\qquad (6.4)$$

where ϕ is the angle of friction of soil. The values of N_q and N_γ for various soil friction angles are given in Table 6.1. The shape and depth factors have been proposed by DeBeer (1970) and Brinch Hanson (1970):

Shape Factors

$$S_q = 1 + \left(\frac{B}{L}\right)\tan \phi \qquad\qquad (6.5)$$

$$S_\gamma = 1 - 0.4\left(\frac{B}{L}\right) \qquad\qquad (6.6)$$

Depth Factors

$$\text{For } \frac{D_f}{B} \leq 1, \quad d_q = 1 + 2\tan\phi(1 - \sin\phi)^2\left(\frac{D_f}{B}\right) \qquad\qquad (6.7)$$

$$d_\gamma = 1 \qquad\qquad (6.8)$$

$$\text{For } \frac{D_f}{B} > 1, \quad d_q = 1 + 2\tan\phi(1 - \sin\phi)^2 \tan^{-1}\left(\frac{D_f}{B}\right) \qquad\qquad (6.9)$$

$$d_\gamma = 1 \qquad\qquad (6.10)$$

Table 6.1   Values[a] of Bearing Capacity Factors, N_q and N_γ

ϕ	N_q	N_γ	ϕ	N_q	N_γ
0	1.00	0.00	26	11.85	12.54
			27	13.20	14.47
1	1.09	0.07	28	14.72	16.72
2	1.20	0.15	29	16.44	19.34
3	1.31	0.24	30	18.40	22.40
4	1.43	0.34			
5	1.57	0.45	31	20.63	25.99
			32	23.18	30.22
6	1.72	0.57	33	26.09	35.19
7	1.88	0.71	34	29.44	41.06
8	2.06	0.86	35	33.30	48.03
9	2.25	1.03			
10	2.47	1.22	36	37.75	56.31
			37	42.92	66.19
11	2.71	1.44	38	48.93	78.03
12	2.97	1.69	39	55.96	92.25
13	3.26	1.97	40	64.20	109.41
14	3.59	2.29			
15	3.94	2.65	41	73.90	130.22
			42	85.38	155.55
16	4.34	3.06	43	99.02	186.54
17	4.77	3.53	44	115.31	224.64
18	5.26	4.07	45	134.88	271.76
19	5.80	4.68			
20	6.40	5.39	46	158.51	330.35
			47	187.21	403.67
21	7.07	6.20	48	222.31	496.01
22	7.82	7.13	49	265.51	613.16
23	8.66	8.20	50	319.07	762.89
24	9.60	9.44			
25	10.66	10.88			

[a] After Vesic (1973).

In Eqs. (6.5)–(6.10), B and L are the width and length of rectangular foundations, respectively. For circular foundations, B is the diameter, and $B = L$.

The preceding equations for static ultimate bearing capacity evaluation are valid for dense sands where the failure surface in the soil extends to the ground surface as shown in Figure 6.1. This is what is referred to as the case of *general shear failure*. For shallow foundations (i.e., $D_f/B \leq 1$), if the relative density of granular soils R_D is less than about 70%, *local* or *punching shear failure* may occur. Hence, for static ultimate bearing capacity calculation, if $0 \leq R_D \leq 0.67$, the values of ϕ in Eqs. (6.3)–(6.10) should be replaced by the modified friction angle

$$\phi' = \tan^{-1}[(0.67 + R_D - 0.75R_D^2)\tan\phi] \qquad (6.11)$$

The facts just described relate to the static bearing capacity of shallow foundations. However, when load is applied rapidly to a foundation to cause failure, the ultimate bearing capacity changes somewhat. This fact has been shown experimentally by Vesic, Banks, and Woodward (1965), who conducted several laboratory model tests with a 4-in. (101.6-mm) diameter rigid rough model footing placed on the surface of a dense river sand (i.e., $D_f = 0$), both dry and saturated. The rate of loading to cause failure was varied in a range of 10^{-5} in./s to over 10 in./s. Hence, the rate was in the range of static (10^{-5} in./s) to impact (10 in./s) loading conditions. All but the four most rapid tests in submerged sand [loading velocity, 0.576–0.790 in./s (14.63–20.07 mm/s)] showed peak failure loads as obtained in the case of general shear failure of soil. The four most rapid tests in *submerged* sand gave the load-displacement plots as obtained in the case of *punching shear* failure, where the failure planes do not extend to the ground surface.

For surface footings ($D_f = 0$) in sand, $q = 0$ and $d_\gamma = 1$. So

$$q_u = \tfrac{1}{2}\gamma B N_\gamma S_\gamma \tag{6.12}$$

or

$$\frac{q_u}{\tfrac{1}{2}\gamma B} = N_\gamma S_\gamma \tag{6.13}$$

The variation of $q_u/\tfrac{1}{2}\gamma B$ with load velocity for the tests of Vesic, Banks, and Woodward (1965) is shown in Figure 6.2. It may be seen that, for any given series

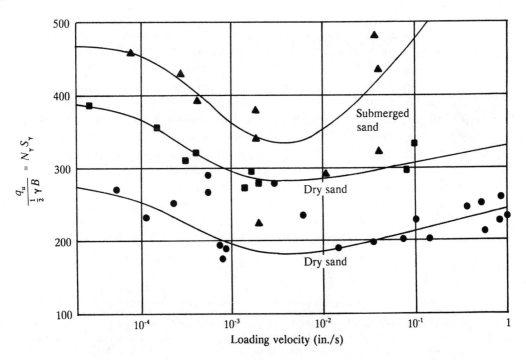

Figure 6.2 Plot of bearing capacity factor versus loading velocity (after Vesic, Banks, and Woodward, 1965)

of tests, the value of $q_u/\frac{1}{2}\gamma B$ gradually decreases with the loading velocity to a minimum value and then continues to increase. This, in effect, corresponds to a decrease in the angle of friction of soil by about 2° when the loading velocity reached a value of about 2×10^{-3} in./s (50.8×10^{-3} mm/s). Such effects of strain rate in reducing the angle of friction of sand has also been observed by Whitman and Healy (1962), as described in Chapter 4.

Based on the experimental results available at this time, the following general conclusions regarding the ultimate dynamic bearing capacity of shallow foundations in sand can be drawn:

1. For a foundation resting on sand and subjected to an acceleration level of $a_{max} \leq 13$ g, it is possible for general shear type of failure to occur in soil (Heller, 1964).

2. For a foundation on sand subjected to an acceleration level of $a_{max} > 13$ g, the nature of soil failure is by punching (Heller, 1964).

3. The difference in the nature of failure in soil is due to the inertial restrain of the soil involved in failure during the dynamic loading. The restrain has almost a similar effect as the overburden pressure as observed during the dynamic loading which causes the punching shear type failure in soil.

4. The *minimum value* of the *ultimate dynamic bearing capacity* of shallow foundations on dense sands obtained between static to impact loading range can be estimated by using a friction angle ϕ_{dy}, such that (Vesic, 1973)

$$\phi_{dy} = \phi - 2° \tag{6.14}$$

The value of ϕ_{dy} can be substituted in place of ϕ in Eqs. (6.2)–(6.10). However, if the soil strength parameters with proper strain rate are known from laboratory testing, they should be used instead of the approximate equation [Eq. (6.14)].

5. The increase of the ultimate bearing capacity at high loading rates as seen in Figure 6.2 is due to the fact that the soil particles in the failure zone do not always follow the *path of least resistance*. This results in a higher shear strength of soil, which leads to a higher bearing capacity.

6. In the case of foundations resting on loose submerged sands, transient liquefaction effects (Chapter 10) may exist (Vesic, 1973). This may result in unreliable prediction of ultimate bearing capacity.

7. The rapid increase of the ultimate bearing capacity in dense saturated sand at fast loading rates is due to the development of negative pore water pressure in the soil.

Example 6.1

A square foundation with dimensions $B \times B$ has to be constructed on a dense sand. Its depth is $D_f = 1$ m. The unit weight and the static angle of friction of the soil can be assigned representative values of 18 kN/m³ and 39°, respectively. The foundation may occasionally be subjected to a maximum dynamic load of 1800 kN increasing at a moderate rate. Determine the size of the foundation using a safety factor of 3.

Solution

Given that $\phi = 39°$ in the absence of any other experimental data, for minimum ultimate dynamic bearing capacity

$$\phi_{dy} = \phi - 2° = 39 - 2 = 37°$$

From Eq. (6.2)

$$q_u = qN_qS_qd_q + \tfrac{1}{2}\gamma BN_\gamma S_\gamma d_\gamma$$

$$q = \gamma D_f = (18)(1) = 18 \text{ kN/m}^2$$

For $\phi_{dy} = 37°$, $N_q = 42.92$ and $N_\gamma = 66.19$.

$$S_q = 1 + \left(\frac{B}{L}\right)\tan\phi = 1 + \tan 37° = 1.754$$

$$S_\gamma = 1 - 0.4\left(\frac{B}{L}\right) = 1 - 0.4 = 0.6$$

$$d_q = 1 + 2\tan\phi(1 - \sin\phi)^2\left(\frac{D_f}{B}\right)$$

$$= 1 + 2\tan 37°(1 - \sin 37°)^2\left(\frac{1}{B}\right) = 1 + \frac{0.239}{B}$$

$$d_\gamma = 1$$

Thus

$$q_u(\text{kN/m}^2) = (18)(42.92)(1.754)\left(1 + \frac{0.239}{B}\right)$$

$$+ \tfrac{1}{2}(18)(B)(66.19)(0.6)(1)$$

$$= 1355 + \frac{323.9}{B} + 357.4B \tag{a}$$

Given

$$q_u = \frac{1800 \times 3}{B^2}\text{kN/m}^2 \tag{b}$$

Combining Eqs. (a) and (b),

$$\frac{5400}{B^2} = 1355 + \frac{323.9}{B} + 357.4B \tag{c}$$

Following is a table to determine the value of B by trial and error. Clearly, $B \approx 1.6$ m.

B (m)	$5400/B^2$ (kN/m^2)	$1355 + 323.9/B + 357.4B$ (kN/m^2)
2	1350	2331.75
1.5	2400	2107
1.6	2109	2133

■

6.3   Bearing Capacity in Clay

For foundations resting on saturated clays ($\phi = 0$ and $c = c_u$; i.e., undrained condition), Eq. (6.1) transforms to the form

$$q_u = c_u N_c S_c d_c + q N_q S_q d_q \tag{6.15}$$

(*Note*: $N_\gamma = 0$ for $\phi = 0$ in Table 6.1.)

$$N_c = 5.14 \tag{6.16}$$

and

$$N_q = 1 \tag{6.17}$$

The values for S_c and S_q (DeBeer, 1970) and d_c and d_q (Brinch Hansen, 1970) are as follows:

$$S_c = 1 + \left(\frac{B}{L}\right)\left(\frac{N_q}{N_c}\right)$$

For $\phi = 0$,

$$S_c = 1 + \left(\frac{B}{L}\right)\left(\frac{1}{5.14}\right) = 1 + 0.1946\left(\frac{B}{L}\right) \tag{6.18}$$

$$S_q = 1 + \tan\phi$$

$$S_q = 1 \tag{6.19}$$

$$d_c = 1 + 0.4\left(\frac{D_f}{B}\right) \qquad \text{for } \frac{D_f}{B} \leq 1 \tag{6.20}$$

$$d_c = 1 + 0.4\tan^{-1}\left(\frac{D_f}{B}\right) \qquad \text{for } \frac{D_f}{B} > 1 \tag{6.21}$$

$$d_q = 1 \tag{6.22}$$

Substituting Eqs. (6.16)–(6.22) into Eq. (6.15),

$$\boxed{q_u = 5.14c_u\left[1 + 0.1946\left(\frac{B}{L}\right)\right]\left[1 + 0.4\left(\frac{D_f}{B}\right)\right] + q \qquad \text{for } \frac{D_f}{B} \leq 1} \tag{6.23}$$

and

$$\boxed{q_u = 5.14c_u\left[1 + 0.1946\left(\frac{B}{L}\right)\right]\left[1 + 0.4\tan^{-1}\left(\frac{D_f}{B}\right)\right] + q \qquad \text{for } \frac{D_f}{B} > 1} \tag{6.24}$$

The ultimate bearing capacity of foundations resting on saturated clay soils can be estimated by using Eqs. (6.23) and (6.24), provided the strain-rate effect due to dynamic loading is taken into consideration in determination of the

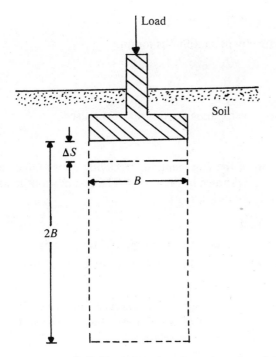

Figure 6.3   Definition of strain rate under a foundation

undrained cohesion. Unlike the case in sand, the undrained cohesion of saturated clays increases with the increase of the strain rate. This fact was discussed in Chapter 4 in relation to the unconsolidated-undrained triaxial tests on *Buckshot* clay. Based on those results, Carroll (1963) suggested that $c_{u(dyn)}/c_{u(stat)}$ may be approximated to be about 1.5.

For a given foundation, the strain rate $\dot{\varepsilon}$ can be approximated as (Figure 6.3)

$$\dot{\varepsilon} = \left(\frac{1}{\Delta t}\right)\left(\frac{\frac{1}{2}\Delta S}{B}\right) \tag{6.25}$$

where B is the width of the foundation.

BEHAVIOR UNDER TRANSIENT LOAD

6.4   **Transient Vertical Load on Foundation (Rotational Mode of Failure)**

Triandafilidis (1965) has presented a solution for dynamic response of continuous footing supported by *saturated cohesive soil* ($\phi = 0$ condition) and subjected to a transient load. Let the transient stress pulse be expressed in the form

$$q_d = q_0 e^{-\beta t} = \lambda q_u e^{-\beta t} \tag{6.26}$$

where

q_d = stress at time t

q_u = *static* bearing capacity of continuous footing

β = decay function

t = time

$q_0 = \lambda q_u$ = *instantaneous* peak intensity of the stress pulse

λ = overload factor

The nature of variation of the exponentially decaying stress pulse represented by Eq. (6.26) with time is shown in Figure 6.4. The assumptions made by Triandafilidis in the analysis are as follows:

1. A condition of $\phi = 0$ exists.

2. The saturated cohesive soil behaves as a rigid plastic material.

3. The failure surface of soil is cylindrical for evaluation of the bearing capacity under static condition.

Such a failure surface is shown in Figure 6.5. The center of rotation of the failure surface (point 0) is located at a height of $0.43B$ above the ground surface; the static bearing capacity can be given by the relation

$$q_u = 5.54c_u \tag{6.27}$$

where c_u = undrained cohesion.

With the stress pulse on the foundation, the equation of motion can be given by (Figure 6.5)

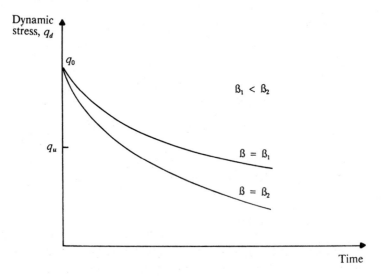

Figure 6.4   Nature of variation of exponentially decaying stress pulse [Eq. (6.26)]

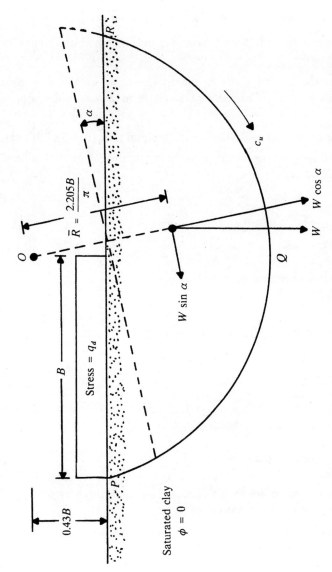

Figure 6.5   Transient vertical load on continuous foundation resting on saturated clay (rotational mode of failure)

moment of the driving forces per unit length about point 0

= moment of the restoring forces per unit length about point 0

$$(6.28)$$

Moment of the *driving force* is due to the pulse load and is equal to $q_d(B \times 1) \times (\frac{1}{2}B) = \frac{1}{2}q_d B^2$. The moment of restoring forces is the sum of the following:

1. Moment of the soil resistance, which is equal to $q_u(B \times 1)(\frac{1}{2}B) = \frac{1}{2}q_u B^2$

2. Soil inertia—resisting moment due to the rigid body motion of the *failed soil mass PQR* (Figure 6.5)

$$M_I = J_0\ddot{\alpha} \tag{6.29}$$

where

$\dot{\alpha}$ = angular acceleration

$$J_0 = \text{polar mass moment of inertia} = \frac{WB^2}{1.36g} \tag{6.30}$$

where

W = weight per unit length of the cylindrical soil mass

$PQR = 0.31\pi\gamma B^2$

$$(6.31)$$

g = acceleration due to gravity

γ = unit weight of soil

Substituting Eq. (6.30) into Eq. (6.29) yields

$$M_I = \left(\frac{WB^2}{1.36g}\right)\ddot{\alpha} \tag{6.32}$$

3. Restoring moment due to the displaced soil mass *PQR*, which is equal to $W\bar{R}\sin\alpha$ (see Figure 6.5 for definition of $\bar{R}$)

Substitution of the preceding moments into Eq. (6.28) gives

$$\frac{1}{2}q_d B^2 = \frac{1}{2}q_u B^2 + \left(\frac{WB^2}{1.36g}\right)\ddot{\alpha} + W\bar{R}\sin\alpha \tag{6.33}$$

Assuming $\sin\alpha \approx \alpha$, rearrangement of the preceding equation yields

$$\ddot{\alpha} + \left(\frac{3g}{\pi B}\right)\alpha = \left(\frac{0.68g}{W}\right)(q_d - q_u) \tag{6.34}$$

Substituting the expression for q_d [Eq. (6.26)] into Eq. (6.34) gives

$$\ddot{\alpha} + \left(\frac{3g}{\pi B}\right)\alpha = \left(\frac{0.68g}{W}\right)q_u(\lambda e^{-\beta t} - 1) \tag{6.35}$$

It may be noted that Eq. (6.35) is of a form similar to that of Eq. (2.23). The natural period of foundations can be given by the relation

$$T = \frac{2\pi}{\sqrt{3g/\pi B}} = 2\pi \sqrt{\frac{\pi B}{3g}}$$

Solution of Eq. (6.35) yields the following relation

$$
\frac{W}{0.68 g q_u}(\alpha) = \frac{T^2}{4\pi^2 + \beta^2 T^2}\left[\left(1 - \lambda + \frac{\beta^2 T^2}{4\pi^2}\right)\cos\left(\frac{2\pi t}{T}\right)\right.
$$
$$
\left. + \frac{\beta \lambda T}{2\pi}\sin\left(\frac{2\pi t}{T}\right) + \lambda e^{-\beta t} - \frac{\beta^2 T^2}{4\pi^2} - 1\right] \tag{6.36}
$$

This relation can be used to trace the history of motion of the foundation. For determination of the maximum angular deflection α, Eq. (6.36) can be differentiated with respect to time. Thus

$$
\frac{W}{0.68 g q_u}\dot{\alpha} = \frac{2\pi T}{4\pi^2 + \beta^2 T^2}\left[\left(\lambda - 1 - \frac{\beta^2 T^2}{4\pi^2}\right)\sin\left(\frac{2\pi t}{T}\right)\right.
$$
$$
\left. + \frac{\beta \lambda T}{2\pi}\cos\left(\frac{2\pi t}{T}\right) - \frac{\beta \lambda T}{2\pi}e^{-\beta t}\right] \tag{6.37}
$$

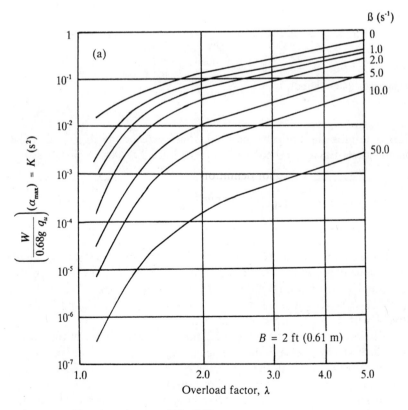

Figure 6.6   (Continued on pp. 270–271)

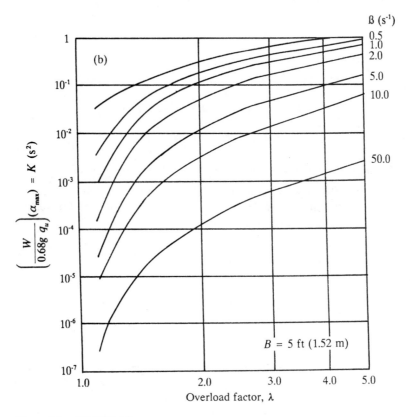

Figure 6.6 (Continued)

For obtaining the critical time $t = t_c$ that corresponds to $\alpha = \alpha_{max}$, the right-hand side of Eq. (6.37) is equated to zero. Since $2\pi T/(4\pi^2 + \beta^2 T^2) \neq 0$,

$$\left(\lambda - 1 - \frac{\beta^2 T^2}{4\pi^2}\right)\sin\left(\frac{2\pi t}{T}\right) + \frac{\beta\lambda T}{2\pi}\cos\left(\frac{2\pi t}{T}\right) - \frac{\beta\lambda T}{2\pi}e^{-\beta t} = 0 \qquad (6.38)$$

By using small increments of time t in Eq. (6.38), the value of t_c can be obtained. This value of $t = t_c$ can then be substituted into Eq. (6.36) with known values of β, λ, and B to obtain $(W/0.68gq_u)\alpha_{max} = K$. Figure 6.6a–c gives the values of $(W/0.68gq_u)\alpha_{max} = K$ (s²) for $B = 2$, 5, and 10 ft (0.61, 1.52, and 3.05 m), respectively, with $\lambda = 1$–5 and $\beta = 0$–50 s⁻¹.

However, it needs to be pointed out that the influence of the *strain rate* on the shear strength of soil and the *dead weight of the foundation* have not been introduced in obtaining the preceding equations. Another factor of concern is the assumption of the nature of failure surface (general shear failure).

It has been pointed out by Triandafilidis that the preceding results should not be applied for rotation $\alpha > 15°$–$20°$.

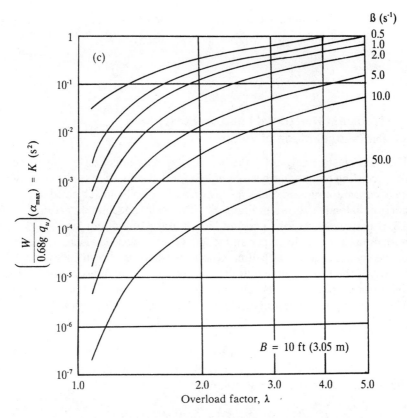

Figure 6.6 Variation of the dynamic load factor, $(W/0.68gq_u)(\alpha_{max})$ with overload factor λ for various values of decay function (after Triandafilidis, 1965)

Example 6.2

A 5-ft-wide strip foundation is subjected to a transient stress pulse, which can be given as $q_d = 10{,}000e^{-\beta t}$ lb/ft^2, $\beta = 10$ s^{-1}. Determine the maximum angular rotation the footing might undergo. The soil supporting the foundation is saturated clay with undrained cohesion $= 900$ lb/ft^2. The saturated unit weight of the soil is 115 lb/ft^3.

Solution

Given $B = 5$ ft, $c_u = 900$ lb/ft^2; thus the static bearing capacity

$$q_u = c_u N_c = 5.54c_u = (5.54)(900) = 4986 \text{ lb/ft}^2$$

Overload factor $= \lambda = \dfrac{q_0}{q_u} = \dfrac{10{,}000}{4986} = 2$

Referring to Figure 6.6b, for $\lambda = 2$, $\beta = 10$ s^{-1};

$$\left(\frac{W}{0.68gq_u}\right)\alpha_{max} \approx 0.003 \text{ s}^2$$

From Eq. (6.31)

$$W = 0.31\pi\gamma B^2 = (0.31)(\pi)(115)(5)^2 = 2800 \text{ lb/ft}$$

$$\alpha_{\text{max}} = \frac{(0.003)(0.68)gq_u}{W} = \frac{(0.003)(0.68)(32.2)(4986)}{2800}$$

$$= 0.117 \text{ rad} = \underline{6.7°} \qquad\blacksquare$$

6.5   Transient Horizontal Load on Foundation (Rotational Mode of Failure)

The rigid plastic analysis for the bearing capacity in cohesive soils presented in the preceding section has been extended for determination of the bearing capacity of continuous foundations resting on a $c - \phi$ soil and subjected to a transient horizontal load as shown in Figure 6.7 (Prakash and Chummar, 1967). In Figure 6.8, Q is the vertical load per unit length of the foundation and $\lambda Q = Q_d$ is the horizontal transient force per unit length (λ is the overload factor). The variation of the dynamic force with time considered here is shown in Figure 6.7. The nature of the failure surface in the soil due to the loadings is assumed to be a logarithmic spiral, with its center located at the corner of the foundation base. The static ultimate bearing capacity of such a foundation can be given by

$$q_u = cN_c + \tfrac{1}{2}\gamma BN_\gamma \qquad (6.39)$$

where c is cohesion, γ is the unit weight of soil, and B is the foundation width. Using such a simplified assumption, Prakash and Chummar have determined the bearing capacity factors as follows:

$$N_c = \frac{(e^{2\pi \tan \phi} - 1)}{\tan \phi} \qquad (6.40)$$

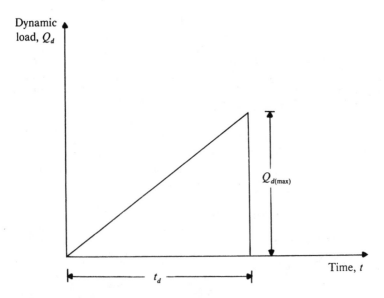

Figure 6.7   Horizontal transient load

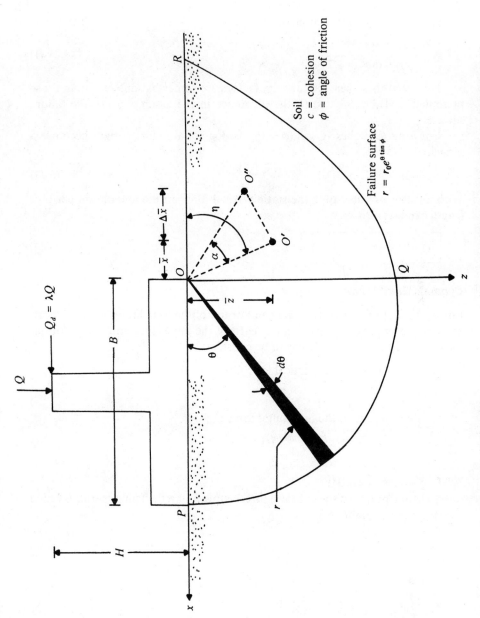

Figure 6.8   Transient horizontal load on a continuous foundation resting on ground surface (rotational mode of failure)

where ϕ is the soil friction angle, and

$$N_\gamma = \frac{4\tan\phi(e^{3\pi\tan\phi} + 1)}{9\tan^2\phi + 1} \qquad (6.41)$$

Note that these bearing capacity factors are somewhat different than those presented in Table 6.1 due to the difference in the assumption of the failure surface.

For a purely cohesive soil ($\phi = 0$), the log-spiral failure surface becomes a semicircle and

$$N_c = 2\pi \qquad (6.42)$$

With a factor of safety of 2, the static vertical force on the foundation per unit length can be given as

$$\boxed{Q = \tfrac{1}{2}B(cN_c + \tfrac{1}{2}\gamma BN_\gamma)} \qquad (6.43)$$

Dynamic Equilibrium

For consideration of the dynamic equilibrium of the foundation with the horizontal transient load, the moment of each of the forces (per unit length) about the center of the log spiral needs to be considered:

1. Moment due to the vertical load Q:

$$M_1 = \tfrac{1}{2}QB \qquad (6.44)$$

2. Moment due to the horizontal force Q_d:

$$M_2 = Q_dH = \frac{Q_{d(max)}Ht}{t_d} = \frac{M_{d(max)}t}{t_d} \qquad (6.45)$$

where $M_{d(max)} = Q_{d(max)}H$.

3. Moment due to the cohesive force acting along the failure surface, which can be given as (Figure 6.8)

$$M_3 = \left(\frac{c}{2\tan\phi}\right)(r_1^2 - r_0^2) \qquad (6.46)$$

In this case, $r_0 = \overline{OP} = B$ and

$$r_1 = r_0 e^{\theta\tan\phi} = \overline{OR} = Be^{\pi\tan\phi}$$

Substitution of these values of r_0 and r_1 into Eq. (6.46) gives

$$M_3 = \psi cB^2 \qquad (6.47)$$

where

$$\psi = \frac{\tfrac{1}{2}(e^{2\pi\tan\phi} - 1)}{\tan\phi} \qquad (6.48)$$

4. Moment due to the frictional resistance along the failure surface: One of the properties of a log spiral is that the radial line at any point on it makes an angle ϕ with the normal at that point. The resultant of the normal and shear forces developed due to friction at any point of a log spiral makes an angle ϕ with the normal at that point; hence, the direction is the same as the radial line which passes through the origin 0. Thus

$$M_4 = 0 \tag{6.49}$$

5. Moment due to the weight of soil mass in the failure wedge PQR:

$$M_5 = \int_0^{\pi} [\tfrac{1}{2}(r\,d\theta)(r)][\tfrac{2}{3}r\cos\theta]\gamma$$

Substituting $r = r_0 e^{\theta\tan\phi} = Be^{\theta\tan\phi}$ and integrating with limits of $\theta = 0$ to $\theta = \pi$ yields

$$M_5 = -\varepsilon\gamma B^3 \tag{6.50}$$

where

$$\varepsilon = \tan\phi(e^{3\pi\tan\phi} + 1)/(9\tan^2\phi + 1) \tag{6.51}$$

The negative sign in Eq. (6.50) indicates that the moment is clockwise. For a purely cohesive soil ($\phi = 0$),

$$M_5 = 0 \tag{6.52}$$

6. Moment of the force due to displacement of the center of gravity of the failure wedge (O' in Figure 6.8) from its initial position:

$$M_6 = W\Delta\bar{x} \tag{6.53}$$

where the weight of the failure wedge is

$$W = \frac{\gamma B^2(e^{2\pi\tan\phi} - 1)}{4\tan\phi} \tag{6.54}$$

$$\Delta\bar{x} = R\cos(\eta - \alpha) - \bar{x} \tag{6.55}$$

and $R = \overline{OO'}$ (Figure 6.8). When α is small, Eq. (6.55) can be written as

$$\Delta\bar{x} = (R\cos\eta)\alpha \tag{6.56}$$

However,

$$R = \sqrt{\bar{x}^2 + \bar{z}^2} \tag{6.57}$$

where

$$\bar{x} = \frac{-4B\tan^2\phi(e^{3\pi\tan\phi} + 1)}{(9\tan^2\phi + 1)(e^{2\pi\tan\phi} - 1)} \tag{6.58}$$

and

$$\bar{z} = \frac{4B\tan\phi(e^{3\pi\tan\phi} + 1)}{3(\sqrt{9\tan^2\phi + 1})(e^{2\pi\tan\phi} - 1)} \tag{6.59}$$

Combining Eqs. (6.53)–(6.59),

$$M_6 = \beta B^3(\sin\eta)\alpha \tag{6.60}$$

where

$$\beta = \frac{e^{3\pi \tan \phi} + 1}{3\sqrt{9 \tan^2 \phi + 1}} \tag{6.61}$$

when

$$\phi = 0, \qquad \beta = \tfrac{2}{3} \tag{6.62}$$

7. Moment due to inertia force of soil wedge:

$$M_7 = \left(\frac{d^2\alpha}{dt^2}\right) J \tag{6.63}$$

where J is the mass moment of inertia of the soil wedge about the axis of rotation:

$$J = \left(\frac{\gamma B^4}{16g \tan \phi}\right)(e^{4\pi \tan \phi} - 1) \tag{6.64}$$

and g is the acceleration due to gravity. Substitution of Eq. (6.64) into Eq. (6.63) yields

$$M_7 = \frac{\mu \gamma B^4}{g} \frac{d^2\alpha}{dt^2} \tag{6.65}$$

where

$$\mu = \frac{e^{4\pi \tan \phi} - 1}{16 \tan \phi} \tag{6.66}$$

For $\phi = 0$,

$$\mu = \tfrac{1}{4}\pi \tag{6.67}$$

Now for the equation of motion,

$$M_1 + M_2 = M_3 + M_4 + M_5 + M_6 + M_7 \tag{6.68}$$

Substitution of the proper terms for the moments in Eq. (6.68) gives

$$\frac{d^2\alpha}{dt^2} + k^2\alpha = A\left(\frac{M_{d(\max)}t}{t_d} + \frac{1}{2}QB - E\right) \tag{6.69}$$

where

$$k = \sqrt{\frac{g\beta \sin \eta}{\mu B}} \tag{6.70}$$

$$A = \frac{g}{\gamma B^4 \mu} \tag{6.71}$$

$$E = \psi c B^2 + \varepsilon \gamma B^3 \tag{6.72}$$

Solution of the differential equation of motion [Eq. (6.69)] with proper boundary conditions yields the following results:

For $t \leq t_d$:

$$
\alpha = \frac{A}{k^2}\left(E - \frac{1}{2}QB\right)\cos(kt) - \frac{A}{k^3}\frac{M_{d(\max)}}{t_d}\sin(kt)
$$
$$
+ \frac{A}{k^2}\left[\frac{M_{d(\max)}t}{t_d} + \frac{1}{2}QB - E\right]
$$

(6.73)

For $t > t_d$:

$$
\alpha = \left(\frac{1}{k}\right)[Gk\cos(kt_d) - H'\sin(kt_d)]\cos(kt)
$$
$$
+ \left(\frac{1}{k}\right)[Gk\sin(kt_d) + H'\cos(kt_d)]\sin(kt) + \left(\frac{A}{k^2}\right)\left(\frac{1}{2}QB - E\right)
$$

(7.74)

where

$$
G = \frac{A}{k^2}\left(E - \frac{1}{2}QB\right)\cos(kt_d) - \frac{A}{k^3}\frac{M_{d(\max)}}{t_d}\sin(kt_d) + \frac{AM_{d(\max)}}{k^2}
$$

(6.75)

and

$$
H' = -\frac{A}{k}\left(E - \frac{1}{2}QB\right)\sin(kt_d) - \frac{A}{k^2}\frac{M_{d(\max)}}{t_d}\cos(kt_d) + \frac{A}{k^2}\frac{M_{d(\max)}}{t_d}
$$

(6.76)

Equations (6.73) and (6.74) can be used to determine the angle of rotation α of a foundation with time. This can be done as follows.

1. Determine the soil parameters, c, ϕ, and γ.

2. Determine Q. Equation (6.43) is based on a factor of safety of 2; however, any other factor of safety can be used.

3. Determine B, H, and t_d.

4. Determine ψ, ε, μ, and β. They are only functions of the friction angle ϕ.

5. Determine $\sin\eta$.

$$
\sin\eta = \frac{\bar{z}}{\sqrt{\bar{x}^2 + \bar{z}^2}}
$$

6. Determine the values of k, A, and E.

7. $M_{d(\max)} = HQ_{d(\max)} = H\lambda Q$

8. Substitute the values of A, k, E, Q, B, and $M_{d(\max)}$ in Eq. (6.73). Note that λ is unknown now, so the expression takes the form

$$
\alpha = f(\lambda, t)
$$

(6.77)

9. Substitute $t = t_d$ in Eq. (6.77) and increase λ gradually starting at zero. The value at which α becomes positive is the critical value of $\lambda = \lambda_{cr}$. For values $\lambda < \lambda_{cr}$, α is negative. This means that the failure surface in soil has not developed.

10. Using $\lambda = \lambda_{cr}$, calculate $M_{d(max)}$, G, and H.

11. Go to Eq. (6.74). For various values of $t > t_d$, calculate α.

Figure 6.9 shows such a calculation as outlined previously for $B = 6.56$ ft (2 m) and $t_d = 0.25$ s for various values of c and ϕ. Note that the maximum value of $\alpha = \alpha_{max}$ occurs at approximately the same time for various combinations of c and ϕ. Figure 6.10 shows a plot of α_{max} versus ϕ for given values of c, B, and t_d. The value of α_{max} decreases with the increase of friction angle. Figure 6.11 shows the variation of α_{max} with the cohesion of soil for given values of ϕ, B, and t_d. Also, the value of the maximum angle of rotation is shown to decrease with increase of the width of the foundation (α_{max} versus B in Figure 6.12) for given values of c, ϕ, and t_d.

It needs to be pointed out that λ_{cr} was about 0.38 in all combinations shown in Figures 6.9, 6.10, 6.11, and 6.12.

6.6   Vertical Transient Load on Strip Foundation (Punching Mode of Failure)

In Sections 6.4 and 6.5, the analyses for strip foundations presented involved a rotational mode of failure. However, it is possible that a foundation may fail by vertically punching into the soil mass due to the application of a vertical transient load. Wallace (1961) has presented a procedure for the estimation of the vertical displacement of a strip foundation with the assumption that the soil behaves as a *rigid plastic material*. In this analysis, the failure surface in the soil mass is assumed to be of similar type as suggested by Terzaghi (1943) for the evaluation of static bearing capacity of strip foundations. This is shown in Figure 6.13. Note that bd is an arc of a logarithmic spiral with its center at 0 which is defined by the equation $r = r_0 e^{\theta \tan \phi}$. In Figure 6.13, r_0 is the distance $0b$.

The static ultimate bearing capacity for such a failure surface can be expressed as

$$q_u = cN_c + qN_q + \tfrac{1}{2}\gamma BN_\gamma \tag{6.78}$$

where B is the foundation width and $q = \gamma D_f$. Comparing Eq. (6.78) with Eq. (6.1), it is apparent that the depth factors (d_c, d_q, and d_γ) have all been assumed to be equal to 1.

If $q = 0$ (i.e., $D_f = 0$) and $\gamma = 0$ (weightless soil),

$$q_u = cN_c \tag{6.79}$$

Again, if $c = 0$ and $\gamma = 0$,

$$q_u = qN_q \tag{6.80}$$

Also, if $c = 0$ and $q = 0$ (i.e., $D_f = 0$),

$$q_u = \tfrac{1}{2}BN_\gamma \tag{6.81}$$

Using this method and assuming O to be the center of the log spiral, Wallace determined the following relations for the bearing capacity factors N_c, N_q, and N_γ.

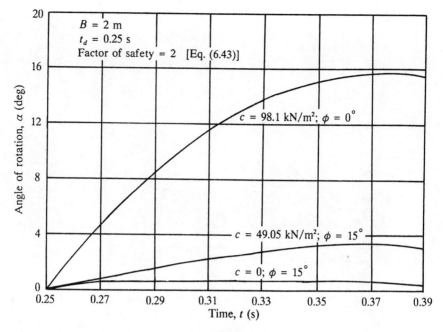

Figure 6.9   Variation of the angle of rotation of continuous foundation due to a horizontal transient load (redrawn after Prakash and Chummar, 1967)

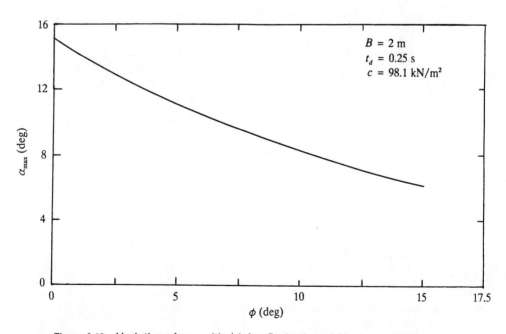

Figure 6.10   Variation of α_{max} with ϕ (after Prakash and Chummar, 1967)

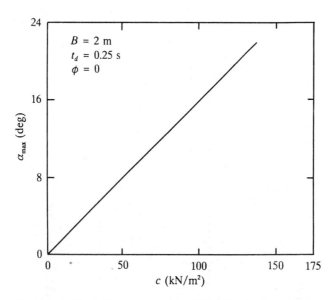

Figure 6.11   Variation of α_{max} with c (after Prakash and Chummar, 1967)

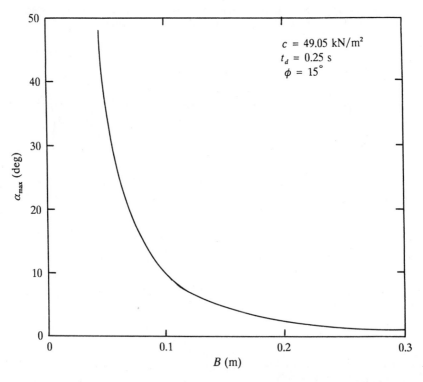

Figure 6.12   Variation of α_{max} with the width of foundation (after Prakash and Chummar, 1967)

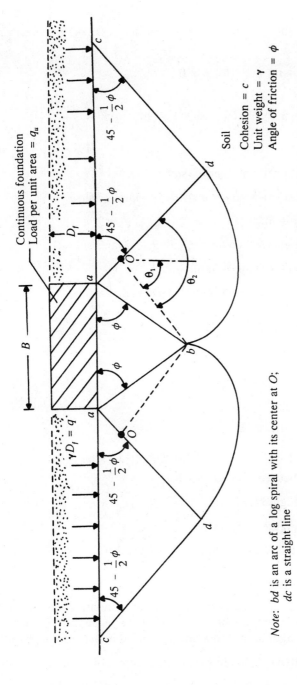

Note: bd is an arc of a log spiral with its center at O; dc is a straight line

Figure 6.13   Failure surface in a soil mass for determination of static bearing capacity of continuous foundation

$$N_c = (\tfrac{1}{2}b^2 + mb^2 \sin \delta)^{-1}\left[(r_2^2 - m^2b^2)(\tfrac{1}{2}\cos \phi) + (2 \tan \phi)^{-1}(r_2^2 - r_0^2)\right.$$

$$\left. -\frac{mb^2}{2\cos \delta}\right] + \tan \phi \qquad \text{for } \phi \neq 0 \tag{6.82}$$

$$N_c = (\tfrac{1}{2}b^2 + mb^2 \sin \delta)^{-1}\left[\tfrac{1}{2}(r_0^2 - m^2b^2) + (\tfrac{3}{4}\pi + \alpha)r_0^2 - \frac{mb^2}{2\cos \delta}\right]$$

$$\text{for } \phi = 0 \tag{6.83}$$

$$N_q = (\tfrac{1}{2}b^2 + mb^2 \sin \delta)^{-1}(r_2^2 - m^2b^2)\sin^2\delta \tag{6.84}$$

$$N_\gamma = (\tfrac{2}{3}b^3 + mb^3 \sin \delta)^{-1}\{\tfrac{1}{3}(r_2 + mb)^2(r_2 - \tfrac{1}{2}mb)\cos \phi \sin \delta$$

$$+ (27 \tan^2\phi + 3)^{-1}$$

$$\times [r_2^3(3 \tan \phi \sin \delta - \cos \delta) + r_0^3(\cos \theta_1 + 3 \tan \phi \sin \theta_1)]$$

$$- \tfrac{1}{6}r_0 \cos(\phi - \alpha)[\tan \phi - \tan(\phi - \alpha)]$$

$$\times [r_0^2 \cos^2(\phi - \alpha) - m^2b^2 \sin^2\delta]\} - \tfrac{1}{2}\tan \phi \tag{6.85}$$

where

$$b = \tfrac{1}{2}B$$

$$\delta = 45 + \tfrac{1}{2}\phi$$

$$\alpha = \angle\, abO$$

$$m = \frac{\text{length } Oa}{b}$$

$\theta_1 =$ internal $\angle$ between r_0 and the vertical line through O

$\theta_2 =$ internal $\angle$ between r_0 and line Od

$r_0 =$ length Ob (Figure 6.13)

$r_2 =$ length Od (Figure 6.13)

The values of N_c, N_q, and N_γ for various values of ϕ and m are given in Table 6.2.

Differential Equation of Motion

In order to obtain the equation of motion for the foundation with a transient load on it, one needs to consider *only one-half of the foundation* (Figure 6.14a). The forces per unit length of the foundation involved in here are as follows:

1. Force due to dynamic loading (refer to Figure 6.14b):

$$\text{Dynamic load} = q_{d(\max)}b\left(\frac{1 - t}{t_d}\right) \qquad \text{for } 0 \leq t \leq t_d \tag{6.86}$$

$$\text{Dynamic load} = 0 \qquad \qquad \text{for } t_d \geq 0 \tag{6.87}$$

2. Force due to soil resistance $= q_u b$.

Table 6.2   Bearing Capacity Factors $(N_c, N_q, N_\gamma, N_I,$ and $N_R)$ [a]

ϕ (deg)	m	N_γ	N_c	N_q	N_I	N_R	$\sqrt{\dfrac{N_R}{N_I}}$
0	-0.05	0.0000	5.7277	1.0	0.0633	2.0125	5.6366
	0.00	0.0000	5.7124	1.0	0.0631	1.9723	5.5887
	$+0.05$	0.0000	5.7258	1.0	0.0633	1.9433	5.5394
5	-0.65	0.1454	79.6255	7.9664	0.3755	8.9076	4.8709
	-0.60	0.1445	29.8163	3.6086	0.2280	6.4362	5.3126
	-0.55	0.1481	18.9958	2.6619	0.1579	5.0332	5.6460
	-0.50	0.1553	14.3469	2.2552	0.1213	4.1699	5.8636
	-0.45	0.1655	11.8179	2.0339	0.1011	3.6088	5.9750
	-0.40	0.1786	10.2699	1.8985	0.0897	3.2299	6.0020
	-0.35	0.1945	9.2580	1.8100	0.0833	2.9674	5.9698
	-0.30	0.2131	8.5723	1.7500	0.0799	2.7828	5.9005
	-0.25	0.2344	8.1007	1.7087	0.0786	2.6523	5.8108
	-0.20	0.2585	7.7778	1.6805	0.0785	2.5604	5.7116
	-0.15	0.2855	7.5629	1.6617	0.0793	2.4969	5.6099
	-0.10	0.3154	7.4291	1.6500	0.0809	2.4547	5.5096
	-0.05	0.3483	7.3580	1.6437	0.0829	2.4288	5.4128
	0.00	0.3843	7.3366	1.6419	0.0853	2.4155	5.3205
	$+0.05$	0.4233	7.3553	1.6435	0.0881	2.4122	5.2330
10	-0.60	0.5700	53.9491	10.5127	0.1120	5.7922	7.1922
	-0.55	0.5588	28.9945	6.1125	0.0935	4.8411	7.1948
	-0.50	0.5645	20.5266	4.6194	0.0833	4.2238	7.1228
	-0.45	0.5832	16.3539	3.8837	0.0779	3.8095	6.9932
	-0.40	0.6127	13.9337	3.4569	0.0757	3.5264	6.8273
	-0.35	0.6521	12.4031	3.1870	0.0755	3.3323	6.6445
	-0.30	0.7008	11.3881	3.0080	0.0767	3.2008	6.4587
	-0.25	0.7586	10.7004	2.8868	0.0790	3.1147	6.2781
	-0.20	0.8253	10.2345	2.8046	0.0821	3.0625	6.1071
	-0.15	0.9012	9.9267	2.7503	0.0858	3.0360	5.9474
	-0.10	0.9863	9.7361	2.7167	0.0901	3.0294	5.7994
	-0.05	1.0807	9.6352	2.6990	0.0948	3.0386	5.6676
	0.00	1.1848	9.6049	2.6936	0.0999	3.0604	5.5360
	$+0.05$	1.2986	9.6313	2.6983	0.1053	3.0923	3.4187
15	-0.55	1.5462	46.5473	13.4724	0.0707	5.2677	8.6324
	-0.50	1.5198	30.2759	9.1124	0.0696	4.7177	8.2310
	-0.45	1.5342	23.2038	7.2175	0.0707	4.3564	7.8481
	-0.40	1.5806	19.3483	6.1844	0.0734	4.1189	7.4903
	-0.35	1.6540	16.9964	5.5542	0.0773	3.9669	7.1622
	-0.30	1.7520	15.4722	5.1458	0.0823	3.8766	6.8645
	-0.25	1.8730	14.4550	4.8732	0.0881	3.8322	6.5961
	-0.20	2.0166	13.7730	4.6905	0.0947	3.8232	6.3542
	-0.15	2.1825	13.3257	4.5706	0.1020	3.8418	6.1361
	-0.10	2.3710	13.0501	4.4968	0.1101	3.8825	5.9388
	-0.05	2.5823	12.9048	4.4579	0.1183	3.9413	5.7596
	0.00	2.8168	12.8613	4.4462	0.1282	4.0149	5.5961
	$+0.05$	3.0750	12.8991	4.4563	0.1383	4.1008	5.4463

Table 6.2   (Continued)

ϕ (deg)	m	N_γ	N_c	N_q	N_I	N_R	$\sqrt{\dfrac{N_R}{N_I}}$
20	−0.50	3.6745	46.2884	17.8477	0.0673	5.6658	9.1768
	−0.45	3.6419	33.8986	13.3381	0.0728	5.3067	8.5380
	−0.40	3.6943	27.6099	11.0492	0.0796	5.0886	7.9941
	−0.35	3.8151	23.9213	9.7067	0.0877	4.9684	7.5267
	−0.30	3.9952	21.5875	8.8572	0.0970	4.9199	7.1214
	−0.25	4.2298	20.0542	8.2992	0.1076	4.9258	6.7672
	−0.20	4.5161	19.0369	7.9289	0.1194	4.9746	6.4552
	−0.15	4.8533	18.3742	7.6877	0.1325	5.0582	6.1783
	−0.10	5.2413	17.9678	7.5398	0.1470	5.1704	5.9309
	−0.05	5.6804	17.7542	7.4620	0.1629	5.3068	5.7084
	0.00	6.1717	17.6903	7.4368	0.1802	5.4638	5.5072
	+0.05	6.7161	17.7457	7.4589	0.1989	5.6486	5.3243
25	−0.50	8.5665	73.8778	35.4499	0.0732	7.2346	9.9384
	−0.45	8.3599	51.2706	24.9079	0.0835	6.8363	9.0503
	−0.40	8.3728	40.7056	19.9814	0.0954	6.6214	8.3291
	−0.35	8.5541	34.7663	17.2119	0.1094	6.5339	7.7297
	−0.30	8.8760	31.1015	15.5029	0.1254	6.5404	7.2223
	−0.25	9.3230	28.7315	14.3977	0.1437	6.6199	6.7864
	−0.20	9.8871	27.1750	13.6720	0.1646	6.7584	6.4075
	−0.15	10.5646	26.1681	13.2024	0.1882	6.9462	6.0748
	−0.10	11.3542	25.5533	12.9157	0.2148	7.1761	5.7803
	−0.05	12.2569	25.2309	12.7654	0.2445	7.4429	5.5178
	0.00	13.2745	25.1345	12.7205	0.2775	7.7423	5.2825
	+0.05	14.4095	25.2180	12.7594	0.3139	8.0710	5.0704
30	−0.45	19.3095	80.8644	47.6872	0.1064	9.3123	9.3540
	−0.40	19.1315	62.4470	37.0539	0.1267	9.0899	8.4705
	−0.35	19.3718	52.5548	31.3426	0.1506	9.0494	7.7518
	−0.30	19.940	46.6067	27.9084	0.1787	9.1446	7.1533
	−0.25	20.8187	42.8208	25.7226	0.2116	9.3473	6.6458
	−0.20	21.9566	40.3597	24.3017	0.2500	9.6392	6.2095
	−0.15	23.3512	38.7778	23.3884	0.2944	10.0081	5.8303
	−0.10	24.9984	37.8159	22.8330	0.3456	10.4452	5.4979
	−0.05	26.8993	37.3127	22.5425	0.4041	10.9441	5.2044
	0.00	29.0580	37.1624	22.4558	0.4706	11.4998	4.9436
	+0.05	31.4810	37.2926	22.5309	0.5457	12.1084	4.7107
35	−0.45	46.2942	134.3023	95.0397	0.1527	13.4981	9.4021
	−0.40	45.4427	100.6609	71.4837	0.1887	13.2639	8.3844
	−0.35	45.6687	83.4477	59.4308	0.2323	13.3114	7.5703
	−0.30	46.7356	73.3676	52.3727	0.2849	13.5708	6.9017
	−0.25	48.5145	67.0529	47.9511	0.3481	14.0015	6.3419
	−0.20	50.9356	62.9887	45.1052	0.4237	14.5786	5.8661
	−0.15	53.9640	60.3926	43.2874	0.5133	15.2859	5.4569
	−0.10	57.5868	58.8199	42.1862	0.6191	16.1127	5.1018
	−0.05	61.8051	57.9989	41.6113	0.7428	17.0515	4.7911
	0.00	66.6296	57.7539	41.4398	0.8868	18.0970	4.5175
	+0.05	72.0773	57.9662	41.5884	1.0529	19.2451	4.2753

Table 6.2   (Continued)

ϕ (deg)	m	N_γ	N_c	N_q	N_I	N_R	$\sqrt{\dfrac{N_R}{N_I}}$
40	−0.40	115.7097	172.8231	146.0161	0.3229	20.8738	8.0404
	−0.35	115.5504	141.1002	119.3973	0.4107	21.1138	7.1701
	−0.30	117.6386	123.0124	104.2199	0.5195	21.7125	6.4650
	−0.25	121.5875	111.8576	94.8599	0.6536	22.6077	5.8817
	−0.20	127.1879	104.7472	88.8935	0.8175	23.7619	5.3914
	−0.15	134.3346	100.2323	85.1051	1.0168	25.1570	4.9741
	−0.10	142.9868	97.5069	82.8181	1.2572	26.7775	4.6152
	−0.05	153.1451	96.0866	81.6263	1.5450	28.6173	4.3038
	0.00	164.839	95.6630	81.2709	1.8870	30.6724	4.0317
	+0.05	178.1176	96.0303	81.5791	2.2904	32.9409	3.7924
45	−0.40	327.6781	322.2748	323.2752	0.6576	36.2961	7.4295
	−0.35	325.4943	259.1345	260.1349	0.8611	37.0113	6.5559
	−0.30	329.9752	224.0769	225.0772	1.1194	38.3965	5.8568
	−0.25	339.8627	202.7837	203.7840	1.4447	40.3468	5.2846
	−0.20	354.4804	189.3358	190.3361	1.8515	42.8070	4.8083
	−0.15	373.4971	180.8450	181.8452	2.3565	45.7496	4.4062
	−0.10	393.7473	175.7358	176.7361	2.9784	49.1634	4.0628
	−0.05	424.2605	173.0775	174.0778	3.7386	53.0475	3.7669
	0.00	456.1177	172.2851	173.2853	4.6607	57.4067	3.5096
	+0.05	492.4763	172.9729	173.9732	5.7709	62.2499	3.2843

[a] After Wallace (1961).

3. Inertia force (IF): With the transient loading, the soil mass in the failure wedge is displaced. The nature of displacement assumed here is shown with broken lines in Figure 6.14a. Note that the vertical displacement of the foundation is equal to Δ. Using energy considerations, Wallace (1961) has stated

$$IF = N_I \gamma b^2 \left(\frac{d^2\Delta}{dt^2}\right) \tag{6.88}$$

The variation of N_I with m and the soil friction angle ϕ is given in Table 6.2.

4. The downward displacement of the foundation results in the movement of soil mass in the failure wedge. This soil mass contributes to a restoring moment about O (Figure 6.14a). The restoring force (RF) on the foundation can be expressed as

$$RF = N_R b \gamma \Delta \tag{6.89}$$

The values of N_R are given in Table 6.2.

Combining the preceding forces, the differential equation of motion can be written as

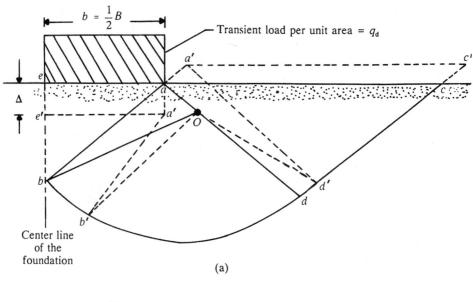

(a)

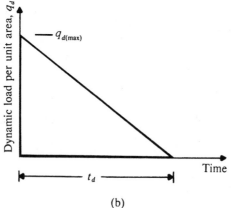

(b)

Figure 6.14 (a) Nature of displacement of the foundation and the soil failure wedge due to transient loading; (b) variation of the transient load on the foundation with time

$$q_u b + N_I \gamma b^2 \left(\frac{d^2\Delta}{dt^2}\right) + N_R \gamma b \Delta = q_{d(\max)} b \left(1 - \frac{t}{t_d}\right) \quad \text{for } 0 \le t \le t_d \quad (6.90)$$

and

$$q_u b + N_I \gamma b^2 \left(\frac{d^2\Delta}{dt^2}\right) + N_R \gamma b \Delta = 0 \quad \text{for } t \ge t_d \quad (6.91)$$

The solutions of these differential equations with proper boundary conditions take the following form.

For $0 \le t \le t_d$:

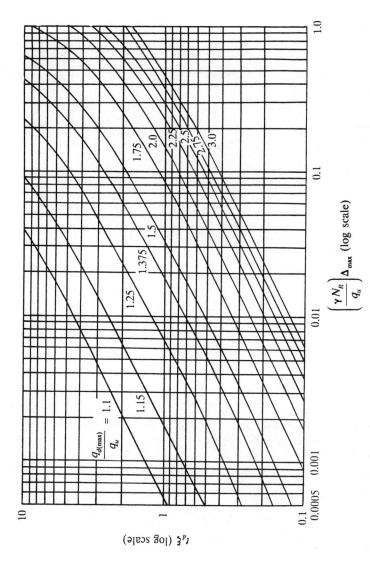

Figure 6.15 Nondimensional plot of the maximum displacement of a continuous foundation due to transient loading (after Wallace, 1961)

$$\left(\frac{N_R\gamma}{q_u}\right)\Delta = \left[\frac{q_{d(\max)}}{q_u} - 1\right]\left[1 - \cos(\zeta t)\right] + \frac{q_{d(\max)}}{q_u t_d \zeta}\left[\sin(\zeta t) - \zeta t\right]$$ (6.92)

For $t \geq t_d$:

$$\left(\frac{N_R\gamma}{q_u}\right)\Delta = \left[1 - \frac{q_{d(\max)}}{q_u} + \frac{q_{d(\max)}}{q_u t_d \zeta}\sin(\zeta t_d)\right]\cos(\zeta t)$$
$$+ \left\{\frac{q_{d(\max)}}{q_u t_d \zeta}\left[1 - \cos(\zeta t_d)\right]\right\}\sin(\zeta t) - 1$$ (6.93)

where

$$\zeta = \sqrt{\frac{2N_R}{N_I B}}$$ (6.94)

The maximum displacements ($\Delta = \Delta_{\max}$) of the foundation evaluated from the last two equations are presented in Figure 6.15 [$t_d \zeta$ versus $(\gamma N_R/q_u)\,\Delta_{\max}$].

The procedure for solving problems for determination of maximum vertical displacements of foundations using preceding is given in Example 6.3.

Example 6.3

A concrete foundation (assumed as continuous) is subjected to a transient load. Given $B = 2\,\text{m}$, $D_f = 1.5\,\text{m}$, $q_{d(\max)} = 1420\,\text{kN/m}^2$, and $t_d = 0.25\,\text{s}$. The properties of the foundation soil are $\gamma = 19\,\text{kN/m}^3$, $\phi = 20°$, and $c = 40\,\text{kN/m}^2$. Calculate the maximum vertical movement of the foundation due to the applied transient load.

Solution

1. Calculate q_u:

$$q_u = cN_c + qN_q + \tfrac{1}{2}\gamma B N_\gamma$$

$$\phi = 20°, \qquad c = 40\,\text{kN/m}^2$$

$$q = \gamma D_f = 19 \times 1.5 = 28.5\,\text{kN/m}^2$$

Referring to Table 6.2 to determine the values of N_c, N_q, and N_γ for $\phi = 20°$, one needs to make several trials for various values of m (see Table 6.3). The minimum value of q_u is obtained at $m = -0.05$, so

$$q_u = 1031\,\text{kN/m}^2 \qquad \text{at } m = -0.05$$

2. For $m = -0.05$ and $\phi = 20°$,

$$N_R = 5.3068, \qquad \sqrt{\frac{N_R}{N_I}} = 5.7084$$

$$\frac{N_R\gamma}{q_u} = \frac{5.3068 \times 19}{1031} = 0.0978$$

Table 6.3 Trials for Various Values of m

$m = -0.35$	$q_u = (40)(23.9213) + (28.5)(9.7067) + \frac{1}{2}(19)(2)(3.8151)$
	$= 1306 \text{ kN/m}^2$
$m = -0.30$	$q_u = (40)(21.5876) + (28.5)(8.8572) + \frac{1}{2}(19)(2)(3.9952)$
	$= 1192 \text{ kN/m}^2$
$m = -0.25$	$q_u = (40)(20.0542) + (28.5)(8.2992) + \frac{1}{2}(19)(2)(4.2298)$
	$= 1119 \text{ kN/m}^2$
$m = -0.20$	$q_u = (40)(19.0369) + (28.5)(7.9289) + \frac{1}{2}(19)(2)(4.5161)$
	$= 1073 \text{ kN/m}^2$
$m = -0.15$	$q_u = (40)(18.3742) + (28.5)(7.6877) + \frac{1}{2}(19)(2)(4.8533)$
	$= 1046 \text{ kN/m}^2$
$m = -0.10$	$q_u = (40)(17.9678) + (28.5)(7.5398) + \frac{1}{2}(19)(2)(5.2413)$
	$= 1033 \text{ kN/m}^2$
$m = -0.05$	$q_u = (40)(17.7542) + (28.5)(7.4620) + \frac{1}{2}(19)(2)(5.6804)$
	$= 1031 \text{ kN/m}^2$
$m = 0.00$	$q_u = (40)(17.6903) + (28.5)(7.4388) + \frac{1}{2}(19)(2)(6.1717)$
	$= 1037 \text{ kN/m}^2$

$$\zeta = \sqrt{\frac{2N_R}{N_I B}} = \sqrt{\frac{N_R}{N_I}}\sqrt{\frac{2}{B}} = (5.7084)\sqrt{\frac{2}{2}} = 5.7084$$

$$t_d \zeta = (0.25)(5.7084) = 1.4271$$

$$\frac{q_{d(max)}}{q_u} = \frac{1420}{1031} = 1.377$$

From Figure 6.15, for $t_d\zeta = 1.4271$ and $q_{d(max)}/q_u = 1.377$,

$$\left(\frac{N_R \gamma}{q_u}\right)\Delta_{max} = 0.04 \quad \text{or} \quad \Delta_{max} = \frac{0.04}{0.0978} = 0.409 \text{ m} \quad \blacksquare$$

6.7   Experimental Observation of Load-Settlement Relationship for Vertical Transient Loading

A limited number of laboratory tests for observation of load-settlement relationships of foundations under transient loading have so far been conducted (Cunny and Sloan, 1961; Shenkman and McKee, 1961; Jackson and Hadala, 1964; Carroll, 1963). The experimental evaluations of these tests are presented in this section.

Load-settlement observations of *square* model footings resting on sand and clay and subjected to transient loads have been presented by Cunny and Sloan (1961). The model footings were of varying sizes from 4.5–9-in. (114.3–228.6-mm) squares and were placed on the surface of the compacted soil layers. The transient loads to which the footings were subjected were of the nature shown in Figure 6.16. The nature of the settlement of footings with time during the application of the dynamic load is also shown in the same figure. In general, during the *rise time* (t_r) of the dynamic load, the settlement of a footing increases rapidly. Once the peak load $[Q_{d(max)}]$ is reached, the rate of settlement with time

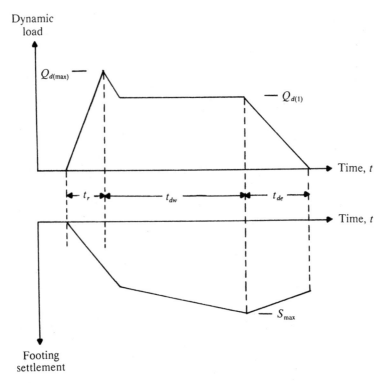

Figure 6.16   Nature of dynamic load applied to laboratory model footings

decreases. However, the total settlement of a footing continues to increase during the *dwell time* of the load (t_{dw}) and reaches a maximum value (S_{max}) at the end of the dwell time. During the *decay period* of the load (t_{de}), the footing rebounds to some degree. The results of the model footing tests on sand obtained by Cunny and Sloan are given in Table 6.4. Also, the results of model tests for square surface footings on clay as reported by Cunny and Sloan are shown in Table 6.5. Based on these results, a few general observations may be made:

1. The settlement of foundations under transient loading is generally uniform. This can be seen by observing the settlements at the three corners of the model footings—both in sand and clay.

2. Footings under dynamic loading may fail by punching type of failure in soil, although general shear failure may be observed for the same footings tested under static conditions.

3. In Table 6.4, the 9-in. (228.6-mm) footing failed at a load of 2590 lb (11.52 kN) under static loading conditions. The total settlement after the failure load application was 2.62 in. (66.55 mm). However, under dynamic loading conditions, when $Q_{d(1)}/Q_u$ was equal to 1.25 (Test 4), the settlement of the footing was about 0.4 in. (10.16 mm). Similarly, in Table 6.5, the static failure load Q_u of the 4.5-in. (114.3-mm) footing was 2460 lb (10.94 kN) with a settlement of 2 in. (50.8 mm). The same footing under dynamic loading with $Q_{d(1)}/Q_u = 1.17$ (Test 2) showed a total settlement of about 0.7 in. (17.78 mm).

Table 6.4 Load-Settlement Relationship of Square Footings on Sand Due to Transient Loading[a]

Test No.	Size of footing (in.)	Q_u (lb)[b]	$Q_{d(max)}$ (lb)	$Q_{d(1)}$ (lb)	$\dfrac{Q_{d(1)}}{Q_u}$ (%)	t_r (ms)	t_{dw} (ms)	t_{de} (ms)	S_{max} (in.)[c]		
									Pot. 1	Pot. 2	Pot. 3
1	6 × 6	770	800	800	104	18	122	110	0.28	0.05	0.11
2	8 × 8	1820	3140	2800	154	8	420	255	—	—	—
3	8 × 8	1820	2275	2175	120	90	280	290	0.83	0.93	0.95
4	9 × 9	2590	3500	3250	125	11	0	350	0.40	0.42	0.40

[a] Compiled from Cunny and Sloan (1961): Compacted dry unit weight of sand = 103.4 lb/ft³ (16.26 kN/m³); relative density of compaction of sand = 96%; triaxial angle of friction of sand = 32°.
[b] Ultimate failure load tested under static conditions.
[c] Settlement of footings measured at three corners of each footing by linear potentiometer.

Table 6.5 Load-Settlement Relationship of Square Footings on Clay Due to Transient Loading[a]

Test No.	Size of footing (in.)	Q_u (lb)[b]	$Q_{d(max)}$ (lb)	$Q_{d(1)}$ (lb)	$\dfrac{Q_{d(1)}}{Q_u}$ (%)	t_r (ms)	t_{dw} (ms)	t_{de} (ms)	S_{max} (in.)[c]		
									Pot. 1	Pot. 2	Pot. 3
1	4.5 × 4.5	2460	2850	2275	93	9	170	350	0.50	0.50	0.48
2	4.5 × 4.5	2460	3100	2820	117	9	0	380	0.66	0.72	0.70
3	4.5 × 4.5	2460	3460	2970	121	10	0	365	1.70	1.68	1.70
4	5 × 5	3040	3580	2950	97	9	0	360	0.58	0.55	0.55

[a] Compiled from Cunny and Sloan (1961): Compacted moist unit weight = 94.1 – 98.4 lb/ft³ (14.79–15.47 kN/m³); moisture content = 22.5 ± 1.7%; c = 2400 lb/ft²; ϕ = 4° (undrained test).
[b] Ultimate failure load tested under static conditions.
[c] Settlement of footings measured at three corners of each footing by linear potentiometer.

These facts show that, for a limiting settlement condition, a foundation can support higher load under dynamic loading conditions than those observed from static tests.

Dynamic Load versus Settlement Prediction in Clayey Soils

Jackson and Hadala (1964) reported several laboratory model tests on 4.5–8-in. (114.3–203.2-mm) square footings resting on highly saturated, compacted, plastic *Buckshot clay*. The tests were similar in nature to those described previously in this section. Based on these results, Jackson and Hadala have shown that there is a unique nondimensional relation between $Q_{d(max)}/B^2 c_u$ and S_{max}/B (c_u = undrained shear strength). This is shown in Figure 6.17. Note that the tests on which Figure 6.17 are based have $t_{dw} = 0$. However, for dynamic loads with $t_{dw} > 0$, the results would not be too different.

The preceding finding is of great practical importance in estimation of the

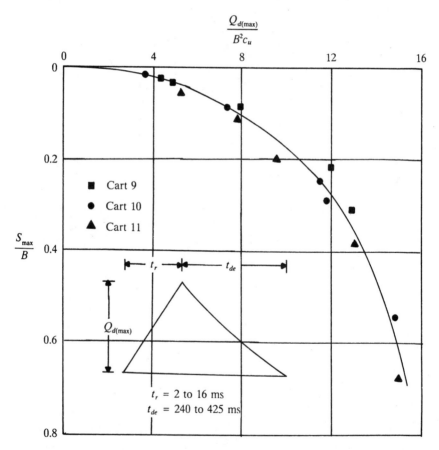

Figure 6.17   Nondimensional relationship of $Q_{d(max)}/B^2 c_u$ and S_{max}/B for model footing tests in Buckshot clay (after Jackson and Hadala, 1964)

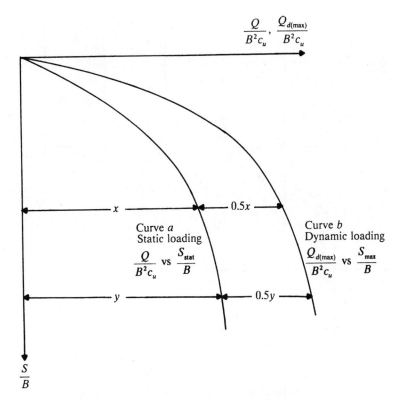

Figure 6.18 Prediction of dynamic load-settlement relationship for foundations on clay

dynamic load-settlement relationships of foundations. Jackson and Hadala have recommended the following procedure for that purpose.

1. Determine the static load Q versus settlement S relationship for a foundation from plate bearing tests in the field.

2. Determine the unconfined compression strength of the soil q_{uc} in the laboratory.

$$q_{uc} = 2c_u$$

3. Plot a graph of Q/B^2c_u versus S_{stat}/B. (See Figure 6.18, curve a.)

4. For any given value of S_{stat}/B, multiply Q/B^2c_u by the strain rate factor (≈ 1.5) and plot it in the same graph. The resulting graph of S_{stat}/B versus $1.5Q/B^2c_u$ will be the predicted relationship between $Q_{d(max)}/B^2c_u$ and S_{max}/B. (See Figure 6.18, curve b.)

Example 6.4

The estimated static plate load bearing test results of a foundation resting on stiff clay and 5 ft in diameter are given next.

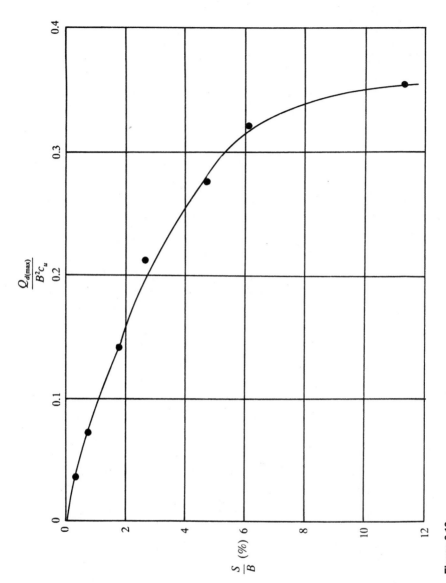

Figure 6.19

Q (lb)	Settlement (in.)	Q (lb)	Settlement (in.)
0	0	6000	1.65
1000	0.25	8000	2.90
2000	0.48	9000	3.70
4000	1.10	10000	6.80

The unconfined compression strength of this clay was 3400 lb/ft².
a. Plot a graph of estimated S_{max}/B versus $Q_{d(max)}/B^2 c_u$ assuming a strain-rate factor of 1.5.
b. Determine the magnitude of the maximum dynamic load $Q_{d(max)}$ that produces a maximum settlement S_{max} of 6 in.

Solution

Given $B = 60$ in. and $c_u = \frac{1}{2}(3400) = 1700$ lb/ft², the following table can be prepared.

Q (lb) (1)	S_{stat} (in.) (2)	S/B (%) (3)	$Q/B^2 c_u$ (4)	$1.5Q/B^2 c_u$ (5)
0	0	0	0	0
1000	0.25	0.417	0.0235	0.035
2000	0.48	0.8	0.0471	0.0707
4000	1.10	1.83	0.094	0.141
6000	1.65	2.75	0.141	0.212
8000	2.90	4.83	0.188	0.282
9000	3.70	6.17	0.212	0.318
10000	6.80	11.33	0.235	0.353

Assuming S/B (Col. 3) to be equal to S_{max}/B and $1.5Q/B^2 c_u$ to be equal to $Q_{d(max)}/B^2 c_u$, a graph can be plotted (Figure 6.19).

For $S_{max} = 6$ in., $\dfrac{S_{max}}{B} = \dfrac{6}{60} = 10\%$

From Figure 6.19, the value of $Q_{d(max)}/B^2 c_u$ corresponding to $S_{max}/B = 10\%$ is about 0.348. Hence

$$Q_{d(max)} = (0.348)(25)(1700) = 14{,}790 \text{ lb} \qquad \blacksquare$$

6.8   Foundation Response to Vertical Transient Load Transmitted through Soil

In most circumstances, design engineers are primarily concerned with the analysis of the behavior of foundations subjected to earthquake-induced forces transmitted from the bedrock. At the present time, limited information is available in regard to the foundation response to the *vertical component* of the earthquake-induced force (Figure 6.20). Varadhi and Saxena (1980) conducted laboratory model studies to evaluate the variation of vertical stress under circu-

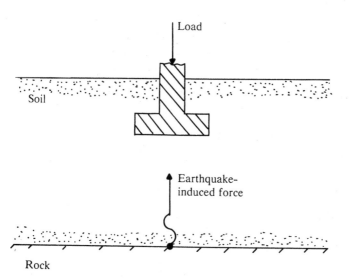

Figure 6.20   Earthquake-induced force on foundation

lar and square model foundations supported by a sand layer. Vertical transient loads were transmitted to the model foundations through the underlying sand. These tests were conducted with a certain bearing pressure, q, on each model foundation. The general findings of these tests are summarized by Figure 6.21, which is for the case of a 6-in.- (152.4-mm-) diameter model surface foundation with $q = 0.55$ lb/in^2 (3.80 kN/m^2). Curve a shows the plot of the overburden pressure (that is, γz) with depth. Curve b is a plot of the variation of vertical stress (σ) due to the bearing pressure (q) below the center of the foundation estimated by using Boussinesq's equation, which may be stated as

$$\sigma = q \left\{ 1 - \frac{1}{[1 + (B/2z)^2]^{3/2}} \right\} \tag{6.95}$$

where B = diameter of the foundation

   z = depth measured from the bottom of the foundation (see insert in Figure 6.21)

For the derivation of the preceding relationship, the reader is referred to any soil mechanics text (for example, Das, 1990). Due to a given transient load transmitted at a certain depth below the foundation, the variation of the peak vertical stress (σ) in soil without the footing on the surface has been measured and shown as curve c. Curve d shows the attenuation of the peak vertical stress with depth below the center of the model foundation. Based on these model test results, the following conclusions can be drawn.

1. The peak dynamic vertical stress intensity may be about 10 times or more as compared to the static bearing pressure.

2. At a depth of about $2B$ (B = diameter of model foundation), the influence of footing mass on the dynamic vertical stress distribution is almost negligible.

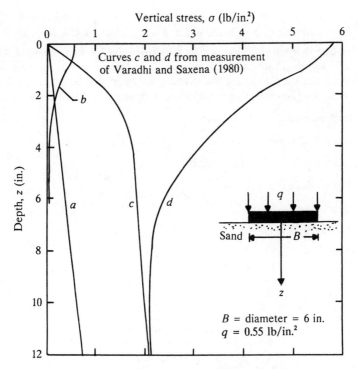

Figure 6.21 Vertical stress under a circular foundation due to a vertical transient load transmitted through the underlying sand

More model and field tests are necessary to establish a proper mathematical relationship for the foundation response to a vertical transient load transmitted through the soil.

PROBLEMS

6.1 A 3-ft-square shallow foundation is supported by dense sand. The relative density of compaction, unit weight, and angle of friction (static) of this sand are 75%, 120 lb/ft³, and 38°, respectively. Given the depth of foundation to be 3 ft, estimate the minimum ultimate bearing capacity of this foundation that might be obtained if the vertical loading velocity on this foundation were varied from static to impact range.

6.2 Redo Problem 6.1 with the depth of foundation as 4.5 ft.

6.3 Redo Problem 6.1 with the following:

$$\text{Foundation width} = 1.6 \text{ m}$$

$$\text{Foundation depth} = 0.75 \text{ m}$$

$$\text{Angle of friction of sand} = 35°$$

$$\text{Unit weight of compacted soil} = 17.4 \text{ kN/m}^3$$

$$\text{Relative density of compaction of sand} = 80\%$$

6.4 A rectangular foundation has a length L of 2.5 m. It is supported by a medium dense sand with a unit weight of 17 kN/m³. The sand has an angle of friction of 36°. The foundation may be subjected to a dynamic load of 735 kN increasing at a moderate rate. Using a factor of safety equal to 2, determine the width of the foundation. Use $D_f = 0.8$ m.

6.5   A foundation 2.25 m square is supported by saturated clay. The unit weight of this clay is 18.6 kN/m³. The depth of the foundation is 1.2 m. Determine the ultimate bearing capacity of this foundation assuming that the load will be applied very rapidly. Given the following for the clay [laboratory unconsolidated-undrained triaxial (static) test results]:

Undrained cohesion, $c_u = 90 \ \text{kN/m}^2$

Strain-rate factor $= 1.4$

6.6   Redo Problem 6.5 with the following changes:

Foundation width $= 1.5$ m

Foundation length $= 2.6$ m

Foundation depth $= 1.75$ m

6.7   Refer to Example 6.2.
   a. Determine the natural period (in seconds) of the foundation for rotational mode of failure due to transient loading.
   b. Using Eq. (6.36), trace the history of motion of the foundation, that is, $\alpha = $ at 0.05, 1, 1.5, 2, 3, and 4 s.

6.8   A 3-m-wide shallow strip foundation on clay is subjected to a vertical transient stress pulse $q_d = 500 \ e^{-\beta t} \ \text{kN/m}^3$. Given $\beta = 5 \ \text{s}^{-1}$, undrained cohesion of clay $= 60 \ \text{kN/m}^2$, and unit weight of soil $= 19.5 \ \text{kN/m}^3$, determine the maximum angular rotation that the footing might undergo.

6.9   Repeat Problem 6.8, given

$$q_d = 1000 \ e^{-\beta t} \ \text{kN/m}^2$$

$$\beta = 10 \ \text{s}^{-1}$$

6.10   A strip foundation 5 ft wide is resting on a soil (as shown in Figure 6.8). Given $\phi = 15°$, $c = 1000 \ \text{lb/ft}^2$, and $\gamma = 115 \ \text{lb/ft}^3$.
   a. Using a factor of safety of 2 and Eqs. (6.39)–(6.41), determine the allowable load the foundation can support per foot length.
   b. Let this foundation be subjected to a horizontal transient load, as shown in Figure 6.7. Given $t_d = 0.2$ s and $H = 10$ ft (Figure 6.8), determine the critical overload factor λ_{cr}.
   c. Using the critical overload factor determined in (b), trace the history of motion of the foundation—that is, the angular rotation α at time $t = 0.2, 0.3, 0.4$, and 0.5 s.

6.11   A continuous foundation 5 ft wide is subjected to a vertical transient load, as shown in Figure 6.14b, with $q_{d(max)} = 60{,}000 \ \text{lb/ft}^2$ and $t_d = 0.2$ s. The foundation is resting on a soil with $c = 0$, $\phi = 35°$, and $\gamma = 115 \ \text{lb/ft}^3$. The depth of the foundation is 4 ft measured from the ground surface. Calculate the maximum vertical movement of the foundation based on the theory described in Section 6.6.

6.12   Repeat Problem 6.11 with the following as given:

Foundation width $= 3$ m

Foundation depth $= 0.4$ m

$$\gamma = 16 \ \text{kN/m}^3, \qquad c = 50 \ \text{kN/m}^2, \qquad \phi = 25°$$

$$q_{d(max)} = 3000 \ \text{kN/m}^2, \qquad t_d = 0.2 \ \text{s}$$

6.13   A clay deposit has an undrained cohesion (static test) of 90 kN/m^2. A static field plate load test was conducted with a plate having a diameter of 0.5 m. When the load per unit area q was 200 kN/m^2, the settlement was 20 mm.

   a. Assume that, for a given value of q, the settlement is proportional to the width of the foundation. Estimate the settlement of a prototype circular foundation in the same clay with a diameter of 3 m (static loading).

   b. The strain-rate factor of the clay is 1.4. If a vertical transient load pulse were applied to the foundation as given in part (a), what would be the maximum transient load (in kilonewtons) that will produce the same maximum settlement (S_{max}) as calculated in part (a)?

REFERENCES

Brinch Hansen, J. (1970). "A Revised and Extended Formula for Bearing Capacity," *Bulletin No. 28*, Danish Geotechnical Institute, Copenhagen, Denmark.

Carroll, W. F. (1963). "Dynamic Bearing Capacity of Soils. Vertical Displacements of Spread Footings on Clay: Static and Impulsive Loadings," *Technical Report No. 3-599*, Report 5, U.S. Army Corps of Engineers, Waterways Experiment Station, Vicksburg, Mississippi.

Caquot, A., and Kerisel, J. (1953). "Sur le Terme de Surface Dans le Calcul des Foundations en Milieu Pulverulent," *Proceedings*, 3rd International Conference on Soil Mechanics and Foundation Engineering, Zurich, Switzerland, Vol. I, pp. 336–337.

Cunny, R. W., and Sloan, R. C. (1961). "Dynamic Loading Machine and Results of Preliminary Small-Scale Footing Tests," *Special Technical Publication No. 305*, American Society for Testing and Materials, pp. 65–77.

Das, B. M. (1990). *Principles of Foundation Engineering*, PWS-KENT, Boston.

DeBeer, E. E. (1970). "Experimental Determination of the Shape Factors and the Bearing Capacity Factors of Sand," *Geotechnique*, Vol. 20, No. 4, pp. 387–411.

Heller, L. W. (1964). "Failure Modes of Impact-Loaded Footings on Dense Sand," *Technical Report R-281*, U.S. Naval Civil Engineering Laboratory, Port Hueneme, California.

Jackson, J. G., Jr., and Hadala, P. F. (1964). "Dynamic Bearing Capacity of Soils. Report 3: The Application of Similitude to Small-Scale Footing Tests," U.S. Army Corps of Engineers, Waterways Experiment Station, Vicksburg, Mississippi.

Prakash, S., and Chummar, A. V. (1967). "Response of Footings to Lateral Loads," *Proceedings*, International Symposium on Wave Propagation and Dynamic Properties of Earth Materials, Ed., G. E. Triandafilidis, University of New Mexico, Albuquerque, New Mexico, pp. 679–691.

Reissner, H. (1924). "Zum Erddrukproblem," *Proceedings*, First International Conference on Applied Mechanics, Delft, The Netherlands, pp. 294–311.

Shenkman, S., and McKee, K. E. (1961). "Bearing Capacity of Dynamically Loaded Footings," *Special Technical Publication No. 305*, American Society for Testing and Materials, pp. 78–90.

Terzaghi, K. (1943). *Theoretical Soil Mechanics*, Wiley, New York.

Triandafilidis, G. E. (1965). "The Dynamic Response of Continuous Footings Supported on Cohesive Soils," *Proceedings*, 6th International Conference on Soil Mechanics and Foundation Engineering, Montreal, Canada, Vol. II, pp. 205–208.

Varadhi, S. N., and Saxena, S. (1980). "Foundation Response to Soil Transmitted Loads," *Journal of the Geotechnical Engineering Division*, ASCE, Vol. 106, No. GT10, pp. 1121–1139.

Vesic, A. S. (1973). "Analysis of Ultimate Loads of Shallow Foundations," *Journal of the Soil Mechanics and Foundations Division*, ASCE, Vol. 99, No. SM1, pp. 45–73.

Vesic, A. S., Banks, D. C., and Woodard, J. M. (1965). "An Experimental Study of Dynamic Bearing Capacity of Footings on Sand," *Proceedings*, 6th International Conference on Soil Mechanics and Foundation Engineering, Montreal, Canada, Vol. II, pp. 209–213.

Wallace, W. F. (1961). "Displacement of Long Footings by Dynamic Loads," *Journal of the Soil Mechanics and Foundations Division*, ASCE, Vol. 87, No. SM5, pp. 45–68.

Whitman, R. V., and Healy, K. A. (1962). "Shear Strength of Sands During Rapid Loading," *Journal of the Soil Mechanics and Foundations Division*, ASCE, Vol. 88, No. SM2, pp. 99–132.

EARTHQUAKE AND
GROUND VIBRATION

7.1   Introduction

The ground vibrations due to earthquakes have resulted in several major structural damages in the past. In the North American continent, earthquakes are believed to originate from the rupture of faults. The ground vibration resulting from an earthquake is due to the upward transmission of the stress waves from rock to the softer soil layer(s). In recent times, several major studies have been performed to study the nature of occurrence of earthquakes and the associated amount of energy released. Also, modern techniques have been developed to analyze and estimate the physical properties of soils under earthquake conditions and to predict the ground motion. These developments are the subjects of discussion in this chapter.

7.2   Definition of Some Earthquake-Related Terms

Focus: The focus of an earthquake is a point below the ground surface where the rupture of a fault first occurs (point F in Figure 7.1a).

Focal Depth: The vertical distance from the ground surface to the focus (EF in Figure 7.1a). The maximum focal depth of all earthquakes recorded so far does not exceed 435 mi (700 km). Based on focal depth, earthquakes may be divided into the following three categories:

1. *Deep-focus earthquakes*: These have focal depths of 185–435 mi (300–700 km). They constitute about 3% of all earthquakes recorded around the world and are mostly located in the Circum-Pacific belt.

2. *Intermediate-focus earthquakes*: These have focal depths of 45–185 mi (70–300 km).

3. *Shallow-focus earthquakes*: The focal depth for these is less than 45 mi (70 km). About 75% of all the earthquakes around the world belong to this category. The California earthquakes have focal depths of about 6–10 mi (10–15 km).

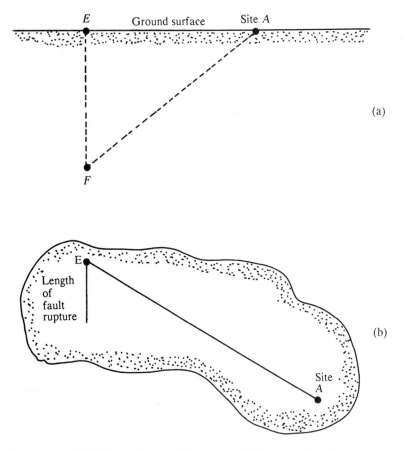

Figure 7.1   Definition of focus and epicenter: (a) section; (b) plan

Epicenter: The point vertically above the focus located on the ground surface (point E in Figure 7.1).

Epicentric Distance: The horizontal distance between the epicenter and a given site (line EA in Figure 7.1).

Hypocentric Distance: The distance between a given site and the focus (line FA in Figure 7.1a).

Effective Distance to Causative Fault: The distance from a fault to a given site for calculation of ground motion (Figure 7.2).

This distance is commonly presumed to be the epicentric distance. This type of assumption, under certain circumstances, may lead to gross errors. It can be explained with reference to Figure 7.2, which shows the plans of two cases of fault rupture. In Figure 7.2a, the length of the fault rupture L is small as compared to the epicentric distance EA. In this case, the effective distance could be taken to be equal to the epicentric distance. However, a better estimate of the effective distance is BA (B is the midpoint of the ruptured fault). Figure 7.2b

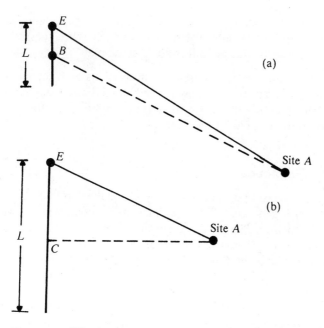

Figure 7.2 Effective distance from a site to the causative fault (*Note: L* is the length of the fault rupture.)

Table 7.1 Abridged Modified Mercalli Intensity Scale[a]

Intensity	Description
I	Detected only by sensitive instruments
II	Felt by a few persons at rest, especially on upper floors; delicate suspended objects may swing
III	Felt noticeably indoors, but not always recognized as a quake; standing autos rock slightly, vibration like passing trucks
IV	Felt indoors by many, outdoors by a few; at night some awaken; dishes, windows, doors disturbed; motor cars rock noticeably
V	Felt by most people; some breakage of dishes, windows and plaster; disturbance of tall objects
VI	Felt by all; many are frightened and run outdoors; falling plaster and chimneys; damage small
VII	Everybody runs outdoors; damage to building varies, depending on quality of construction; noticed by drivers of autos
VIII	Panel walls thrown out of frames; fall of walls, monuments, chimneys; sand and mud ejected; drivers of autos disturbed
IX	Buildings shifted off foundations, cracked, thrown out of plumb; ground cracked; underground pipes broken
X	Most masonry and frame structures destroyed; ground cracked; rails bent; landslides
XI	New structures remain standing; bridges destroyed; fissures in ground; pipes broken; landslides; rails bent
XII	Damage total; waves seen on ground surface; lines of sight and level distorted; objects thrown up into air

[a] After Wiegel, R. W. (1970).

shows the case where the length of the fault rupture is large. In such circumstances, the length AC is the effective distance, which is the perpendicular distance from the site to the line of fault rupture in the plan.

Intensity: An arbitrary scale developed to measure its destructiveness. The *Modified Mercalli Scale* is presently in use in the United States for that purpose, divided into 12 degrees of intensity. An abridged version of the *Modified Mercalli Scale* is given in Table 7.1 on p. 303.

7.3   Earthquake Magnitude

Magnitude: Is a measure of the size of an earthquake, based on the amplitude of elastic waves it generates. The *magnitude scale* presently in use was first developed by C. F. Richter. The historical developments of the magnitude scale have been summarized by Richter himself (1958).

Richter's earthquake magnitude is defined by the equation

$$\log_{10} E = 11.4 + 1.5M \tag{7.1}$$

where E is the energy released (in ergs) and M is magnitude. Båth (1966) slightly modified the constants given in Eq. (7.1) and presented it in the form

$$\log_{10} E = 12.24 + 1.44M \tag{7.2}$$

From Eq. (7.2), it can be seen that the increase of M by one unit will generally correspond to about a 30-fold increase of the energy released (E) due to the earthquake. A comparison of the magnitude M of an earthquake with the *maximum intensity* of the Modified Mercalli Scale is given in Table 7.2. Table 7.3 gives a list of some of the past major earthquakes around the world with their magnitudes.

As mentioned previously, the main cause of earthquakes is the rupture of faults. In general, the greater the length of fault rupture, the greater the magnitude of an earthquake. Several relations for the magnitude of the earthquake and the length of fault rupture have been presented by various investigators

Table 7.2 Comparison of the Richter Scale Magnitude with the Modified Mercalli Scale

Richter scale magnitude M	Maximum intensity, Modified Mercalli Scale
1	—
2	I, II
3	III
4	IV
5	VI, VII
6	VIII
7	IX, X
8	XI

Table 7.3   Some Past Major Earthquakes

Name	Epicenter Location	Date	Magnitude
Alaska	61.1° N, 147.5° W	March 27, 1964	8.4
Chile (South America)	38° S, 73.5° W	May 22, 1960	8.4
Colombia (South America)	1° N, 82° W	January 31, 1906	8.6
Peru (South America)	9.2° S, 78.8° W	May 31, 1970	7.8
San Francisco, California	38° N, 123° W	April 18, 1906	8.3
Kern County, California	35° N, 119° W	July 21, 1952	7.7
Dixie Valley, Nevada	39.8° N, 118.1° W	December 16, 1954	6.8
Hebgen Lake, Montana	44.8° N, 111.1° W	August 17, 1959	7.1

(Tocher, 1958; Bonilla, 1967; Housner, 1969). Tocher (1958), based on observations of some earthquakes in the area of California and Nevada, suggested the relationship

$$\log L = 1.02M - 5.77 \tag{7.3}$$

where L is the length of fault rupture (kilometers).

Based on Eq. (7.3), it can be seen that for an earthquake of magnitude 6, the length of fault rupture is about 1.44 mi (2.3 km). However, when the magnitude is increased to 8, the length of fault rupture associated is about 153 mi (250 km).

7.4   Characteristics of Rock Motion During an Earthquake

The ground motion near the surface of a soil deposit is mostly attributed to the upward propagation of shear waves from the underlying rock or "rocklike" layers. The term rocklike implies that the shear wave velocity in the material is similar to that associated with soft rocks. The typical range of shear wave velocities in hard rocks such as granite is about 10,000–12,000 ft/s ($\approx$ 3050–3660 m/s). Shear wave velocities associated with soft rocks can be in the low range of 2500–3000 ft/s (762–915 m/s). However, the rocklike material may not exhibit the characteristics associated with hard base rocks (Seed, Idriss, and Kiefer, 1969). Hence, for arriving at a solution of the nature of ground motion at or near the ground surface, one needs to know some aspects of the earthquake-induced motion in the rock or rocklike materials. The most important of these are

duration of the earthquake,
predominant *period* of acceleration, and
maximum *amplitude* of motion.

Each of these factors has been well summarized by Seed, Idriss, and Kiefer.

Duration of Earthquake

In general, it can be assumed that the duration of an earthquake will be somewhat similar to that of the fault rupture. The rate of propagation of fault rupture has been estimated by Housner (1965) to be about 2 mi/s (3.2 km/s). Based on this, Housner has estimated the following variation of the duration of fault rupture with the magnitude of an earthquake.

Magnitude of earthquake (Richter scale)	Duration of fault break (s)
5	5
6	15
7	25–30

It may be noted that the approximate duration of fault rupture can be estimated from Eq. (7.3). Once the length of rupture L for a given magnitude of earthquake is estimated, the duration can be given by $L/$(velocity of rupture).

Predominant Period of Rock Acceleration

Gutenberg and Richter (1956) have given an estimate of the predominant periods of *accelerations* developed in rock for *California* earthquakes. Similar results for earthquakes of magnitude $M > 7$ have been reported by Figueroa (1960). Using these results, Seed, Idriss, and Kiefer (1969) developed a chart for the *average predominant periods of accelerations* for various earthquake magni-

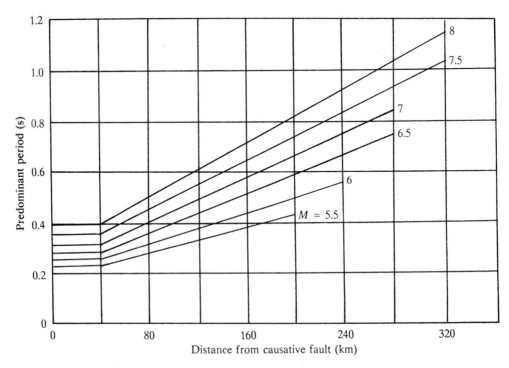

Figure 7.3 Predominant period for maximum rock acceleration (after Seed, Idriss, and Kiefer, 1969)

tudes. This is shown in Figure 7.3. Note that in this figure the predominant periods are plotted against the distance from the causative fault. This distance is approximately the epicentric distance for the case where the length of fault rupture is small. However, where the length of fault rupture is large, it is the perpendicular distance from the site to the ruptured fault line in the plan (Figure 7.2).

Maximum Amplitude of Acceleration

The maximum amplitude of acceleration in rock in the epicentric region for shallow earthquakes [focal depth about 10 m (16 km)] can be approximated as (Gutenberg and Richter, 1956)

$$\log a_0 = -2.1 + 0.81M - 0.027M^2 \tag{7.4}$$

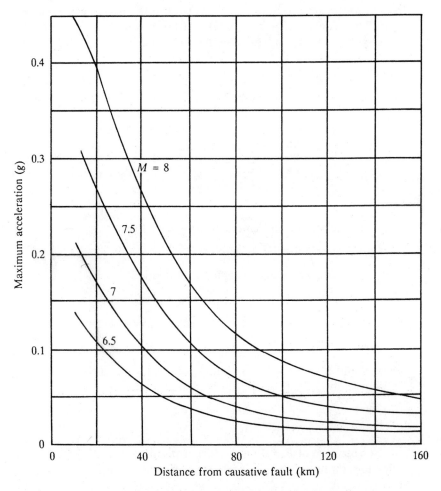

Figure 7.4 Variation of maximum acceleration with earthquake magnitude and distance from causative fault (after Seed, Idriss, and Kiefer, 1969)

$$\log a_0 = -2.1 + 0.81M - 0.027M^2 \qquad (7.4)$$

where a_0 is the maximum amplitude of acceleration.

At any other point away from the epicenter, the magnitude of the maximum amplitude of acceleration decreases. Relations for the attenuation factor of the maximum acceleration have been given by Gutenberg and Richter (1956), Banioff (1962), Esteva and Rosenblueth (1963), Kanai (1966) and Blume (1965). Based on these studies, Seed, Idriss, and Kiefer (1969) have given the average values of maximum acceleration for various magnitudes of earthquakes and distances from the causative faults. These are given in Figure 7.4.

7.5   Vibration of Horizontal Soil Layers with Linearly Elastic Properties

As stated before, the vibration of the soil layers due to an earthquake is due to the upward propagation of shear waves from the underlying rock or rocklike layer. The reponse of a horizontal soil layer with *linearly elastic properties*, developed by Idriss and Seed (1968), is presented in this section.

Homogeneous Soil Layer

Figure 7.5 shows a horizontal soil layer of thickness H underlain by a rock or rocklike material. Let the underlying rock layer be subjected to a seismic motion u_g that is a function of time t. Considering a soil column of unit cross-sectional area, the equation of motion can be written as

$$\rho(y)\frac{\partial^2 u}{\partial t^2} + c(y)\frac{\partial u}{\partial t} - \frac{\partial}{\partial y}\left[G(y)\frac{\partial u}{\partial y}\right] = -\rho(y)\frac{\partial^2 u_g}{\partial t^2} \qquad (7.5)$$

where

$\quad u$ = relative displacement at depth y and time t

$\quad G(y)$ = shear modulus at depth y

$\quad c(y)$ = viscous damping coefficient at depth y

$\quad \rho(y)$ = density of soil at depth y

The shear modulus can be given by the equation (see the discussion in Chapter 4)

$$G(y) = Ay^B \qquad (7.6)$$

where A and B are constant depending on the nature of the soil.

Substituting Eq. (7.6) into Eq. (7.5), we obtain

$$\rho\frac{\partial^2 u}{\partial t^2} + c\frac{\partial u}{\partial t} - \frac{\partial}{\partial y}\left[Ay^B\frac{\partial u}{\partial y}\right] = -\rho\frac{\partial^2 u_g}{\partial t^2} \qquad (7.7)$$

For the case of $B \neq 0$ (but <0.5), using the method of separation of variables, the solution to Eq. (7.7) can be given in the form

$$u(y,t) = \sum_{n=1}^{n=\infty} Y_n(y)X_n(t) \qquad (7.8)$$

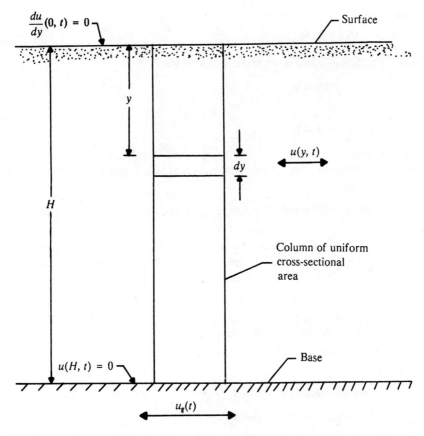

Figure 7.5   Cross section and boundary conditions of a semi-infinite soil layer subjected to a horizontal seismic motion at its base

where

$$Y_n(y) = \left(\frac{1}{2}\beta_n\right)^b \Gamma(1-b)\left(\frac{y}{H}\right)^{b/\theta} J_{-b}\left[\beta_n\left(\frac{y}{H}\right)^{1/\theta}\right] \tag{7.9}$$

and

$$\ddot{X}_n + 2D_n\omega_n\dot{X}_n + \omega_n^2 X_n = -R_n\ddot{u}_g \tag{7.10}$$

J_{-b} is the Bessel function of first kind of order $-b$, β_n represents the roots of $J_{-b}(\beta_n) = 0$, $n = 1, 2, 3\ldots$, and the circular natural frequency of nth mode of vibration is

$$\omega_n = \frac{\beta_n\sqrt{A/\rho}}{\theta H^{1/\theta}} \tag{7.11}$$

The damping ratio in the nth mode is

$$D_n = \frac{\frac{1}{2}c}{\rho\omega_n} \tag{7.12}$$

and Γ is the gamma function,

$$R_n = [(\tfrac{1}{2}\beta_n)^{1+b}\Gamma(1-b)J_{1-b}(\beta_n)]^{-1} \tag{7.13}$$

The terms b and θ are related as follows:

$$B\theta - \theta + 2b = 0 \tag{7.14}$$

and

$$B\theta - 2\theta + 2 = 0 \tag{7.15}$$

For detailed derivations, see Idriss and Seed (1967).

For obtaining the relative displacement at a depth y, the general procedure is as follows:

1. Determine the system shape $Y_n(y)$ during the nth mode of vibration [Eq. (7.9)].

2. Determine $X_n(t)$ from Eq. (7.10). This can be done by direct numerical step-by-step procedure (Berg and Housner, 1961; Wilson and Clough, 1962) or the iterative procedure as proposed by Newmark (1962).

3. Determine $u(y, t)$ from Eq. (7.8).

4. The relative velocity $[\dot{u}(y, t)]$, relative acceleration $[\ddot{u}(y, t)]$, and strain $\partial u/\partial y$ can be obtained by differentiation of Eq. (7.8).

5. The values of total acceleration, velocity, and displacement can be obtained as

$$\text{Total acceleration} = \ddot{u} + \ddot{u}_g$$

$$\text{Total velocity} = \dot{u} + \dot{u}_g$$

$$\text{Total displacement} = u + u_g$$

The values of $\dot{u}_g$ and u_g can be obtained by integration of the acceleration record $[\ddot{u}_g(t)]$.

Special Cases

Cohesionless Soils: In the case of cohesionless soils, the shear modulus [Eq. (7.6)] can be approximated as

$$G(y) = Ay^{1/2} \quad \text{or} \quad G(y) = Ay^{1/3}$$

Assuming the latter to be representative (i.e., $B = \tfrac{1}{3}$), Eqs. (7.14) and (7.15) can be solved, yielding

$$b = 0.4 \quad \text{and} \quad \theta = 1.2$$

Hence, Eqs. (7.9)–(7.11) take the following form:

$$Y_n(y) = (\tfrac{1}{2}\beta_n)^{0.4}\Gamma(0.6)\left(\frac{y}{H}\right)^{1/3} J_{-0.4}\left[\beta_n\left(\frac{y}{H}\right)^{5/6}\right] \tag{7.16}$$

$$\ddot{X}_n + 2D_n\omega_n\dot{X}_n + \omega_n^2 X_n = -\ddot{u}_g\left[\left(\frac{\beta_n}{2}\right)^{1.4}\Gamma(0.6)J_{0.6}(\beta_n)\right]^{-1} \tag{7.17}$$

and

$$\omega_n = \frac{\beta_n \sqrt{A/\rho}}{1.2H^{5/6}} \tag{7.18}$$

(*Note:* $\beta_1 = 1.7510$, $\beta_2 = 4.8785$, $\beta_3 = 8.0166$, $\beta_4 = 11.1570\ldots$.)

Cohesive Soils: In cohesive soils, the shear modulus may be considered to be approximately constant with depth; so, in Eq. (7.6), $B = 0$ and

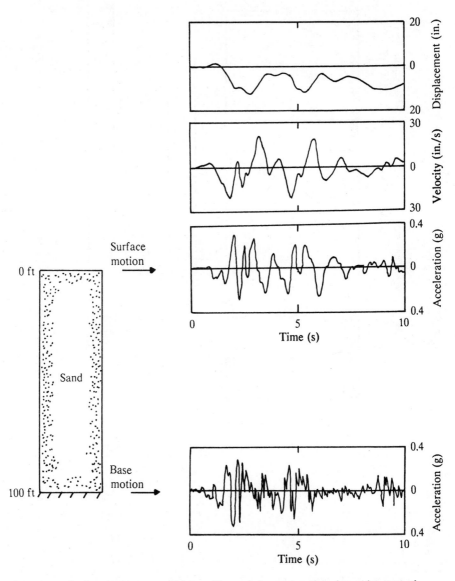

Figure 7.6 Surface response of layer with modulus proportional to cube root of depth (after Idriss and Seed, 1968)

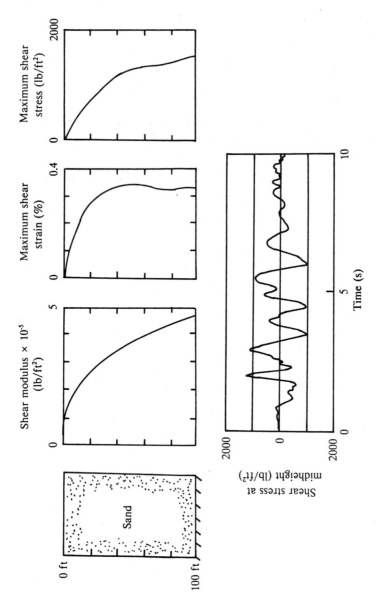

Figure 7.7   Stress and strain developed within the layer of soil shown in Figure 7.6 (after Idriss and Seed, 1968)

$$G(y) = A \tag{7.19}$$

With this assumption, Eqs. (7.9)–(7.11) are simplified as

$$Y_n(y) = \cos\left[\frac{1}{2}(2n - 1)\left(\frac{y}{H}\right)\right] \tag{7.20}$$

$$\ddot{X}_n + 2D_n\omega_n\dot{X}_n + \omega_n^2 X_n = (-1)^n\left[\frac{4}{(2n - 1)\pi}\right]\ddot{u}_g \tag{7.21}$$

and

$$\omega_n = \left[\frac{(2n - 1)\pi}{2H}\right]\sqrt{\frac{G}{\rho}} \tag{7.22}$$

Computer programs for determination of acceleration, velocity, and displacement of soil profiles for these two special cases can be found in Idriss and Seed (1967, Appendix C).

An example of a solution for cohesionless (granular) soil is given in Figure 7.6. Figure 7.7 shows the variation of shear modulus, maximum shear strain, and maximum shear stress with depth for the same soil layer shown in Figure 7.6. For this example,

$$H = 100 \text{ ft (30.49 m)}$$

$$\text{Total unit weight of soil} = \gamma = 125 \text{ lb/ft}^3 \quad (19.65 \text{ kN/m}^3)$$

$$\text{Effective unit weight of soil} = \gamma' = 60 \text{ lb/ft}^3 \quad (9.43 \text{ kN/m}^3)$$

$$\text{Shear modulus of soil} = 1 \times 10^5 y^{1/3} \text{ lb/ft}^2$$

$$D = 0.2 \quad (\text{for all modes, } n = 1, 2, \ldots, \infty)$$

For a discussion on the damping coefficient of soil under earthquake conditions, see Chapter 4.

Layered Soils

If a soil profile consists of several layers of varying properties that are linearly elastic, a lumped mass type of approach can be taken (Idriss and Seed, 1968). These lumped masses $(m_1, m_2, \ldots, m_N)$ are shown in Figure 7.8. Note

$$m_1 = \frac{\gamma_1 h_1}{g} \tag{7.23}$$

where m_1 is a lumped mass placed at the top of soil layer 1, γ_1 is the unit weight of soil in layer 1, h_1 is the half thickness of soil layer 1, and

$$m_i = \frac{\gamma_{i-1}h_{i-1} + \gamma_i h_i}{g}, \qquad i = 2, 3, \ldots, N \tag{7.24}$$

These masses are connected by springs which resist lateral deformation. The

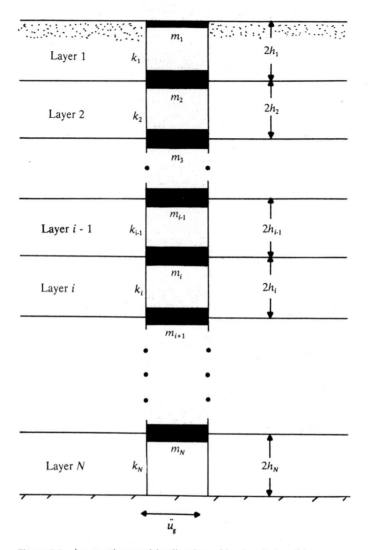

Figure 7.8 Lumped mass idealization of horizontal soil layers

spring constants can be given by

$$k_i = \frac{G_i}{2h_i}, \qquad i = 1, 2, \ldots, N \tag{7.25}$$

where k_i is the spring constant of the spring connecting the masses m_i and m_{i+1}, and G_i is the shear modulus of layer i.

The equation of motion of the system can be given by the expression

$$[M]\{\ddot{u}\} + [C]\{\dot{u}\} + [K]\{u\} = \{R(t)\} \tag{7.26}$$

where $[M]$ is a matrix for mass, $[C]$ is a matrix for viscous damping, $[K]$ is the stiffness matrix, and $\{u\}$, $\{\dot{u}\}$, and $\{\ddot{u}\}$ are relative displacement, relative velocity, and relative acceleration vectors, respectively. The matrices $[M]$, $[C]$, and $[K]$ are of the order N (the number of layers considered). The matrix $[M]$ is a diagonal matrix such that

$$\text{diag}[M] = (m_1, m_2, m_3, \ldots, m_N) \tag{7.27}$$

The matrix $[K]$ is tridiagonal and symmetric and

$$K_{11} = k_1$$
$$K_{ij} = k_{i-1} + k_i \qquad \text{for } i = j$$
$$K_{ij} = -k_i \qquad \text{for } i = j - 1$$
$$K_{ij} = -k_j \qquad \text{for } i = j + 1$$

All other K_{ij} are equal to zero.

The load vector $\{R(t)\}$ is

$$\{R(t)\} = -\text{col}(m_1, m_2, \ldots, m_N)\ddot{u}_g \tag{7.28}$$

A computer program for solution of Eq. (7.26) is given in Idriss and Seed (1967, Appendix C). The general outline of the solution is as follows:

1. The number of layers of soil (N) and the mass and stiffness matrices are first obtained.

2. The mode shapes and frequencies are obtained from the characteristic value problem as

$$[K]\{\phi^n\} = \omega_n^2[M]\{\phi^n\} \tag{7.29}$$

where ϕ_i^n is the mode shape at the ith level during the nth mode of vibration and ω_n is the circular frequency at the nth mode of vibration.

3. Eq. (7.26) is then reduced to a set of uncoupled normal equations. The normal equations are solved for the response of each mode at each instant of time. The relative displacement at level i can then be expressed as

$$u_i(t) = \sum_{n=1}^{N} \phi_i^n X_n(t) \tag{7.30}$$

where $X_n(t)$ is the normal coordinate for the nth mode and $u_i(t)$ is the relative displacement at the ith level at time t.

4. The relative velocity $[\dot{u}_i(t)]$ and the relative acceleration $[\ddot{u}_i(t)]$ can be obtained by differentiation of Eq. (7.30), or

$$\dot{u}_i(t) = \sum_{n=1}^{N} \phi_i^n \dot{X}_n(t) \tag{7.31}$$

$$\ddot{u}_i(t) = \sum_{n=1}^{N} \phi_i^n \ddot{X}_n(t) \tag{7.32}$$

5. The total acceleration, velocity, and displacement at level i and time t can be given as follows:

Total acceleration $= \ddot{u}_i(t) + \ddot{u}_g$

Total velocity $= \dot{u}_i(t) + \dot{u}_g$

Total displacement $= u_i(t) + u_g$

6. The shear strain between level i and $i + 1$ can be expressed as

Shear strain $[u_i(t) - u_{i+1}(t)]/2h_i$ (7.33)

7. The shear stress between level i and $i + 1$ can now be obtained as

$$\tau_i(t) = (\text{shear strain})G \tag{7.34}$$

Degree of Accuracy and Stability of the Analysis

The degree of accuracy of the lumped mass solution depends on the number of layers of soils used in an analysis. (*Note:* The value of the shear modulus for each layer is assumed to be constant.) In order to select a reasonable number of layers N with a tolerable degree of accuracy, Idriss and Seed (1968) prepared the graph shown in Figure 7.9, where ERS means the percentage of error in the lumped mass representation. The use of this figure can be explained as follows.

Let the height, shear modulus, and unit weight of the ith layer of soil be H_i, G_i, and γ_i, respectively. The fundamental frequency of this layer can be obtained

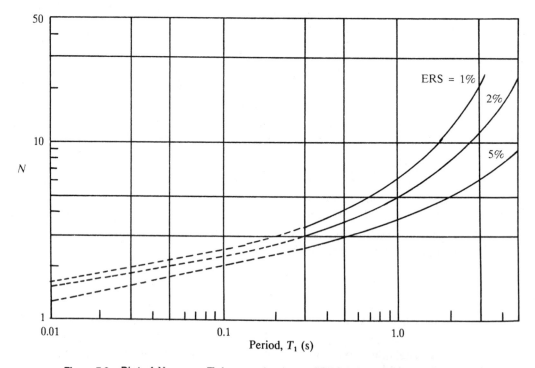

Figure 7.9   Plot of N versus T_1 for equal values of ERS (after Idriss and Seed, 1968)

from Eq. (7.22) as

$$\omega_{n(i)} = \left[\frac{(2n-1)\pi}{2H_i}\right]\sqrt{\frac{G_i}{\rho_i}}$$

$$= \left(\frac{\pi}{2H_i}\right)\sqrt{\frac{G_i}{\rho_i}} \quad \text{for } n = 1$$

Hence, the fundamental period can be given by

$$T_{1(i)} = \frac{2\pi}{\omega_{1(i)}} = \frac{4H_i}{\sqrt{G_i g/\gamma_i}} \tag{7.35}$$

where g is acceleration due to gravity.

Using the value of this $T_{1(i)}$ and a given value of ERS, the value of N_i can be obtained from Figure 7.9; this is the number of layers into which the ith layer has to be divided for the analysis of the ground vibration. Since this needs to be done for each layer of soil,

$$N = \sum N_i \tag{7.36}$$

For the stability of the lumped mass solution, Idriss and Seed (1978) have suggested the following condition.

For the step-by-step solution (Berg and Housner, 1961; Wilson and Clough, 1962):

$$T_{NN} \geq 2\,\Delta t \tag{7.37}$$

For Newmark's iterative solution (1962)

$$T_{NN} \geq 5\,\Delta t \tag{7.38}$$

where Δt is the time interval used for integrating the normal equations and T_{NN} is the lowest period included in the analysis. Note that this corresponds to the highest mode of vibration.

General Remarks for Ground Vibration Analysis

First of all, it should be kept in mind that soil deposits, in general, tend to amplify the underlying rock motion to some degree.

Secondly, for appropriate analysis of ground motion due to an earthquake, it is necessary that an earthquake acceleration–time record be available at the level of the *bedrock* or *bedrocklike material* for a given site. The design accelerogram can be obtained by selecting an actual motion, which has been recorded in the past, of a somewhat similar magnitude and fault distance as the design conditions. This accelerogram is then modified by taking into account the differences between the recorded and design conditions. This modification can be better explained by the following example.

Let the design earthquake be of magnitude 7 and the site be located at a distance of 80 km. Hence, its predominant period at *bedrock* or *bedrocklike material* is 0.4 s (Figure 7.3) and the maximum acceleration is of the order of

0.04g (Figure 7.4). The estimated duration of this earthquake is about 30 sec (equal to the duration of the fault break; Section 7.4). Also, let the recorded earthquake have a predominant period of 0.45 s, maximum acceleration of 0.05g, and a duration of 40 s. The recorded earthquake may now be modified by reducing the ordinates (i.e., magnitudes of acceleration) by $0.04/0.05 = \frac{4}{5}$ and by compressing the time scale by $0.40/0.45 = \frac{8}{9}$. This results in a maximum acceleration of 0.04g with a predominant period of 0.4 s and a duration of 35.5 s. The first 30 s of this accelerogram can now be taken for the analysis of ground motion.

Appropriate parts of an accelerogram could be repeated to obtain the desired period of predicted significant motion.

Example 7.1

In a soil deposit, a clay layer has a thickness of 16 m. The unit weight and the shear modulus of the clay soil deposit are 17.8 kN/m³ and 24,000 kN/m², respectively. Determine the number of layers into which this should be divided so that the ERS in the lumped mass solution does not exceed 5%.

Solution

Given that $H_i = 16$ m and $G_i = 24,000$ kN/m²,

$$T_{1(i)} = \frac{4H_i}{\sqrt{G_i g / \gamma_i}} = \frac{(4)(16)}{\sqrt{(24,000 \times 9.81)/17.8}} = 0.556 \text{ s}$$

From Figure 7.9, with $T_{1(i)} = 0.556$ s and ERS = 5%, the value of N_i is equal to 3. Thus, this clay layer should be subdivided into at least 3 layers with thicknesses of 5.33 m each. ∎

7.6   Other Studies for Vibration of Soil Layers Due to Earthquakes

In the preceding section, for the evaluation of the ground vibration, it was assumed that

1. the soil layer(s) possess linearly elastic properties, and

2. the soil layer(s) are horizontal.

Under strong ground-shaking conditions, the stress-strain relationships may be of the nature shown in Figure 7.10a, and not linearly elastic. This type of stress-strain relationship can be approximated to a bilinear system as shown in Figure 7.10b and the analysis of ground vibration can then be carried out. The lumped mass type of solution using bilinear stress-strain relationships of horizontally layered soils (Figure 7.11) have been presented by Parmelee et al. (1964) and Idriss and Seed (1967, 1968), whose works may be examined for further details.

Studies of the vibration of soils with sloping boundaries have also been made by Idriss, Dezfulian, and Seed (1969) and Dezfulian and Seed (1970). This involves a finite element method of analysis. For a computer program of such an analysis, refer to Idriss, Dezfulian, and Seed.

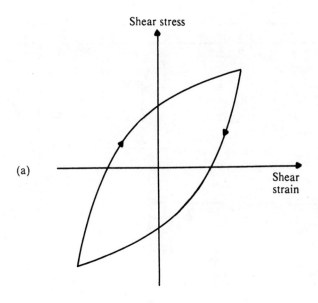

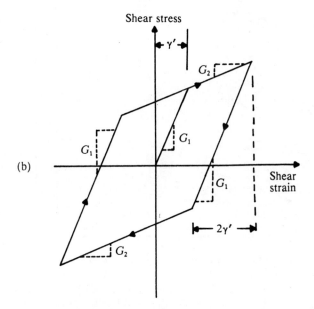

Figure 7.10   Shear stress-strain characteristics of soil: (a) stress-strain curve; (b) bilinear idealization

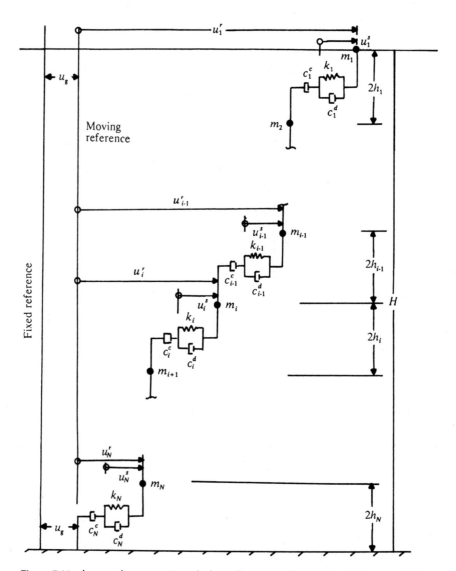

Figure 7.11 Lumped parameter solution of a semi-infinite layer: bilinear solution (after Idriss and Seed, 1968)

7.7   Equivalent Number of Significant Uniform Stress Cycles for Earthquakes

In the study of soil liquefaction of granular soils (Chapter 10), it becomes necessary to determine the equivalent number of significant uniform stress cycles for an earthquake that has *irregular stress–time history*. This is explained with the aid of Figure 7.12. Figure 7.12a shows the irregular pattern of shear stress on a soil deposit with time for an earthquake. The maximum shear stress induced is τ_{max}. This irregular stress–time history may be *equivalent* to uniformly intense

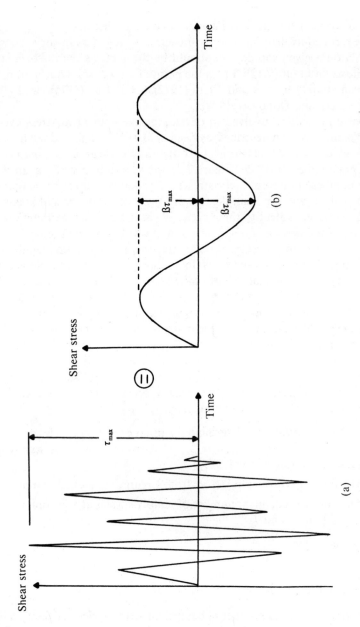

Figure 7.12 Equivalent uniform stress cycles: (a) irregular stress–time history; (b) equivalent uniform stress–time history

N number of cyclic shear stresses of maximum magnitude equal to $\beta\tau_{max}$ (Figure 7.12b). The term equivalent means that the effect of the stress history shown in Figure 7.12a on a given soil deposit should be the same as the uniform stress cycles as shown in Figure 7.12b. From the point of view of soil liquefaction, this fact has been studied by Lee and Chan (1972), Seed et al. (1975), Seed (1976, 1979), and Valera and Donovan (1977).

The basic procedure involved in developing the equivalent stress cycle is fairly simple and has been described by Seed et al. (1975). This is done by using the results of the soil liquefaction study by simple shear tests obtained by DeAlba, Chan, and Seed (1975). Figure 7.13 shows a plot of τ/τ_{max} against the equivalent number of uniform cyclic stresses N at a maximum stress magnitude of $0.65\tau_{max}$. This means, for example, that *one cycle* of shear stress of maximum magnitude τ_{max} is equivalent to three cycles of shear stress of maximum magnitude $0.65\tau_{max}$. Similarly, *one cycle* of shear stress with maximum magnitude of $0.75\tau_{max}$ is equivalent to 1.4 cycles of shear stress with a maximum magnitude of $0.65\tau_{max}$. Figure 7.13 can be used to evaluate the values of N for various earthquakes for a maximum magnitude of uniform cyclic shear stress level equaling $0.65\tau_{max}$. (*Note:* $\beta = 0.65$.) This can be most effectively explained by a numerical example. While doing this, one must recognize that, within the top 20 ft (≈ 6–7 m) of a given soil deposit, the cyclic shear stress–time history of an earthquake is similar in form to the acceleration–time history at the ground surface. The acceleration–time history for the San Jose earthquake (1955) is shown in Figure 7.14. Note that the maximum acceleration in this case is $0.106g$. Hence, τ_{max} is proportional to $0.106g$. In order to determine N, one needs to prepare Table 7.4. This can be done in the following manner.

1. Looking at Figure 7.14, determine the number of stress cycles at various stress levels such as τ_{max}, $0.95\tau_{max}$, $0.9\tau_{max}$, ... above the horizontal axis (col. 2) and below the horizontal axis (col. 5).

2. Determine the conversion factors from Figure 7.13 (cols. 3 and 6).

3. Determine the equivalent number of uniform cycles at a maximum stress level of $0.65\tau_{max}$ (cols. 4 and 7).

col. 2 × col. 3 = col. 4

and

col. 5 × col. 6 = col. 7

4. Determine the total number of equivalent stress cycles at $0.65\tau_{max}$ above and below the horizontal axis.

5. $N = \frac{1}{2}$(equivalent no. of cycles above the horizontal
        + equivalent no. of cycles below the horizontal)

Equivalent numbers of uniform stress cycles (at a maximum level of $0.65\tau_{max}$) for several earthquakes with magnitudes of 5.3–7.7 analyzed in the preceding manner are shown in Figure 7.15. These are for the strongest component of the ground motion recorded. The mean and the mean ± 1 standard deviation (i.e., 16, 50, and 84 percentile) are also shown. This helps the designer

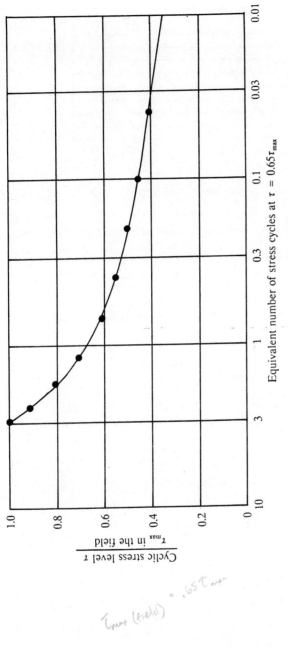

Figure 7.13   Plot of τ/τ_{max} versus N at $\tau = 0.65\tau_{max}$ (after Seed et al. 1975)

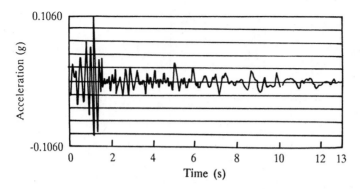

Figure 7.14   San Jose earthquake record, 1955 (after Seed et al., 1975)

Table 7.4   Example of Determination of Equivalent Uniform Cyclic Stress Series from Figure 7.14[a]

	Above horizontal axis			Below horizontal axis		
Stress level $(\otimes \tau_{max})$ (1)	no. of stress cycles (2)	Conversion factor (3)	Equivalent no. of cycles at $0.65\tau_{max}$ (4)	No. of stress cycles (5)	Conversion factor (6)	Equivalent no. of cycles at $0.65\tau_{max}$ (7)
1.00	1	3.00	3.00			
0.95						
0.90	—	—	—			
0.85	—	—	—	1	2.05	2.05
0.80	—	—	—	1	1.70	1.70
0.75	—	—	—			
0.70	—	—	—			
0.65	—	—	—			
0.60	1	0.70	0.70			
0.55	1	0.40	0.40	1	0.40	0.40
0.50						
0.45						
0.40	1	0.04	0.04	1	0.04	0.04
0.35	2	0.02	0.02	1	0.02	0.02
		Total	4.2		Total	4.2

Average number of cycles of $0.65\tau_{max} \approx 4.2$

[a] Seed et al. (1975).

choose the proper value of the equivalent uniform stress cycles depending on the degree of conservation required.

Using a similar procedure, Lee and Chan (1972) have given the variation of N with the earthquake magnitude for maximum uniform cyclic stress levels of $0.65\tau_{max}$, $0.75\tau_{max}$, and $0.85\tau_{max}$.

A cumulative damage approach has also been described by Valera and Donovan (1977) for determination of N. This approach is based on Miner's law

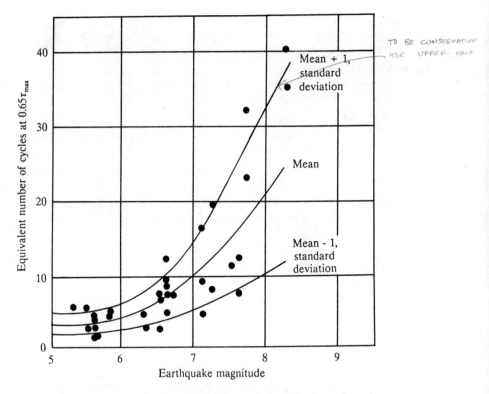

Figure 7.15 Equivalent numbers of uniform stress cycles based on strong component of ground motion (after Seed et al. 1975)

and involves the natural period of the soil deposit and the duration of earthquake shaking.

REFERENCES

Banioff, H. (1962). Unpublished report to A. R. Golze, Chief Engineer, Department of Water Resources, Sacramento, California, by Department of Water Resources Consulting Board for Earthquake Analysis.

Båth, M. (1966). "Earthquake Seismology," *Earth Science Reviews*, Vol. 1, p. 69.

Berg, G. V., and Housner, G. W. (1961). "Integrated Velocity and Displacement of Strong Earthquake Gound Motion," *Bulletin*, Seismological Society of America, Vol. 51, No. 2, pp. 175–189.

Blume, J. A. (1965). "Earthquake Ground Motion and Engineering Procedures for Important Installations Near Active Faults," *Proceedings*, 3rd World Conference on Earthquake Engineering, New Zealand, Vol. 3.

Bonilla, M. G. (1967). "Historic Surface Faulting in Continental United States and Adjacent Parts of Mexico," Interagency Report, U.S. Department of the Interior, Geological Survey.

DeAlba, P., Chan, C., and Seed, H. B. (1975). "Determination of Soil Liquefaction Characteristics by Large-Scale Laboratory Tests," Earthquake Engineering Research Center, *Report EERC 75-14*, University of California, Berkeley.

Dezfulian, H., and Seed, H. B. (1970). "Seismic Response of Soil Deposits Underlain by Sloping Rock Boundaries," *Journal of the Soil Mechanics and Foundations Division*, ASCE, Vol. 96, No. SM6, pp. 1893–1916.

Esteva, L., and Rosenblueth, E. (1963). "Espectros de Temblores a Distancias Moderadas y Grandes," *Proceedings*, Chilean Conference on Seismology and Earthquake Engineering, Vol. 1, University of Chile.

Figueroa, J. J. (1960). "Some Considerations About the Effect of Mexican Earthquakes," *Proceedings*, 2nd World Conference on Earthquake Engineering, Japan, Vol. III.

Gutenberg, B., and Richter, C. F. (1956). "Earthquake Magnitude, Intensity, Energy and Acceleration," *Bulletin*, Seismological Society of America, Vol. 46, No. 2, pp. 105–146.

Housner, G. W. (1965). "Intensity of Earthquake Ground Shaking Near the Causative Fault," *Proceedings*, 3rd World Conference on Earthquake Engineering, New Zealand, Vol. I.

Housner, G. W. (1969). "Engineering Estimate of Ground Shaking and Maximum Earthquake Magnitude," *Proceedings*, 4th World Conference on Earthquake Engineering, Santiago, Chile.

Housner, G. W. (1970). "Design Spectrum," in *Earthquake Engineering*, ed., R. W. Wiegel, Prentice-Hall, Englewood Cliffs, New Jersey, pp. 97–106.

Idriss, I. M., Dezfulian, H., and Seed, H. B. (1969). "Computer Programs for Evaluating the Seismic Response of Soil Deposits with Nonlinear Characteristics Using Equivalent Linear Procedures," *Research Report*, Earthquake Engineering Research Center, College of Engineering, University of California, Berkeley.

Idriss, I. M., and Seed, H. B. (1967). "Response of Horizontal Soil Layers During Earthquakes," *Research Report*, Soil Mechanics and Bituminous Materials Laboratory, University of California, Berkeley.

Idriss, I. M., and Seed, H. B. (1968). "Seismic Response of Horizontal Soil Layers," *Journal of the Soil Mechanics and Foundations Division*, ASCE, Vol. 94, No. SM4, pp. 1003–1031.

Kanai, K. (1966). "Improved Empirical Formula for the Characteristics of Strong Earthquake Motions," *Proceedings*, Japan Earthquake Engineering Symposium, Tokyo, pp. 1–4.

Lee, K. L., and Chan, K. (1972). "Number of Equivalent Significant Cycles in Strong Motion Earthquakes," *Proceedings*, International Conference on Microzonation, Seattle, Washington, Vol. 2, pp. 609–627.

Newmark, N. M. (1962). "A Method of Computations for Structural Dynamics," *Transactions*, ASCE, Vol. 127, Part I, pp. 1406–1435.

Parmelee, R., Penzien, J., Scheffey, C. F., Seed, H. B., and Thiers, G. R. (1964). "Seismic Effects on Structures Supported on Piles Extending Through Deep Sensitive Clays," *Report No. 64-2*, Institute of Engineering Research, University of California, Berkeley.

Richter, C. F. (1958). *Elementary Seismology*, W. H. Freeman, San Francisco, California.

Seed, H. B. (1976). "Evaluation of Soil Liquefaction Effects on Level Ground During Earthquakes," *Preprint No. 2752*, ASCE National Convention, Sept. 27–Oct. 1, pp. 1–104.

Seed, H. B. (1979). "Soil Liquefaction and Cyclic Mobility Evaluation for Level Ground During Earthquakes," *Journal of the Geotechnical Engineering Division*, ASCE, Vol. 105, No. GT2, pp. 102–155.

Seed, H. B., Idriss, I. M., and Kiefer, F. W. (1969). "Characteristics of Rock Motion During Earthquakes," *Journal of the Soil Mechanics and Foundations Division*, ASCE, Vol. 95, No. SM5, pp. 1199–1218.

Seed, H. B., Idriss, I. M., Makdisi, F., and Banerjee, N. (1975). "Representation of Irregular Stress–Time Histories by Equivalent Uniform Stress Series in Liquefaction Analyses," *Report No. EERC 75-29*, Earthquake Engineering Research Center, University of California, Berkeley.

Tocher, D. (1958). "Earthquake Energy and Ground Breakage," *Bulletin*, Seismological Society of America, Vol. 48, No. 2, pp. 147–153.

Valera, J. E. and Donovan, N. C. (1977). "Soil Liquefaction Procedures—A Review," *Journal of the Geotechnical Engineering Division*, ASCE, Vol. 103, No. GT6, pp. 607–625.

Wiegel, R. W. (ed.) (1970). *Earthquake Engineering*, Prentice-Hall, Englewood Cliffs, New Jersey.

Wilson, E. L., and Clough, R. W. (1962). "Dynamic Response by Step-by-Step Matrix Analysis," *Proceedings*, Symposium on the Use of Computers in Civil Engineering, Lisbon, Portugal.

Chapter 8

LATERAL EARTH PRESSURE ON RETAINING WALLS

8.1   Introduction

Excessive dynamic lateral earth pressure on retaining structures resulting from earthquakes has caused several major damages in the past. The increase of lateral earth pressure during earthquakes induces sliding and/or tilting to the retaining structures. The majority of case histories of failures reported in the literature until now concern waterfront structures such as quay walls and bridge abutments. Some of the examples of failures and lateral movements of quay walls due to earthquakes are given in Table 8.1. Seed and Whitman (1970) have suggested that some of these failures may have been due to several reasons, such as

1. increase of lateral earth pressure behind the wall,
2. reduction of water pressure at the front of the wall, and
3. liquefaction of the backfill material (see Chapter 10).

Nazarian and Hadjan (1979) have given a comprehensive review of the dynamic lateral earth pressure studies advanced so far. Based on this study, the theories can be divided into three broad categories, such as

1. fully plastic (static or pseudostatic) solution,
2. solutions based on elastic wave theory, and
3. solutions based on elastoplastic and nonlinear theory.

In this chapter, the lateral earth pressure theory, based on the fully plastic solution which is widely used by most of the design engineers, is developed.

8.2   Mononobe–Okabe Active Earth Pressure Theory

In 1776, Coulomb derived an equation for active earth pressure on a retaining wall due to a dry cohesionless backfill (Figure 8.1), which is of the form

$$P_A = \tfrac{1}{2}\gamma H^2 K_A \qquad\qquad (8.1)$$

Table 8.1   Failures and Movements of Quay Walls[a]

Earthquake	Date	Magnitude	Harbor	Distance from Epicenter	Damage	Approximate Movement
Kitaizu	25 November 1930	7.1	Shimizu	30 mi (48 km)	Failure of gravity walls[b]	26 ft (7.93 m)
Shizuoka	11 July 1935		Shimizu		Retaining wall collapse[b]	16 ft (4.88 m)
Tonankai	7 December 1944	8.2	Shimizu	110 mi (175 km)	Sliding of retaining wall[b]	
			Nagoya	80 mi (128 km)	Outward movement of bulkhead with relieving platform[b]	10–13 ft (3.05–3.96 m)
			Yokkaichi	90 mi (144 km)	Outward movement of pile-supported deck[b]	12 ft (3.66 m)
Nankai	21 December 1946	8.1	Nagoya		Outward movement of bulkhead with relieving platform[b]	13 ft (3.96 m)
			Osaka	125–190 mi	Failure of retaining wall above relieving platform[b]	14 ft (4.27 m)
			Yokkaichi	(200–304 km)	Outward movement of pile-supported deck[b]	12 ft (3.66 m)
			Uno			
Tokachioki	4 March 1952	7.8	Kushiro	90 mi (144 km)	Outward movement of gravity wall[b]	2 ft (0.61 m)
Chile	22 May 1960	8.4	Puerto Montt	70 mi (112 km)	Outward movement of gravity wall[b]	18 ft (5.49 m)
					Complete overturning of gravity walls[c]	>15 ft (4.57 m)
					Outward movement of anchored bulkheads[c]	2–3 ft (0.61–0.915 m)
Niigata	16 June 1964	7.5	Niigata	32 mi (51.2 km)	Tilting of gravity wall[d]	10 ft (3.05 m)
					Outward movement of anchored bulkheads[d]	1–7 ft (0.305–2.13 m)

[a] After Seed and Whitman (1970).
[b] Reported by Amano, Azuma, and Ishii (1956).
[c] Reported by Duke and Leeds (1963).
[d] Reported by Hayashi, Kubo, and Nakase (1966).

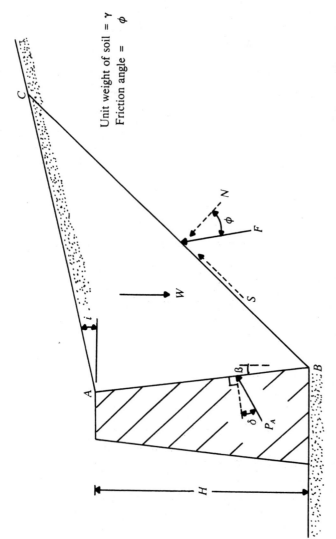

Unit weight of soil = γ
Friction angle = ϕ

Figure 8.1 Coulomb's active earth pressure (*Note*: *BC* is the failure plane; W = weight of the wedge *ABC*; *S* and *N* = shear and normal forces on the plane *BC*; F = resultant of *S* and *N*.)

where

P_A = active force per unit length of the wall

γ = unit weight of soil

H = height of the retaining wall

K_A = active earth pressure coefficient

$$= \frac{\cos^2(\phi - \beta)}{\cos^2\beta \cos(\delta + \beta)\left[1 + \left\{\dfrac{\sin(\delta + \phi)\sin(\phi - i)}{\cos(\delta + \beta)\cos(\beta - i)}\right\}^{1/2}\right]^2} \qquad (8.2)$$

where

ϕ = soil friction angle

δ = angle of friction between the wall and the soil

β = slope of the back of the wall with respect to the vertical

i = slope of the backfill with respect to the horizontal

The values of K_A for $\beta = 0°$ and various values of ϕ and δ are given in Table 8.2.

In the actual design of retaining walls, the value of the wall friction δ is assumed to be between $\phi/2$ and $\frac{2}{3}\phi$. The active earth pressure coefficients for various values of ϕ, i, and β with $\delta = \frac{2}{3}\phi$ are given in Table 8.3 on p. 332. This is a very useful table for design considerations.

Coulomb's active earth pressure equation can be modified to take into account the vertical and horizontal coefficients of acceleration induced by an earthquake. This is generally referred to as the *Mononobe–Okabe analysis* (Mononobe, 1929; Okabe, 1926). The Mononobe–Okabe solution is based on the following assumptions:

1. The failure in soil takes place along a plane such as *BC* shown in Figure 8.2 on p. 333.

Table 8.2 Values of K_A [Eq. (8.2)] for $\beta = 0°$ and $i = 0°$

ϕ (deg)	δ (deg)					
	0	5	10	15	20	25
28	0.3610	0.3448	0.3330	0.3251	0.3203	0.3186
30	0.3333	0.3189	0.3085	0.3014	0.2973	0.2956
32	0.3073	0.2945	0.2853	0.2791	0.2755	0.2745
34	0.2827	0.2714	0.2633	0.2579	0.2549	0.2542
36	0.2596	0.2497	0.2426	0.2379	0.2354	0.2350
38	0.2379	0.2292	0.2230	0.2190	0.2169	0.2167
40	0.2174	0.2098	0.2045	0.2011	0.1994	0.1995
42	0.1982	0.1916	0.1870	0.1841	0.1828	0.1831

Table 8.3   Values of K_A [Eq. (8.2)] (*Note:* $\delta = \frac{2}{3}\phi$ in all cases)

i (deg)	ϕ (deg)	β (deg)					
		0	5	10	15	20	25
0	28	0.3213	0.3588	0.4007	0.4481	0.5026	0.5662
	30	0.2973	0.3349	0.3769	0.4245	0.4794	0.5435
	32	0.2750	0.3125	0.3545	0.4023	0.4574	0.5220
	34	0.2543	0.2916	0.3335	0.3813	0.4367	0.5017
	36	0.2349	0.2719	0.3137	0.3615	0.4170	0.4825
	38	0.2168	0.2535	0.2950	0.3428	0.3984	0.4642
	40	0.1999	0.2361	0.2774	0.3250	0.3806	0.4468
	42	0.1840	0.2197	0.2607	0.3081	0.3638	0.4303
5	28	0.3431	0.3845	0.4311	0.4843	0.5461	0.6191
	30	0.3165	0.3578	0.4043	0.4575	0.5194	0.5926
	32	0.2919	0.3329	0.3793	0.4324	0.4943	0.5678
	34	0.2691	0.3097	0.3558	0.4088	0.4707	0.5443
	36	0.2479	0.2881	0.3338	0.3866	0.4484	0.5222
	38	0.2282	0.2679	0.3132	0.3656	0.4273	0.5012
	40	0.2098	0.2489	0.2937	0.3458	0.4074	0.4814
	42	0.1927	0.2311	0.2753	0.3271	0.3885	0.4626
10	28	0.3702	0.4164	0.4686	0.5287	0.5992	0.6834
	30	0.3400	0.3857	0.4376	0.4974	0.5676	0.6516
	32	0.3123	0.3575	0.4089	0.4683	0.5382	0.6220
	34	0.2868	0.3314	0.3822	0.4412	0.5107	0.5942
	36	0.2633	0.3072	0.3574	0.4158	0.4849	0.5682
	38	0.2415	0.2846	0.3342	0.3921	0.4607	0.5438
	40	0.2214	0.2637	0.3125	0.3697	0.4379	0.5208
	42	0.2027	0.2441	0.2921	0.3487	0.4164	0.4990
15	28	0.4065	0.4585	0.5179	0.5869	0.6685	0.7671
	30	0.3707	0.4219	0.4804	0.5484	0.6291	0.7266
	32	0.3384	0.3887	0.4462	0.5134	0.5930	0.6895
	34	0.3091	0.3584	0.4150	0.4811	0.5599	0.6554
	36	0.2823	0.3306	0.3862	0.4514	0.5295	0.6239
	38	0.2578	0.3050	0.3596	0.4238	0.5006	0.5949
	40	0.2353	0.2813	0.3349	0.3981	0.4740	0.5672
	42	0.2146	0.2595	0.3119	0.3740	0.4491	0.5416
20	28	0.4602	0.5205	0.5900	0.6715	0.7690	0.8810
	30	0.4142	0.4728	0.5403	0.6196	0.7144	0.8303
	32	0.3742	0.4311	0.4968	0.5741	0.6667	0.7800
	34	0.3388	0.3941	0.4581	0.5336	0.6241	0.7352
	36	0.3071	0.3609	0.4233	0.4970	0.5857	0.6948
	38	0.2787	0.3308	0.3916	0.4637	0.5587	0.6580
	40	0.2529	0.3035	0.3627	0.4331	0.5185	0.6243
	42	0.2294	0.2784	0.3360	0.4050	0.4889	0.5931

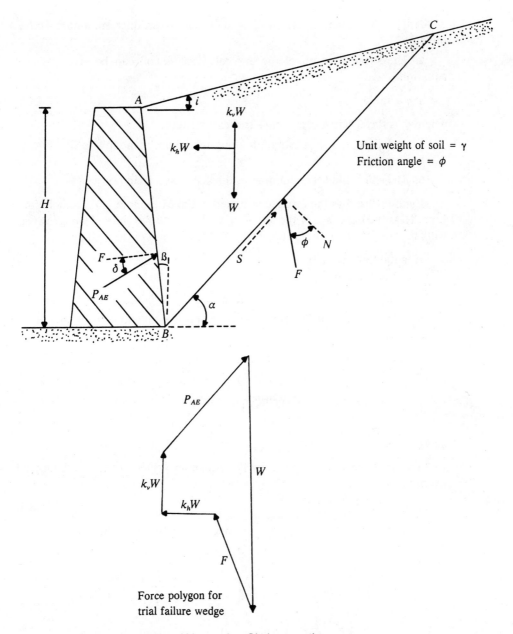

Force polygon for
trial failure wedge

Figure 8.2   Derivation of Mononobe–Okabe equation

2. The movement of the wall is sufficient to produce minimum active pressure.

3. The shear strength of the dry cohesionless soil can be given by the equation

$$s = \sigma' \tan \phi \tag{8.3}$$

where σ' is the effective stress and s is shear strength.

4. At failure, full shear strength along the failure plane (plane BC, Figure 8.2) is mobilized.

5. The soil behind the retaining wall behaves as a rigid body.

Figure 8.2 shows the forces considered in the Mononobe–Okabe solution. Line AB is the back face of the retaining wall and ABC is the soil wedge which will fail. The forces on the failure wedge per unit length of the wall are

a. weight of wedge W,

b. active force P_{AE},

c. resultant of shear and normal forces along the failure plane F, and

d. $k_h W$ and $k_v W$, the inertia forces in the horizontal and vertical directions, respectively, where

$$k_h = \frac{\text{horiz. component of earthquake accel.}}{g}$$

$$k_v = \frac{\text{vert. component of earthquake accel.}}{g}$$

and g is acceleration due to gravity.

The active force determined by the wedge analysis described here may be expressed as

$$P_{AE} = \tfrac{1}{2}\gamma H^2 (1 - k_v) K_{AE} \tag{8.4}$$

where K_{AE} is the active earth pressure coefficient with earthquake effect:

$$K_{AE} = \frac{\cos^2(\phi - \theta - \beta)}{\cos\theta \cos^2\beta \cos(\delta + \beta + \theta)\left[1 + \sqrt{\dfrac{\sin(\phi + \delta)\sin(\phi - \theta - i)}{\cos(\delta + \beta + \theta)\cos(i - \beta)}}\right]^2} \tag{8.5}$$

$$\theta = \tan^{-1}\left(\frac{k_h}{1 - k_v}\right) \tag{8.6}$$

Equation (8.4) is generally referred to as the *Mononobe–Okabe active earth pressure equation*. For the active force condition (P_{AE}), the angle α that the soil wedge ABC located behind the retaining wall (Figure 8.2) makes with the horizontal (for $k_v = 0°$, $\beta = 0°$, $i = 0°$, $\phi = 30°$, and $\delta = 0°$ and $20°$) is shown in Figure 8.3.

Table 8.4 gives the values of K_{AE} [Eq. (8.5)] for various values of ϕ, δ, i, and k_h with $k_v = 0$ and $\beta = 0°$.

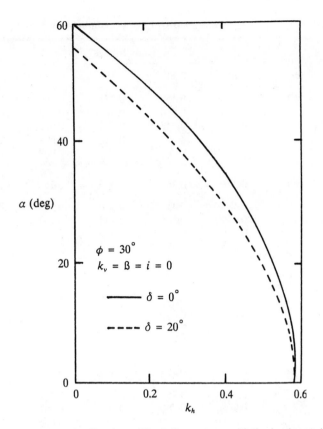

Figure 8.3 Inclination of the failure plane with the horizontal (after Davies, Richards, and Chen, 1986)

Table 8.4 Values of K_{AE} [Eq. (8.5)] with $k_v = 0$ and $\beta = 0°$

			ϕ (deg)				
k_h	δ (deg)	i (deg)	28	30	35	40	45
0.1	0	0	0.427	0.397	0.328	0.268	0.217
0.2			0.508	0.473	0.396	0.382	0.270
0.3			0.611	0.569	0.478	0.400	0.334
0.4			0.753	0.697	0.581	0.488	0.409
0.5			1.005	0.890	0.716	0.596	0.500
0.1	0	5	0.457	0.423	0.347	0.282	0.227
0.2			0.554	0.514	0.424	0.349	0.285
0.3			0.690	0.635	0.522	0.431	0.356
0.4			0.942	0.825	0.653	0.535	0.442
0.5			—	—	0.855	0.673	0.551

(continued)

Table 8.4 Values of K_{AE} [Eq. (8.5)] with $k_v = 0$ and $\beta = 0°$ (Continued)

k_h	δ (deg)	i (deg)	ϕ (deg) 28	30	35	40	45
0.1	0	10	0.497	0.457	0.371	0.299	0.238
0.2			0.623	0.570	0.461	0.375	0.303
0.3			0.856	0.748	0.585	0.472	0.383
0.4			—	—	0.780	0.604	0.486
0.5			—	—	—	0.809	0.624
0.1	$\frac{\phi}{2}$	0	0.396	0.368	0.306	0.253	0.207
0.2			0.485	0.452	0.380	0.319	0.267
0.3			0.604	0.563	0.474	0.402	0.340
0.4			0.778	0.718	0.599	0.508	0.433
0.5			1.115	0.972	0.774	0.648	0.552
0.1	$\frac{\phi}{2}$	5	0.428	0.396	0.326	0.268	0.218
0.2			0.537	0.497	0.412	0.342	0.283
0.3			0.699	0.640	0.526	0.438	0.367
0.4			1.025	0.881	0.690	0.568	0.475
0.5			—	—	0.962	0.752	0.620
0.1	$\frac{\phi}{2}$	10	0.472	0.433	0.352	0.285	0.230
0.2			0.616	0.562	0.454	0.371	0.303
0.3			0.908	0.780	0.602	0.487	0.400
0.4			—	—	0.857	0.656	0.531
0.5			—	—	—	0.944	0.722
0.1	$\frac{2}{3}\phi$	0	0.393	0.366	0.306	0.256	0.212
0.2			0.486	0.454	0.384	0.326	0.276
0.3			0.612	0.572	0.486	0.416	0.357
0.4			0.801	0.740	0.622	0.533	0.462
0.5			1.177	1.023	0.819	0.693	0.600
0.1	$\frac{2}{3}\phi$	5	0.427	0.395	0.327	0.271	0.224
0.2			0.541	0.501	0.418	0.350	0.294
0.3			0.714	0.655	0.541	0.455	0.386
0.4			1.073	0.921	0.722	0.600	0.509
0.5			—	—	1.034	0.812	0.679
0.1	$\frac{2}{3}\phi$	10	0.472	0.434	0.354	0.290	0.237
0.2			0.625	0.570	0.463	0.381	0.317
0.3			0.942	0.807	0.624	0.509	0.423
0.4			—	—	0.909	0.699	0.573
0.5			—	—	—	1.037	0.800

8.3   Some Comments on the Active Force Equation

Considering the active force relation given by Eqs. (8.4)–(8.6), the term $\sin(\phi - \theta - i)$ in Eq. (8.5) has some important implications.

First, if $\phi - \theta - i < 0$ (i.e., negative), no real solution of K_{AE} is possible. Physically it implies that an *equilibrium condition will not exist*. Hence, for stability, the limiting slope of the backfill may be given by

$$i \leq \phi - \theta \tag{8.7}$$

For no earthquake condition, $\theta = 0$; for stability, Eq. (8.7) gives the familiar relation

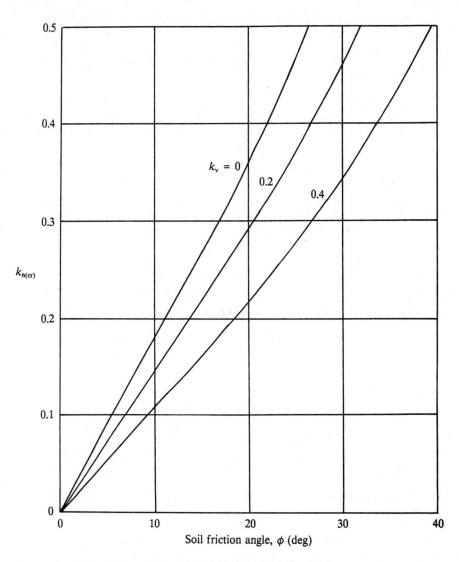

Figure 8.4   Critical values of horizontal acceleration (Eq. 8.11)

$$i \leq \phi \tag{8.8}$$

Secondly, for horizontal backfill, $i = 0$; for stability,

$$\theta \leq \phi \tag{8.9}$$

Since $\theta = \tan^{-1}[k_h/(1 - k_v)]$, for stability, combining Eqs. (8.6) and (8.9) results in

$$k_h \leq (1 - k_v) \tan \phi \tag{8.10}$$

Hence, the critical value of the horizontal acceleration can be defined as

$$k_{h(cr)} = (1 - k_v) \tan \phi \tag{8.11}$$

where $k_{h(cr)}$ = critical value of horizontal acceleration (Figure 8.4).

8.4   Procedure for Obtaining P_{AE} Using Standard Charts of K_A

Since the values of K_A are available in most standard handbooks and textbooks, I. Arango (1969) developed a simple procedure for obtaining the values of K_{AE} from the standard charts of K_A. This procedure has been described by Seed and Whitman (1970). Referring to Eq. (8.1),

$$P_A = \tfrac{1}{2}\gamma H^2 K_A = \tfrac{1}{2}\gamma H^2 A_c (\cos^2 \beta)^{-1} \tag{8.12}$$

where

$$A_c = K_A \cos^2 \beta$$
$$= \frac{\cos^2(\phi - \beta)}{\cos(\delta + \beta)\left[1 + \left\{\dfrac{\sin(\delta + \phi)\sin(\phi - i)}{\cos(\delta + \beta)\cos(\beta - i)}\right\}^{1/2}\right]^2} \tag{8.13}$$

In a similar manner, from Eq. (8.4)

$$P_{AE} = \tfrac{1}{2}\gamma H^2(1 - k_v)K_{AE} = \tfrac{1}{2}\gamma H^2(1 - k_v)(\cos \theta \cos^2 \beta)^{-1}(A_m) \tag{8.14}$$

where

$$A_m = K_{AE} \cos \theta \cos^2 \beta$$
$$= \frac{\cos^2(\phi - \beta - \theta)}{\cos(\delta + \beta + \theta)\left[1 + \left\{\dfrac{\sin(\phi + \delta)\sin(\phi - i - \theta)}{\cos(\delta + \beta + \theta)\cos(i - \beta)}\right\}^{1/2}\right]^2} \tag{8.15}$$

Now let

$$i' = i + \theta \tag{8.16}$$

and

$$\beta' = \beta + \theta \tag{8.17}$$

Substitution of Eqs. (8.16) and (8.17) into Eq. (8.15) yields

$$A_m = \frac{\cos^2(\phi - \beta')}{\cos(\delta + \beta')\left[1 + \left\{\dfrac{\sin(\phi + \delta)\sin(\phi - i')}{\cos(\delta + \beta')\cos(\beta' - i')}\right\}^{1/2}\right]^2}$$ (8.18)

The preceding equation is similar to Eq. (8.13) except for the fact that i' and β' are used in place of i and β. Thus, it can be said that

$$A_m = A_c(i', \beta') = K_A(i', \beta')\cos^2\beta'$$

The active earth pressure P_{AE} can now be expressed as

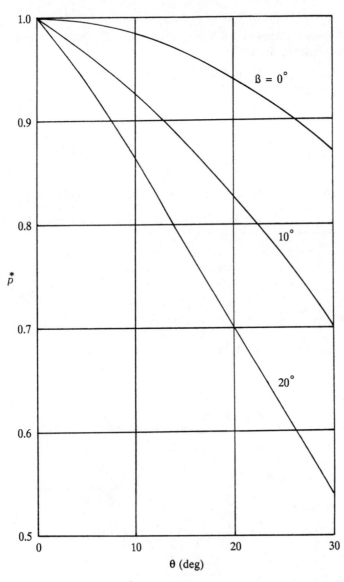

Figure 8.5   Variation of $\overset{*}{p}$ and θ

$$P_{AE} = \tfrac{1}{2}\gamma H^2 (1 - k_v)\left(\frac{\cos^2 \beta'}{\cos \theta \cos^2 \beta}\right)K_A(i', \beta')$$

$$= P_A(i', \beta')(1 - k_v)(\overset{*}{p}) \tag{8.19}$$

where

$$\overset{*}{p} = \left(\frac{\cos^2 \beta'}{\cos \theta \cos^2 \beta}\right) \tag{8.20}$$

In order to calculate P_{AE} by using Eq. (8.19), one needs to follow these steps:

1. Calculate i' [Eq. (8.16)].

2. Calculate β' [Eq. (8.17)].

3. With known values of ϕ, δ, i', and β', calculate K_A (from Tables 8.2, Table 8.3, or other available charts).

4. Calculate P_A as equal to $\tfrac{1}{2}\gamma H^2 K_A$ (K_A from Step 3).

5. Calculate $(1 - k_v)$.

6. Calculate $\overset{*}{p}$ [Eq. (8.20)].

7. Calculate

$$P_{AE} = \underset{\text{(Step 4)}}{P_A(i', \beta')}\underset{\text{(Step 5)}}{(1 - k_v)}\ \underset{\text{(Step 6)}}{(\overset{*}{p})}$$

For convenience, some typical values of $\overset{*}{p}$ are plotted in Figure 8.5.

Example 8.1

Refer to Figure 8.2. If $\beta = 0°$, $i = 0°$, $\phi = 36°$, $\delta = 18°$, $H = 15$ ft, $\gamma = 110$ lb/ft³, $k_v = 0.2$, and $k_h = 0.3$, determine the active force per unit length of the wall.

Solution

$$\theta = \tan^{-1}\left(\frac{k_h}{1 - k_v}\right) = \tan^{-1}\left(\frac{0.3}{1 - 0.2}\right) = 20.56°$$

$$i' = i + \theta = 0 + 20.56° = 20.56°$$

$$\beta' = \beta + \theta = 0 + 20.56° = 20.56°$$

$$K_A(i', \beta') = \frac{\cos^2(\phi - \beta')}{\cos^2 \beta' \cos(\delta + \beta')\left[1 + \left\{\dfrac{\sin(\delta + \phi)\sin(\phi - i')}{\cos(\delta + \beta')\cos(\beta' - i')}\right\}^{1/2}\right]^2}$$

$$= \frac{\cos^2(15.44)}{(\cos^2 20.56)(\cos 38.56)\left[1 + \left\{\dfrac{(\sin 54)(\sin 15.44)}{(\cos 38.56)(\cos 0)}\right\}^{1/2}\right]^2}$$

$$= 0.583$$

$$P_A(i', \beta') = \frac{1}{2}\gamma H^2 K_A(i', \beta') = \frac{1}{2}(110)(15)^2(0.583) = 7214.62 \text{ lb/ft}$$

$$\overset{*}{p} = \left(\frac{\cos^2 \beta'}{\cos \theta \cos^2 \beta}\right) = \frac{\cos^2 20.56}{(\cos 20.56)(\cos 0)}$$

$$= 0.9363$$

Hence, from Eq. (8.19),

$$P_{AE} = P_A(i', \beta')(1 - k_v)(\overset{*}{p}) = (7214.62)(1 - 0.2)(0.9363) = \underline{5404 \text{ lb/ft}} \qquad \blacksquare$$

8.5   Effect of Various Parameters on the Value of the Active Earth Pressure Coefficient

Parameters such as the angle of wall friction, angle of friction of soil, and slope of the backfill influence the magnitude of the active earth pressure coefficient K_{AE} to varying degrees. The effect of each of these factors is considered briefly.

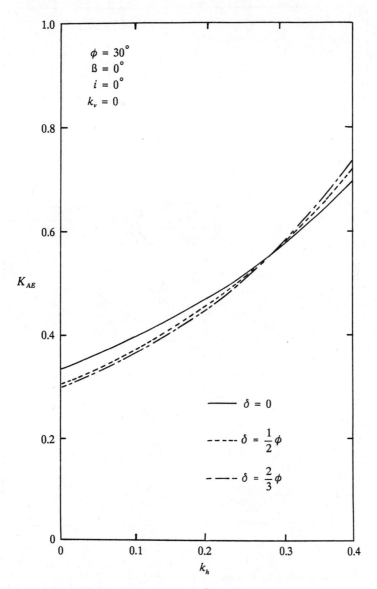

Figure 8.6   Influence of wall friction, δ, on K_{AE}

A. Effect of Wall Friction Angle δ

Figure 8.6 shows the variation of the active earth pressure coefficient K_{AE} with k_h for $\phi = 30°$ with $\delta = 0°$, $\phi/2$, and $\frac{2}{3}\phi$ ($k_v = 0$, $\beta = 0°$, and $i = 0°$). It can be seen from the plot that for $0 \leq \delta \leq \frac{2}{3}\phi$, the effect of wall friction on the active earth pressure coefficient is rather small.

B. Effect of Soil Friction Angle φ

Figure 8.7 shows the plot of $K_{AE} \cos \delta$ (that is, the horizontal component of the active earth pressure coefficient) for a vertical retaining wall with horizontal

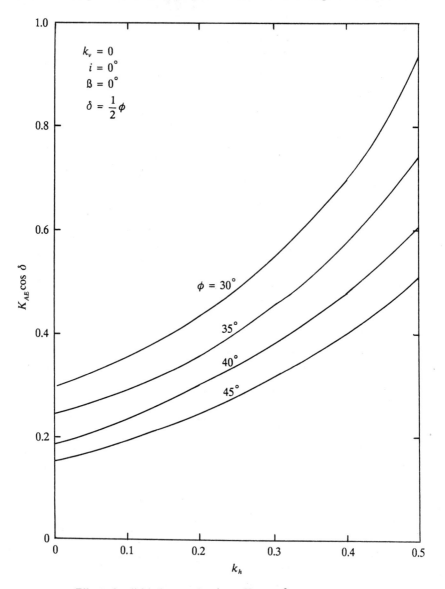

Figure 8.7   Effect of soil friction angle, ϕ, on $K_{AE} \cos \delta$

backfill ($\beta = 0°$ and $i = 0°$). In this plot, it has been assumed that $\delta = \phi/2$. From the plot, it may be seen that, for $k_v = 0$, $k_h = 0$ and $\delta = \frac{1}{2}\phi$, $K_{AE(\phi=30°)}$ is about 35% higher than $K_{AE(\phi=40°)}$. Hence, a small error in the assumption of the soil friction angle could lead to a large error in the estimation of P_{AE}.

C. Effect of Slope of the Backfill i

Figure 8.8 shows the variation of the value of $K_{AE} \cos \delta$ with i for a wall with $\beta = 0$, $\delta = \frac{2}{3}\phi$, $\phi = 30°$, and $k_v = 0$. Note that the value of $K_{AE} \cos \delta$ sharply increases with the increase of the slope of the backfill.

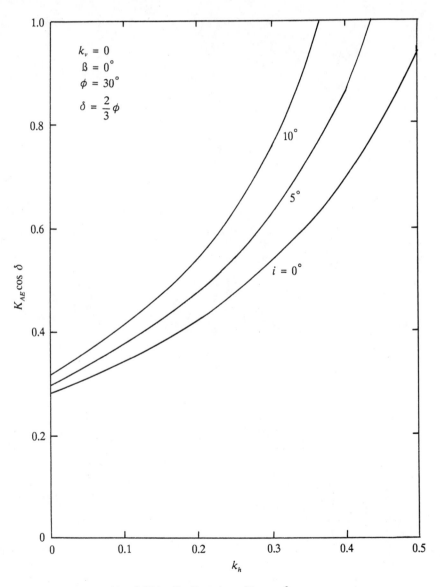

Figure 8.8   Effect of backfill inclination, i, on $K_{AE} \cos \delta$

8.6 Graphical Construction for Determination of Active Force, P_{AE}

Culmann (1875) developed a graphical method for determination of the active force P_A [Eq. (8.1)] developed behind a retaining wall. A modified form of Culmann's graphical construction for determination of the active force P_{AE} per unit length of a retaining wall has been proposed by Kapila (1962). In order to understand this, consider the force polygon for the wedge ABC shown in Figure 8.2. For convenience, this has been replotted in Figure 8.9a. The force polygon can be reduced to a force triangle with forces P_{AE}, F, and $W\sqrt{(1 - k_v)^2 + k_h^2}$

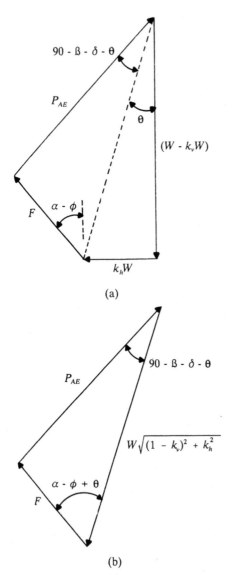

(a)

(b)

Figure 8.9   (Continued)

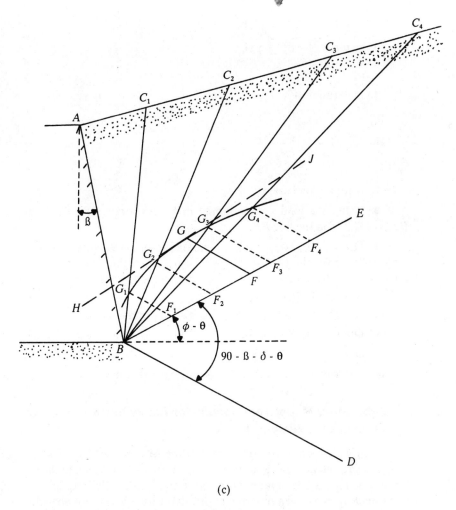

(c)

Figure 8.9   Modified Culmann construction

(Figure 8.9b). Note that in Figure 8.9a, b, α is the angle that the failure wedge makes with the horizontal.

The idea behind this graphical construction is to determine the *maximum* value of P_{AE} by considering several trial wedges. With reference to Figure 8.9c, following are steps for the graphical construction:

1. Draw line *BE*, which makes an angle $\phi - \theta$ with the horizontal.
2. Draw a line *BD*, which makes an angle $90° - \beta - \delta - \theta$ with the line *BE*.
3. Draw BC_1, BC_2, BC_3, ..., which are the trial failure surfaces.
4. Determine k_h and k_v and then $\sqrt{(1 - k_v)^2 + k_h^2}$.
5. Determine the weights W_1, W_2, W_3, ... of trial failure wedges ABC_1, ABC_2, ABC_3, ..., respectively (per unit length at right angle to the cross section shown). Note

$$W_1 = (\text{area of } ABC_1) \cdot \gamma \cdot 1$$

$$W_2 = (\text{area of } ABC_2) \cdot \gamma \cdot 1$$

$$\vdots$$

6. Determine $W_1', W_2', W_3', \ldots$ as

$$W_1' = \sqrt{(1 - k_v)^2 + k_h^2}\, W_1$$

$$W_2' = \sqrt{(1 - k_v)^2 + k_h^2}\, W_2$$

$$\vdots$$

7. Adopt a load scale.

8. Using the load scale adopted in Step 7, draw $BF_1 = W_1'$, $BF_2 = W_2'$, $BF_3 = W_3', \ldots$ on the line BE.

9. Draw $F_1 G_1$, $F_2 G_2$, $F_3 G_3, \ldots$ parallel to line BD. Note that $BF_1 G_1$ is the force triangle for the trial wedge ABC_1 similar to that shown in Figure 8.9b. Similarly, $BF_2 G_2$, $BF_3 G_3, \ldots$, are the force triangles for the trial wedges ABC_2, $ABC_3, \ldots$, respectively.

10. Join the points $G_1, G_2, G_3, \ldots$, by a smooth curve.

11. Draw a line HJ parallel to line BE. Let G be the point of tangency.

12. Draw line GF parallel to BD.

13. Determine active force P_{AE} as $GF \times$ (load scale)

8.7   Laboratory Model Test Results for Active Earth Pressure Coefficient, K_{AE}

In the early stages of the development of the Mononobe–Okabe solution [Eq. (8.4)], several small-scale laboratory model test results relating to the determination of the magnitude of lateral force on a rigid wall with dry granular backfill, and thus K_{AE}, have been reported in the literature (e.g., Mononobe and Matsuo, 1929; Jacobsen, 1939). More recently, Sherif, Ishibashi, and Lee (1982), Sherif and Fang (1984), and Ishibashi and Fang (1987) have published results of lateral earth pressure measurement behind a heavily instrumented rigid retaining wall. For all the preceding tests, the height of the retaining wall was 1 m. The retaining wall was resting on a shaking table with a granular backfill. A sinusoidal input motion with a $3\frac{1}{2}$-Hz frequency and maximum acceleration up to 0.5 g was applied to the shaking table during the experiments. The results of these tests are very instructive and will be summarized here.

The nature of distribution of active earth pressure and thus the magnitude of the active force on a retaining wall is very much dependent on the nature of yielding of the wall itself. Figure 8.10 shows the three possible modes of wall yielding for the development of an active state:

a. Rotation about the bottom (Figure 8.10a)

b. Translation (Figure 8.10b)

c. Rotation about the top (Figure 8.10c)

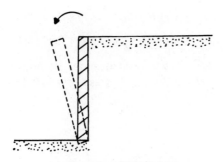

(a) Rotation about bottom

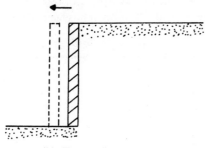

(b) Translation

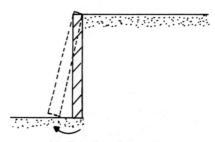

(c) Rotation about top

Figure 8.10   Modes of wall rotation for active pressure

Model test results relating to each of the three modes of wall yielding are described next.

A. Rotation About the Bottom

Ishibashi and Fang (1987) measured the dynamic active earth pressure distribution behind the model rigid retaining wall of 1-m height ($\beta = 0°$) described in the first paragraph of this section. For these tests, dry sand was used as a backfill material. The surface of the backfill was kept horizontal (that is, $i = 0$; Figure 8.2). The properties of the sand backfill were:

Dry unit weight of compaction of the backfill: 15.94–16.11 kN/m^3

Relative density of the backfill: 49.5–57.6%

Angle of friction of the soil: 38.5–40.1°

For these tests, the model retaining wall was rotated about its bottom. The magnitude of k_h was varied from 0 to about 0.6, and k_v was equal to 0. From Eq. (8.4) with $k_v = 0$,

$$K_{AE} = \frac{P_{AE}}{\frac{1}{2}\gamma H^2} \tag{8.21}$$

Figure 8.11 shows the variation of the experimental values of $K_{AE}\cos\delta$ obtained from the tests of Ishibashi and Fang (1987). Also plotted in Figure 8.11 is the theoretical variation of $K_{AE}\cos\delta$ obtained from Eq. (8.5) with $k_v = 0$, $\beta = 0°$, and $i = 0°$. In plotting this theoretical variation, it has been assumed that

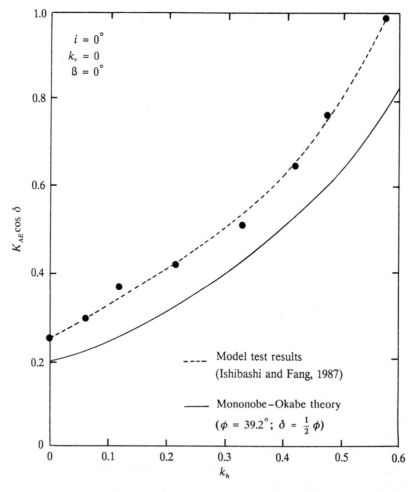

Figure 8.11 Wall rotation about the bottom for active pressure—comparison of theory with model test results

$\phi = 39.2°$ and $\delta = \phi/2$. The comparison between the Mononobe–Okabe theoretical curve and the experimental curve shows that

$$P_{AE(\text{measured})} \approx 1.23 \text{ to } 1.43P_{AE(\text{theory})}$$

B. Translation of the Wall

Dynamic active earth pressure measurement behind a vertical rigid model retaining wall undergoing translation was reported by Sherif, Ishibashi, and Lee (1982). The details of the test conditions are as follows:
Retaining wall:

Height $= 1$ m

$\beta = 0°$

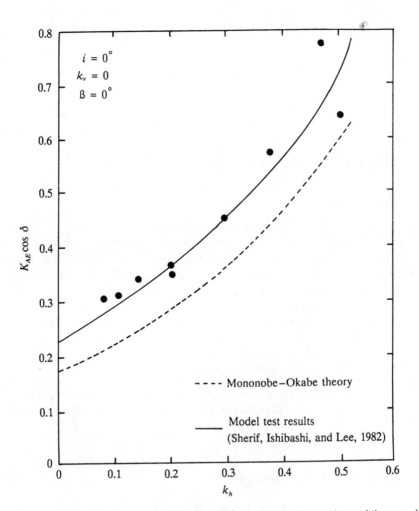

Figure 8.12   Translation of wall for active pressure—comparison of theory with model test results

Average properties of backfill (sand):

Unit weight $= 16.28$ kN/m^3

Angle of friction, $\phi = 40.9°$

Angle of wall friction, $\delta = 23.9°$

Slope of the backfill, $i = 0°$

For these tests the magnitude of k_h was varied from 0 to 0.5 and k_v was 0.

Figure 8.12 shows the experimental variation of $K_{AE} \cos \delta$ obtained from these model tests. Also shown in this figure is the variation of $K_{AE} \cos \delta$ obtained from the Mononobe–Okabe theory [Eq. (8.5)]. Based on this plot it appears that the experimental values of P_{AE} are about 30% higher than those obtained from Eqs. (8.4) and (8.5).

Sherif, Ishibashi, and Lee (1982) also developed an empirical relationship for the magnitude of wall translation for development of the active state, which can be given as

$$\Delta = H(7 - 0.13\phi)10^{-4} \tag{8.22}$$

where

$\Delta =$ lateral translation of the wall

$H =$ height of the wall

In Eq. (8.22), the value of ϕ is in degrees.

C. Rotation of the Wall about the Top

Sherif and Fang (1984) reported the dynamic earth pressure distribution behind a 1-m high rigid vertical retaining wall ($\beta = 0°$) undergoing rotation about its top. A sand with an average unit weight of 15.99 kN/m^3 was used as a backfill. The surface of the backfill was horizontal (that is, $i = 0°$). The nature of variation of the maximum active horizontal earth pressure distribution ($p_{AE} \cos \delta$, where $p =$ active earth pressure at a given depth) obtained from these tests is shown in Figure 8.13. Also plotted in this figure are the theoretical variations of $p_{AE} \cos \delta$ obtained from the Mononobe–Okabe solution (with $\beta = 0°$, $i = 0$, and $k_v = 0$) for various values of k_h. From the comparison of the theoretical and experimental plots, the following general conclusions can be drawn.

1. The nature of variation of dynamic earth pressure for wall rotation about the top is very much different than that predicted by the Mononobe–Okabe theory.

2. For a given value of k_h,

$$P_{AE} \cos \delta = \int (p_{AE} \cos \delta) \, dy \tag{8.23}$$

where $y =$ depth measured from the top of the wall.

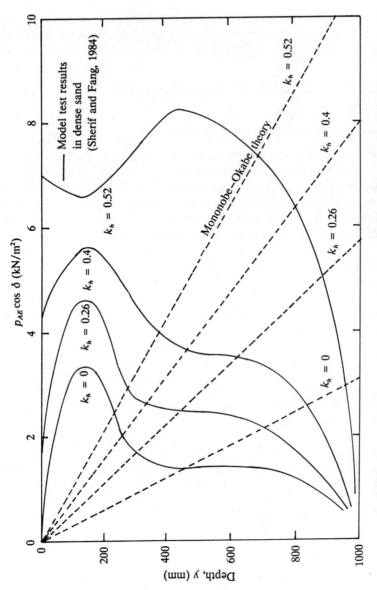

Figure 8.13 Rotation of wall about the top for active pressure—comparison of theory with model test results ($i = 0°$, $\beta = 0°$, $k_v = 0$)

3. For a given value of k_h, the horizontal component of the lateral force, $P_{AE} \cos \delta$, calculated from the experimental curves by using Eq. (8.23), is about 15% to 20% higher than that predicted by the Mononobe–Okabe theory.

8.8   Point of Application of the Resultant Active Force, P_{AE}

A. Rotation about the Bottom of the Wall

The original Mononobe–Okabe solution for the active force on retaining structures implied that the resultant force will act at a distance of $\frac{1}{3}H$ measured from the bottom of the wall (H = height of the wall) similar to that in the static case ($k_h = k_v = 0$). However, all the laboratory tests that have been conducted so far indicate that the resultant pressure P_{AE} acts at a distance $\bar{H}$, which is somewhat greater than $\frac{1}{3}H$ measured from the bottom of the wall. This is shown in Figure 8.14.

Prakash and Basavanna (1969) have made a theoretical evaluation for determination of $\bar{H}$. Based on the force-equilibrium analysis, their study shows that $\bar{H}$ increases from $\frac{1}{3}H$ for $k_h = 0$ to about $\frac{1}{2}H$ for $k_h = 0.3$ (for $\phi = 30°$, $\delta = 7.5°$, $k_v = 0$, $i = \beta = 0$). For similar conditions, the moment-equilibrium analysis gave a value of $\bar{H} = \frac{1}{3}H$ and $k_h = 0$, which increases to a value of $\bar{H} \approx H/1.9$ at $k_h = 0.3$.

For practical design considerations, Seed and Whitman (1969) have proposed the following procedure for determination of the line of action of P_{AE}.

1. Calculate P_A [Eq. (8.1)].

2. Calculate P_{AE} [Eq. (8.4)].

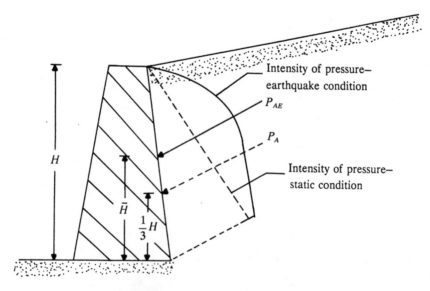

Intensity of pressure—earthquake condition

P_{AE}

P_A

Intensity of pressure—static condition

H

$\bar{H}$

$\frac{1}{3}H$

Figure 8.14   Point of application of resultant active earth pressure

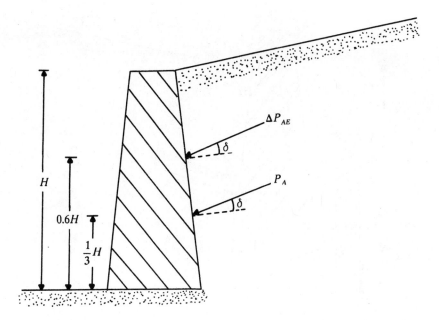

Figure 8.15

3. Calculate $\Delta P_{AE} = P_{AE} - P_A$. The term ΔP_{AE} is the incremental force due to earthquake condition.

4. Assume that P_A acts at a distance of $\frac{1}{3}H$ from the bottom of the wall (Figure 8.15).

5. Assume that ΔP_{AE} acts at a distance of $0.6H$ from the bottom of the wall (Figure 8.15); then

$$\overline{H} = \frac{(P_A)(\frac{1}{3}H) + (\Delta P_{AE})(0.6H)}{P_{AE}}$$

B. Translation of the Wall

Sherif, Ishibashi, and Lee (1982) suggested that, for wall translation, the following procedure can be used to estimate the location of the line of action of the active force, P_{AE}.

1. Calculate P_A [Eq. (8.1)].

2. Calculate P_{AE} [Eq. (8.4)].

3. Calculate $\Delta P_{AE} = P_{AE} - P_A$.

4. Referring to Figure 8.16, calculate

$$\overline{H} = \frac{(P_A)(0.42H) + (\Delta P_{AE})(0.48H)}{P_{AE}}$$

C. Rotation about the Top of the Wall

For rotation of the wall about its top (Figure 8.17), $\overline{H}$ is about $0.55H$ (Sherif and Fang, 1984).

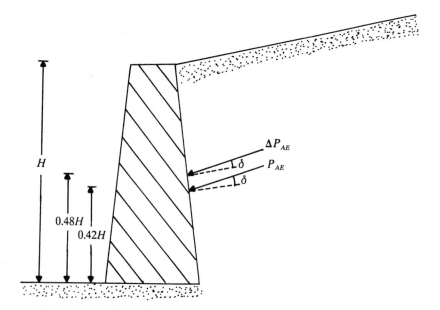

Figure 8.16

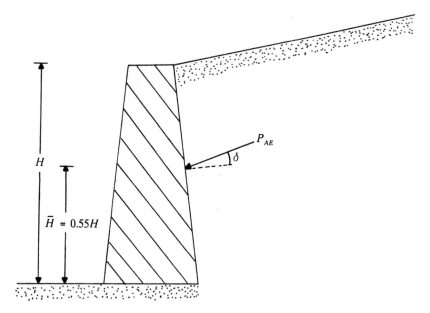

Figure 8.17

Example 8.2

Referring to Example 8.1, determine the location of the line of action for P_{AE}. Assume rotation of the wall about its bottom.

Solution

The value of P_{AE} in Example 8.1 has been determined to be 5404 lb/ft.

$$P_A = \tfrac{1}{2}\gamma H^2 K_A$$

For $\phi = 36°$, $\delta = 18°$, $K_A = 0.236$ [Eq. (8.2)]. Thus,

$$P_A = \tfrac{1}{2}(110)(15)^2(0.236) = 2920.5 \text{ lb/ft}$$

This acts at a distance equal to $\frac{15}{3} = 5$ ft from the bottom of the wall. Again,

$$\Delta P_{AE} = 5404 - 2920.5 = 2483.5 \text{ lb/ft}$$

The line of action of ΔP_{AE} intersects the wall at a distance of $0.6H = 9$ ft measured from the bottom, so

$$\bar{H} = \frac{(5)(2920.5) + (9)(2483.5)}{5404} = 6.83 \text{ ft} \qquad \blacksquare$$

8.9   Design of Gravity Retaining Walls Based on Limited Displacement

Richards and Elms (1979) have proposed a procedure for design of gravity retaining walls based on limited displacement. In their study, they have taken into consideration the wall inertia effect and concluded that there is some lateral movement of the wall even for mild earthquakes. In order to develop this procedure, consider a gravity retaining wall as shown in Figure 8.18, along with the forces acting on it during an earthquake. For stability, summing the forces in the vertical direction,

$$N = W_w - k_v W_w + P_{AE} \sin(\delta + \beta) \qquad (8.24)$$

where N is the vertical component of the reaction at the base of the wall and W_w is the weight of the wall. Similarly, summing the forces in the horizontal direction,

$$S = k_h W_w + P_{AE} \cos(\delta + \beta) \qquad (8.25)$$

where S is the horizontal component of the reaction at the base of the wall. At sliding,

$$S = N \tan \phi_b \qquad (8.26)$$

where ϕ_b is the soil-wall friction angle at the base of the wall.
Substituting Eqs. (8.24) and (8.25) into Eq. (8.26), one obtains

$$k_h W_w + P_{AE} \cos(\delta + \beta) = [W_w(1 - k_v) + P_{AE} \sin(\delta + \beta)] \tan \phi_b$$

or

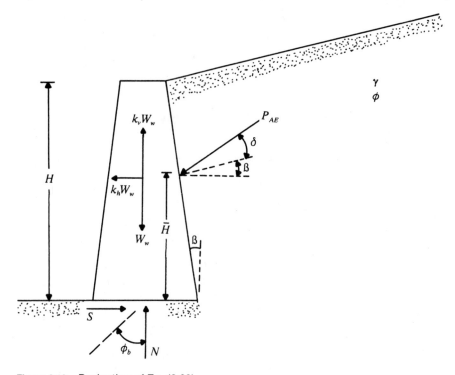

Figure 8.18   Derivation of Eq. (8.28)

$$W_w[(1 - k_v)\tan\phi_b - k_h] = P_{AE}[\cos(\delta + \beta) - \sin(\delta + \beta)\tan\phi_b]$$

$$W_w = \frac{P_{AE}[\cos(\delta + \beta) - \sin(\delta + \beta)\tan\phi_b]}{(1 - k_v)\tan\phi_b - k_h} \tag{8.27}$$

From Eq. (8.4), $P_{AE} = \frac{1}{2}\gamma H^2(1 - k_v)K_{AE}$. Substitution of this equation into Eq. (8.27) yields

$$W_w = \frac{\frac{1}{2}\gamma H^2 K_{AE}[\cos(\delta + \beta) - \sin(\delta + \beta)\tan\phi_b]}{(\tan\phi_b - \tan\theta)} \tag{8.28}$$

where $\tan\theta = k_h/(1 - k_v)$.

It may be noted that, in Eq. (8.28), W_w is equal to infinity if

$$\tan\phi_b = \tan\theta \tag{8.29}$$

This implies that infinite mass of the wall is required to prevent motion. The critical value of $k_h = k_{h(cr)}$ can thus be given by the relation

$$\tan\theta = \frac{k_{h(cr)}}{1 - k_v} = \tan\phi_b$$

or

$$k_{h(cr)} = (1 - k_v)\tan\phi_b \tag{8.30}$$

Equation (8.27) can also be written in the form

$$W_w = [\tfrac{1}{2}\gamma H^2(1 - k_v)K_{AE}]C_{IE} \tag{8.31}$$

where

$$C_{IE} = \frac{\cos(\delta + \beta) - \sin(\delta + \beta)\tan\phi_b}{(1 - k_v)(\tan\phi_b - \tan\theta)} \tag{8.32}$$

Figure 8.19 shows the variation of C_{IE} with k_h for various values of k_v ($\phi = \phi_b = 35°$, $\delta = \tfrac{1}{2}\phi$, $i = \beta = 0$). Also, Figure 8.20 shows the variation of C_{IE}

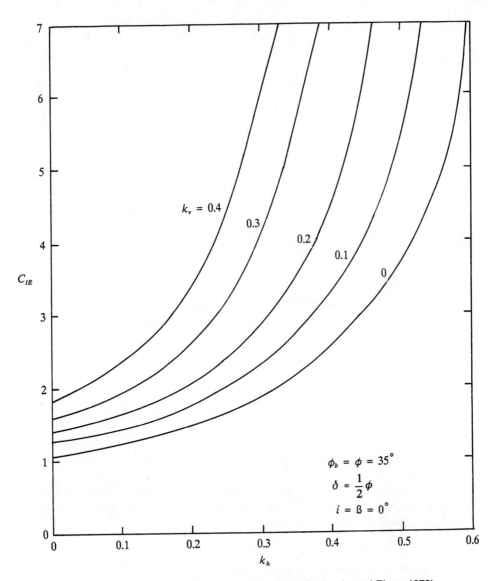

Figure 8.19   Effect of k_v on the value of C_{IE} (after Richards and Elms, 1979)

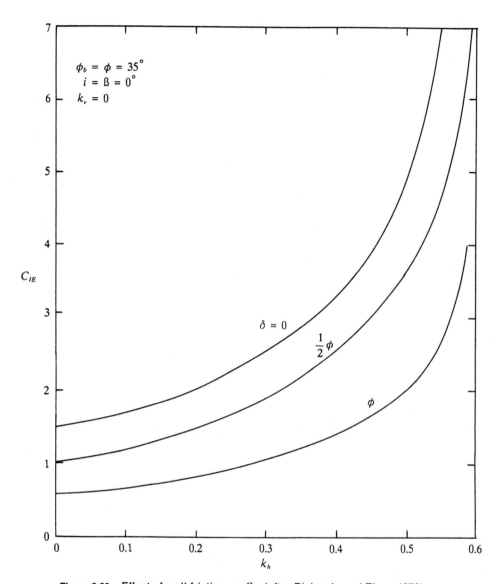

Figure 8.20   Effect of wall friction on C_{IE} (after Richards and Elms, 1979)

with k_h for various values of wall friction angle, δ ($\phi = \phi_b = 35°$, $i = \beta = 0$, $k_v = 0$).

Note that Eq. (8.31) is for the limiting equilibrium condition for sliding with earthquake effects taken into consideration. For the static condition (i.e., $k_h = k_v = 0$), Eq. (8.31) becomes

$$W = \tfrac{1}{2}\gamma H^2 K_A C_I \tag{8.33}$$

where $W = W_w$ (for static condition) and

$$C_I = \frac{\cos(\delta + \beta) - \sin(\delta + \beta)\tan\phi_b}{\tan\phi_b} \tag{8.34}$$

Thus, comparing Eqs. (8.31) and (8.33), we can write that

$$\frac{W_w}{W} = F_T F_I = F_W \tag{8.35}$$

where

$$F_T = \frac{K_{AE}(1 - k_v)}{K_A} = \text{soil thrust factor}$$

$$F_I = \frac{C_{IE}}{C_I} = \text{wall inertia factor}$$

and F_w is a factor of safety applied to the weight of the wall to take into account the effects of soil pressure and wall inertia.

Figure 8.21 shows a plot of F_T, F_I, and F_W for various values of k_h ($\phi = \phi_b = 35°$, $\delta = \frac{1}{2}\phi$, $k_v = 0$, $\beta = i = 0$). Richards and Elms (1979) have ex-

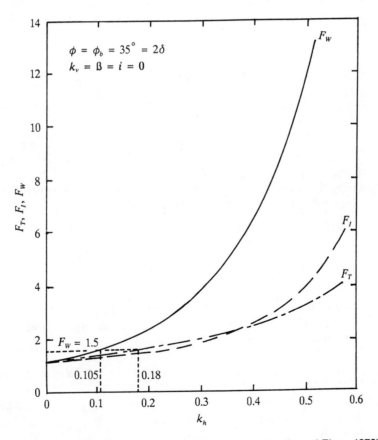

Figure 8.21 Variation of F_T, F_I, and F_W (after Richards and Elms, 1979)

plained the importance of the inertia factors given in Eq. (8.35). Referring to Figure 8.21, suppose that one neglects the wall inertia factor (which is not considered in the design procedure outlined in Sections 8.2 and 8.3; i.e., $F_I = 1$). In such a case,

$$F_W = F_T = \frac{W_w}{W}$$

For a value of $F_W = 1.5$, the critical horizontal acceleration is equal to 0.18. However, if the wall inertial factor is considered, the critical horizontal acceleration corresponding to $F_W = 1.5$ is equal to 0.105. In other words, if a gravity retaining wall is designed such that $W_w = 1.5W$, the wall will start to move laterally at a value of $k_h = 0.105$. Based on the procedure described in Section 8.2, if $W_w = 1.5W$, it is assumed that the wall will not move laterally until a value of $k_h = 0.18$ is reached.

These considerations show that, for no lateral movement, the weight of the wall has to be increased by a considerable amount over the static condition, which may prove to be very expensive. Thus, for actual design with reasonable cost, one has to assume some lateral displacement of the wall will take place during an earthquake; the procedure for determination of the wall weight (W_w) is then as follows:

1. Determine an acceptable displacement d of the wall.

2. Determine a design value of k_h from the equation

$$k_h = A_a \left(\frac{0.2A_v^2}{A_a d} \right)^{1/4} \tag{8.36}$$

where A_a and A_v are effective acceleration coefficients and displacement d is in inches. The values of A_a and A_v for a given region in the United States are given by the Applied Technology Council (1978).

Equation (8.36) has been suggested by Richards and Elms (1979), and is based on the study of Newmark (1965) and Franklin and Chang (1977).

3. Using the above value of k_h, and assuming $k_v = 0$, determine the value of K_{AE}.

4. Determine the weight of the wall W_w from Eq. (8.31).

5. Apply a factor of safety to W_w obtained in Step 4.

A slight modification of this design procedure was proposed by Nadim and Whitman (1983). This modification is intended primarily to account for the amplification of the ground motion in the backfill.

Example 8.3

Determine the weight of a retaining wall 4 m high (given $\beta = 0$, $i = 0$, $\gamma = 17.29 \text{ kN/m}^3$, $\phi_b = \phi = 34°$, $\delta = \frac{1}{2}\phi$, $A_v = 0.2$, $A_a = 0.2$, factor of safety = 1.5)
a. for static condition,
b. for zero displacement condition under earthquake loading, and
c. for a displacement of 50.8 mm under earthquake loading.

Solution

a. From Eq. (8.33),

$$W = \tfrac{1}{2}\gamma H^2 K_A C_I$$

From Table 8.2, $K_A = 0.256$ (for $\phi = 34°$, $\delta = 17°$, $i = 0$, $\beta = 0$).

$$C_I = \frac{\cos(\delta + \beta) - \sin(\delta + \beta)\tan\phi_b}{\tan\phi_b} = \frac{\cos 17 - \sin 17(\tan 34)}{\tan 34}$$

$$= 1.125$$

Thus

$$W = \tfrac{1}{2}(17.29)(4)^2(0.256)(1.125) = 39.84 \text{ kN/m}$$

With a factor of safety of 1.5, the weight of the wall is equal to $(1.5)(39.84) = \underline{59.76 \text{ kN/m}}$.

b. From Eq. (8.32),

$$W_w = \tfrac{1}{2}\gamma H^2(1 - k_v)K_{AE}C_{IE}$$

Assume $k_v = 0$.

$$C_{IE} = \frac{\cos(\delta + \beta) - \sin(\delta + \beta)\tan\phi_b}{(1 - k_v)(\tan\phi_b - \tan\theta)}$$

$$\tan\theta = \frac{k_h}{1 - k_v} = \frac{0.2}{1} = 0.2; \qquad \theta = 11.31°$$

$$C_{IE} = \frac{\cos 17 - \sin 17(\tan 34)}{\tan 34 - 0.2} = 1.6$$

Again, from Eq. (8.5),

$$K_{AE} = \frac{\cos^2(34 - 11.31)}{\cos(11.31)[\cos(17 + 11.31)]\left[1 + \sqrt{\dfrac{\sin(34 + 17)\sin(34 - 11.31)}{\cos(17 + 11.31)}}\right]^2}$$

$$= 0.393$$

$$W_w = \tfrac{1}{2}(17.29)(4)^2(1 - 0)(0.393)(1.6) = 86.98 \text{ kN/m}$$

With a factor of safety of 1.5, the weight of wall $= (86.98)(1.5) = \underline{130.47 \text{ kN/m}}$.

c. From Eq. (8.36),

$$k_h = A_a\left[\frac{0.2A_v^2}{A_a d}\right]^{1/4} = 0.2\left[\frac{(0.2)(0.2)^2}{(0.2)(2)}\right]^{1/4} = 0.075$$

$$\tan\theta = \frac{k_h}{1 - k_v} = \frac{0.075}{1 - 0} = 0.075$$

or

$$\theta = 4.29°$$

$$C_{IE} = \frac{\cos 17 - \sin 17(\tan 34)}{\tan 34 - 0.075} = \frac{0.7591}{0.5995} = 1.27$$

Using Eq. (8.5)

$$K_{AE} = \frac{\cos^2(34 - 4.29)}{\cos(4.29)[\cos(17 + 4.29)]\left[1 + \sqrt{\dfrac{\sin(34 + 17)\sin(34 - 4.29)}{\cos(17 + 4.29)}}\right]^2}$$

$$= 0.3$$

Thus, with a factor of safety of 1.5,

$$W_w = (1.5)(\tfrac{1}{2})(17.29)(4)^2(0.3)(1.27) = \underline{79.05 \text{ kN/m}} \qquad \blacksquare$$

8.10   Hydrodynamic Effects of Pore Water

The lateral earth pressure theory developed in the preceding sections of this chapter has been for retaining walls with dry soil backfills. However, for quay walls (Figure 8.22), the hydrodynamic effect of the water also has to be taken into consideration. This is usually done according to the Westergaard theory (1933) which was derived to obtain the dynamic water pressure on the face of a concrete dam. Based on this theory, the water pressure due to an earthquake at a depth y (Figure 8.22) may be expressed as

$$p_1 = \tfrac{7}{8}k_h\gamma_w h^{1/2}y^{1/2} \qquad (8.37)$$

where p_1 is the intensity of pressure on the seaward side, γ_w is the unit weight of water, and h is the total depth of water. Hence, the total dynamic water force on the seaward side per unit length of the wall $[P_{1(w)}]$ can be obtained by integration as

$$P_{1(w)} = \int p_1 \, dy = \int_0^h \tfrac{7}{8}k_h\gamma_w h^{1/2}y^{1/2} \, dy = \frac{7}{12}k_h\gamma_w h^2 \qquad (8.38)$$

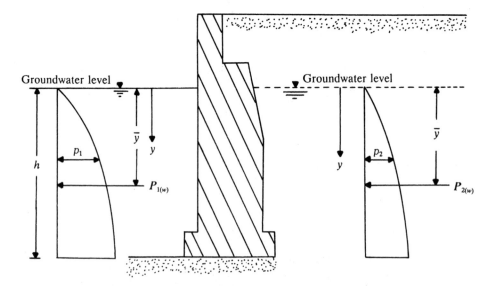

Figure 8.22 Hydrodynamic effects on a quay wall

The location of the resultant water pressure is

$$\bar{y} = \frac{1}{P_{1(w)}} \int_0^h (p_1\, dy) y = \frac{1}{P_{1(w)}} (\tfrac{7}{8} k_h \gamma_w h^{1/2}) \int_0^h (y^{1/2})(y)\, dy$$

$$= \frac{1}{P_{1(w)}} (\tfrac{7}{8} k_h \gamma_w h^{1/2})(h^{5/2})\tfrac{2}{5} = \frac{1}{P_{1(w)}} (\tfrac{7}{20} k_h \gamma_w h^3)$$

or

$$\bar{y} = (\tfrac{7}{12} k_h \gamma_w h^2)^{-1} (\tfrac{7}{20} k_h \gamma_w h^3) = 0.6h \tag{8.39}$$

Matsuo and O'Hara (1960) have suggested that the increase of the pore water pressure on the landward side is approximately 70% of that on the seaward side. Thus,

$$p_2 = 0.7(\tfrac{7}{8} k_h \gamma_w h^{1/2} y^{1/2}) = 0.6125 k_h \gamma_w h^{1/2} y^{1/2} \tag{8.40}$$

where p_2 is the dynamic pore water pressure on the landward side at a depth y. The total dynamic pore water force increase $[P_{2(w)}]$ per unit length of the wall is

$$P_{2(w)} = 0.7(\tfrac{7}{12} k_h \gamma_w h^2) = 0.4083 k_h \gamma_w h^2 \tag{8.41}$$

During an earthquake, the force on the wall per unit length on the seaward side will be reduced by $P_{1(w)}$ and that on the landward side will be increased by $P_{2(w)}$. Thus, the total increase of the force per unit length of the wall is equal to

$$P_w = P_{1(w)} + P_{2(w]} = 1.7(\tfrac{7}{12} k_h \gamma_w h^2) = 0.9917 k_h \gamma_w h^2 \tag{8.42}$$

Example 8.4

Refer to Figure 8.22. For the quay wall, $h = 10$ m. Determine the total dynamic force increase due to water for $k_h = 0.2$.

Solution

From Eq. (8.42)

$$P_w = 0.9917 k_h \gamma_w h^2 = 0.9917(0.2)(9.81)(10)^2 = \underline{194.6 \text{ kN/m}} \qquad \blacksquare$$

8.11   Dynamic Passive Force on Retaining Wall

Figure 8.23 shows a retaining wall having a granular soil as the backfill material. If the wall is pushed toward the soil mass, at a certain stage failure in the soil will occur along a plane BC. At failure the force, P_{PE}, per unit length of the retaining wall is the *dynamic passive force*. The force per unit length of the wall that needs to be considered for equilibrium of the soil wedge is shown in Figure 8.23. The notations W, ϕ, δ, γ, k_h, and k_v have the same meaning as described in Figure 8.2 (Section 8.2). Using the basic assumptions for the soil given in Section 8.2, the passive force (P_{PE}) may also be derived as (Kapila, 1962)

$$P_{PE} = \tfrac{1}{2} \gamma H^2 (1 - k_v) K_{PE} \tag{8.43}$$

where

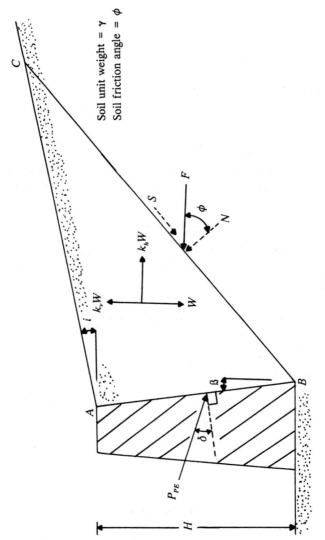

Soil unit weight = γ
Soil friction angle = ϕ

Figure 8.23 Passive force, P_{PE}, on a retaining wall

$$K_{PE} = \frac{\cos^2(\phi + \beta - \theta)}{\cos\theta\cos^2\beta\cos(\delta - \beta + \theta)\left[1 - \left\{\dfrac{\sin(\phi + \delta)\sin(\phi + i - \theta)}{\cos(i - \beta)\cos(\delta - \beta + \theta)}\right\}^{1/2}\right]^2}$$

$$(8.44)$$

and $\theta = \tan^{-1}[k_h/1 - k_v]$.

Note that Eq. (8.43) has been derived for dry cohesionless backfill. Kapila has also developed a graphical procedure for determination of P_{PE}.

Figure 8.24 shows the variation of K_{PE} for various values of soil friction angle ϕ and k_h (with $k_v = i = \beta = \delta = 0$). From the figure it can be seen that, with other parameters remaining the same, the magnitude of K_{PE} increases with the increase of soil friction angle ϕ.

Figure 8.25 shows the influence of the backfill slope angle on K_{PE}. Other factors remaining constant, the magnitude of K_{PE} increases with the increase of i.

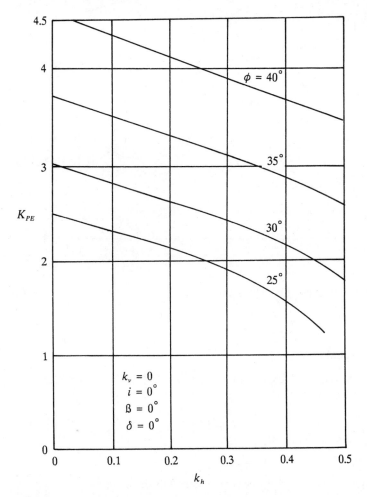

Figure 8.24 Variation of K_{PE} with soil friction angle and k_h (after Davies, Richards, and Chen, 1986)

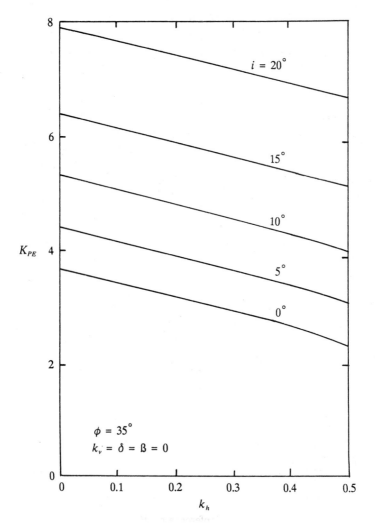

Figure 8.25 Influence of backfill slope on K_{PE} (after Davies, Richards, and Chen, 1986)

PROBLEMS

8.1 A retaining wall is 18 ft high with a vertical back ($\beta = 0°$). It has a horizontal cohesionless soil (dry) as backfill. Given:

Unit weight of soil $= 95 \ \text{lb/ft}^3$

Angle of friction $\phi = 30°$

$k_h = 0.35 \qquad k_v = 0 \qquad \delta = 15°$

Determine the active force P_{AE} per unit length of the retaining wall.

8.2 Refer to Problem 8.1. Determine the location of the point of intersection of the resultant force P_{AE} with the back face of the retaining wall. Assume

a. the wall is rotating about the bottom, and
b. the wall translates.

8.3 Refer to Figure 8.2. Given:

$$H = 3\,\text{m} \qquad \phi = 40° \qquad k_h = 0.3$$

$$\beta = 10° \qquad \delta = 13° \qquad k_v = 0.1$$

$$i = 10° \qquad \gamma = 15 \cdot 72\ \text{kN/m}^3$$

Determine the active force per unit length P_{AE} and the location of the resultant. Assume that the wall is rotating about its bottom.

8.4 Refer to Problem 8.3. Where would be the location of the resultant if the wall is rotating about its top?

8.5 Redo Example 8.1 using the modified Culmann graphical solution procedure outlined in Section 8.6.

8.6 Redo Problem 8.1 using the modified Culmann graphical solution procedure.

8.7 Redo Problem 8.3 using the modified Culmann graphical solution procedure.

8.8 For the retaining wall and the backfill given in Problem 8.1, determine the passive force P_{PE} per unit length.

8.9 For the retaining wall and the backfill given in Problem 8.3, determine the passive force P_{PE} per unit length.

8.10 Consider a 12-ft-high vertical retaining wall ($\beta = 0°$) with a horizontal backfill ($i = 0°$). Given for the soil are $\phi = 32°$, $\gamma = 120\ \text{lb/ft}^3$, and $\delta = 0$.
a. Calculate P_{AE} and the location of the resultant with $k_v = 0.1$ and $k_h = 0.15$.
b. For the results of (a), what should be the weight of the wall per foot length for no lateral movement? The factor of safety against sliding is 1.4.
c. What should be the weight of the wall for an allowable lateral displacement of 1 in? Given $A_v = A_a = 0.15$; the factor of safety against sliding is 1.4.

REFERENCES

Amano, R., Azuma, H., and Ishii, Y. (1956). "Aseismic Design of Quay Walls in Japan," *Proceedings*, 1st World Conference on Earthquake Engineering, Berkeley, California.

Arango, I. (1969). Personal communication with Seed, H. B. and Whitman, R. V. (1970).

Applied Technology Council (1978). "Tentative Provisions for the Development of Seismic Regulations for Buildings," *Publication ATC 3-06*, Palo Alto, California.

Coulomb, C. A. (1776). "Essai sur une Application des Règles de Maximis et Minimis à quelques Problèmes de Statique, relatifs a l'Architecture," *Mem. Roy. des Sciences*, Paris, Vol 3, p. 38.

Culmann, C. (1875). *Die graphische Statik*, Meyer and Zeller, Zurich.

Davies, T. G., Richards, R., and Chen, K. H. (1986). "Passive Pressure During Seismic Loading," *Journal of Geotechnical Engineering*, ASCE, Vol. 112, No. GT4, pp. 479–484.

Duke, C. M., and Leeds, D. J. (1963). "Response of Soils, Foundations, and Earth Structures," *Bulletin of the Seismological Society of America*, Vol. 53, No. 2, pp. 309–357.

Franklin, A. G., and Chang, F. K. (1977). "Earthquake Resistance of Earth and Rockfill Dams," *Report 5, Miscellaneous Paper S71-17*, Soils and Pavement Laboratory, U.S. Army Engineer Waterways Experiment Station, Vicksburg, Mississippi.

Hayashi, S., Kubo, K., and Nakase, A. (1966). "Damage to Harbor Structures in the Nigata Earthquake," *Soils and Foundations*, Vol. 6, No. 1, pp. 26–32.

Ishibashi, I., and Fang, Y. S. (1987). "Dynamic Earth Pressures with Different Wall Movement Modes," *Soils and Foundations*, Vol. 27, No. 4, pp. 11–22.

Jacobsen, L. S. (1939). Described in Appendix D of "The Kentucky Project," *Technical Report No. 13*, Tennessee Valley Authority, 1951.

Kapila, J. P. (1962). "Earthquake Resistant Design of Retaining Walls," *Proceedings*, 2nd Earthquake Symposium, University of Roorkee, Roorkee, India.

Matsuo, H., and O'Hara, S. (1960). "Lateral Earth Pressures and Stability of Quay Walls During Earthquakes," *Proceedings*, 2nd World Conference on Earthquake Engineering, Japan, Vol. 1.

Mononobe, N. (1929). "Earthquake-Proof Construction of Masonry Dams," *Proceedings*, World Engineering Conference, Vol. 9, pp. 274–280.

Mononobe, N., and Matsuo, H. (1929). "On the Determination of Earth Pressures During Earthquakes," *Proceedings*, World Engineering Conference, Vol. 9, pp. 176–182.

Nadim, F., and Whitman, R. V. (1983). "Seismically Induced Movement of Retaining Walls," *Journal of Geotechnical Engineering*, ASCE, Vol. 109, No. GT7, pp. 915–931.

Nazarian, H. N., and Hadjan, A. H. (1979). "Earthquake-Induced Lateral Soil Pressure on Structures," *Journal of the Geotechnical Engineering Division*, ASCE, Vol. 105, No. GT9, pp. 1049–1066.

Newmark, N. M. (1965). "Effect of Earthquakes on Dams and Embankments," *Geotechnique*, Vol. 15, No. 2, pp. 139–160.

Okabe, S. (1926). "General Theory of Earth Pressure," *Journal of the Japanese Society of Civil Engineers*, Vol. 12, No. 1.

Prakash, S., and Basavanna, B. M. (1969). "Earth Pressure Distribution Behind Retaining Wall During Earthquake," *Proceedings*, 4th World Conference on Earthquake Engineering, Santiago, Chile.

Richards, R., and Elms, D. G. (1979). "Seismic Behavior of Gravity Retaining Walls," *Journal of the Geotechnical Engineering Division*, ASCE, Vol. 105, No. GT4, pp. 449–464.

Seed, H. B., and Whitman, R. V. (1970). "Design of Earth Retaining Structures for Dynamic Loads," *Proceedings*, Specialty Conference on Lateral Stresses in the Ground and Design of Earth Retaining Structures, ASCE, pp. 103–147.

Sherif, M. A., and Fang, Y. S. (1984). "Dynamic Earth Pressures on Walls Rotating About the Top," *Soils and Foundations*, Vol. 24, No. 4, pp. 109–117.

Sherif, M. A., Ishibashi, I., and Lee, C. D. (1982). "Earth Pressure Against Rigid Retaining Walls," *Journal of the Geotechnical Engineering Division*, ASCE, Vol. 108, No. GT5, pp. 679–696.

Westergaard, H. M. (1933). "Water Pressures on Dams During Earthquakes," *Transactions*, ASCE, Vol. 98, pp. 418–433.

Chapter 9

COMPRESSIBILITY OF SOILS UNDER DYNAMIC LOADS

9.1   Introduction

Permanent settlements under vibratory machine foundations can generally be placed under two categories:

1. Elastic and consolidation settlement due to the static weight
2. Settlement due to vibratory compaction of the foundation soil

Permanent settlement in soils can also be induced due to the vibration caused by an earthquake. The elastic and consolidation settlement due to static loads is not discussed here since the conventional methods of calculation can be found in most standard soil mechanics texts. In this chapter, the present available methods of evaluation of permanent settlement due to dynamic loading conditions are presented.

9.2   Compaction of Granular Soils: Effect of Vertical Stress and Vertical Acceleration

The fact that granular soils can be compacted by vibration is well known. Dry granular soils are likely to exhibit more compaction due to vibration as compared to moist soils. This is because of the surface tension effect in moist soils, which offers a resistance for the soil particles to roll and slide and arrange themselves into a denser state.

Laboratory studies have been made in the past to evaluate the effect of cycling *controlled vertical stress* at low frequencies, i.e., at low acceleration levels on confined granular soils (D'Appolonia, 1970). Such laboratory tests can be performed by taking a granular soil specimen in a mold, as shown in Figure 9.1a. A confining vertical air pressure σ_z is first applied to the specimen, after which a vertical dynamic stress of amplitude σ_d is applied repeatedly. The permanent compressions of the specimen are recorded after the elapse of several cycles of dynamic stress application.

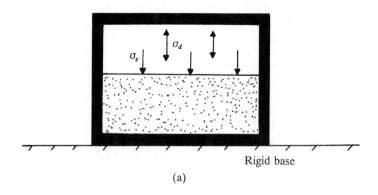

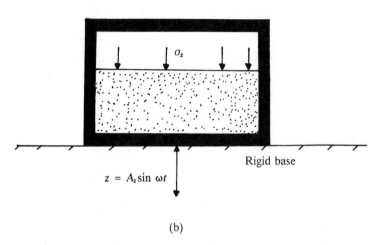

Figure 9.1   Compaction of granular soil by (a) controlled vertical stress; (b) controlled vertical acceleration

Also, several investigations on confined dry granular soils have been conducted (e.g., see D'Appolonia and D'Appolonia, 1967; Ortigosa and Whitman, 1968) in which a *controlled vertical acceleration* is imposed on the specimen, which produces small dynamic stress changes. For these tests, the specimen is placed in a mold fixed to a vibrating table (Figure 9.1b). Then a vertical confining air pressure σ_z is applied to the specimen. After that, the specimen is subjected to a vertical vibration for a period of time. Note that, for a vertical vibration,

$$z = A_z \sin \omega t$$

where A_z is the amplitude of the vertical vibration. The magnitude of the *peak acceleration* is equal to $A_z \omega^2 = A_z (2\pi f)^2$. Thus, the peak acceleration is controlled by the *amplitude of displacement* and *the frequency of vibration*. For a

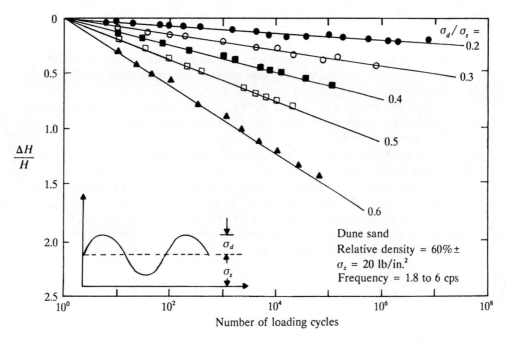

Figure 9.2 Compression of a dune sand under controlled vertical stress condition (after D'Appolonia, 1970)

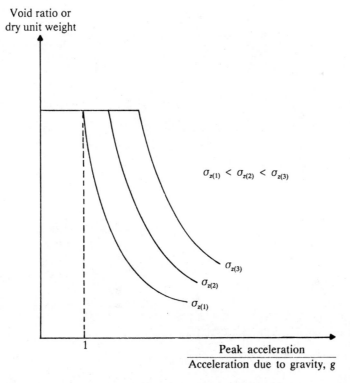

Figure 9.3 Nature of variation of void ratio or dry unit weight of dry sand in controlled vertical acceleration tests

constant peak acceleration of vibration, the drive mechanism has to be adjusted for A_z and f. The vertical compression of the specimen can be determined at the end of a test.

Thus the first type of test described above is run with repeated stresses with *negligible acceleration*; the second type is for repeated acceleration with *small dynamic stress on soils*.

Figure 9.2 shows the results of a number of tests conducted on a dune sand for controlled vertical stress condition. For all tests, the sand specimens were compacted to an initial relative density of about 60%. The frequencies of load application were in a range of 1.8–6 cps. Along the ordinate are plotted the vertical strain, which is equal to $\Delta H/H$ (where H is the initial height of the specimen and ΔH is the vertical compression of the specimen after a given number of load cycles). It may be seen that, for a given value of σ_d/σ_z,

$$\frac{\Delta H}{H} \propto \log N$$

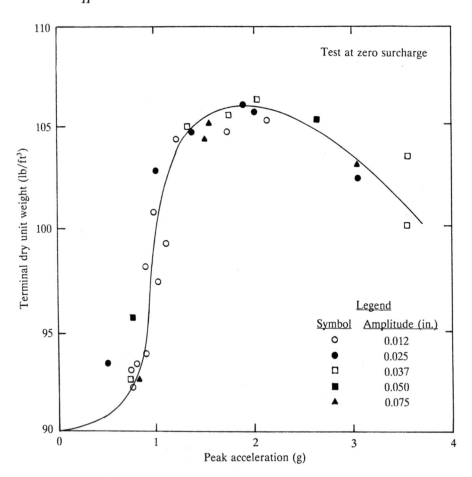

Figure 9.4 Correlation between terminal unit weight and peak vertical acceleration for a dune sand (redrawn after D'Appolonia, 1970)

Table 9.1   Details of the Specimens Used in the Tests by Krizek and Fernandez (1971)

| Soil | Percent of mix | Moisture content | | Modified Proctor dry unit weight | | Optimum moisture content, modified Proctor test (%) |
		Air dry (%)	Damp (%)	(lb/ft³)	(kN/m³)	
Ottawa sand	—	0.6	4.4 ± 0.5	107.6	16.92	11.0
Grundite	—	2.42	Not tested	101.8	16.00	18.5
Mix-10	90% Ottawa sand + 10% grundite	0.26	5 ± 0.5	114.4	17.99	8
Mix-20	80% Ottawa sand + 20% grundite	0.51	4.5 ± 0.5	120.6	18.96	9
Mix-30	70% Ottawa sand + 30% grundite	0.72	5 ± 0.3	124.0	19.49	9.5

where N is the number of load cycle applications. Also note that, for a given number of load cycles, the vertical strain increases with increasing values of σ_d/σ_z.

Figure 9.3 shows the nature of the results obtained from controlled vertical acceleration tests on dry sand by Ortigosa and Whitman (1968). Note that, even at zero confining pressure, no vertical strain is induced up to a peak acceleration of about 1 g (1 × acceleration due to gravity). A similar test result of D'Appolonia (1970) is shown in Figure 9.4, for which $\sigma_z = 0$. The terminal dry unit weight shown in Figure 9.4 is the unit weight of sand at the end of the test.

Krizek and Fernandez (1971) also conducted several laboratory tests with *controlled vertical acceleration* to study the densification of damp clayey sand. Tests were conducted with air-dry and damp specimens of Ottawa sand, grundite, and three mixtures of Ottawa sand and grundite: 90%–10%, 80%–20%, 70%–30%. Table 9.1 gives the details of the specimens used for the tests.

For conducting the tests, approximately 0.6 ft³ (0.017 m³) of soil samples— air dry and moist—were placed in a loose condition in a cylindrical mold 18 in. (0.457 m) high and 12 in. (0.305 m) in diameter. They were subjected to vertical vibrations for a period of time under various vertical pressures (σ_z). The range of time for vibratory compaction for the specimens was varied. Maximum vertical accelerations up to a value of about 6 g were used. Figure 9.5 shows the time rate of vibratory compaction of air-dry and moist sand–grundite mixtures.

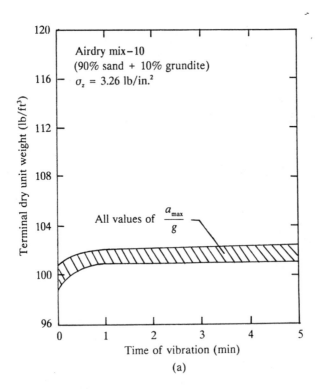

Figure 9.5   (Continued)

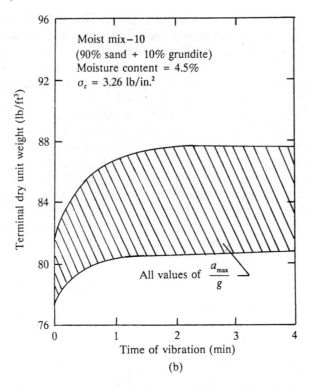

Figure 9.5   Time rate of vibratory compaction for air dry and moist sand–grundite mixtures (redrawn after Krizek and Fernandez, 1971)

It needs to be pointed out that very few tests were conducted for $a_{max}/g < 1$ (a_{max} = peak acceleration). However, from this study the following general conclusions may be drawn.

1. Significant vibratory densification does not occur with peak acceleration levels of less than 1 g.

2. The terminal vibratory dry unit weight of *air-dry* soils slightly decreases for $a_{max}/g > 2$. This is true only for zero confining pressure ($\sigma_z = 0$).

3. An increase of the clay percentage in soils has a tendency to reduce $\gamma_{d(termin - vibrat)}/\gamma_{d(modif\ max\ Proctor)}$.

4. Increase of moisture content has a significant influence in reducing $\gamma_{d(termin - vibrat)}/\gamma_{d(modif\ max\ Proctor)}$.

9.3 ___ **Settlement of Strip Foundation on Granular Soil under the Effect of Controlled Cyclic Vertical Stress**

In Section 9.2, some laboratory experimental observations of settlement of *laterally confined* sand specimens were presented. In these cases, the loads have been applied over the full surface area. However, in the field, the load covers only a small area and settlements in these cases include those caused by the induced shear strains. In the case of foundations, the shear strains increase with the increase of σ_d/q_u (where σ_d is the amplitude of dynamic load and q_u is the ultimate bearing capacity). In this section, some developments on settlements of strip footings under the effect of controlled cycling vertical stress applied *at low frequencies* (i.e., negligible acceleration) are discussed.

Raymond and Komos (1978) conducted laboratory model tests with strip footings with widths of 2.95 in. (75 mm) and 8.98 in. (228 mm) resting on 20-30 Ottawa sand in a large box. The cyclic loads on model strip footings were applied by a Bellofram loading piston activated by an air pressure system. The loadings approximated a rectangular wave form as shown in Figure 9.6a with a frequency of 1 cps. The settlements of the footings were measured by a dial gauge together with a DVDT activating a strip chart recorder. For conducting the tests, the ultimate static bearing capacities (q_u) were first experimentally determined. The footings were then subjected to various magnitudes of cyclic load ($\sigma_d/q_u = 13.5\%$–90%, where $\sigma_d = Q_0/A$, and A is the area of the model footing). The load settlement relationships obtained from the tests for the 228-mm footing are shown in Figure 9.6b. In this figure, S_N is the permanent settlement of the footing and N is the number of cycles of load application. Such plots may be given by an empirical relation as

$$\frac{S_N}{\log N} = a + bS_N \tag{9.1}$$

where a and b are two constants.

The experimental values of a and b for these two footings may be approximated by the following equations.

For 2.95-in.- (75-mm-) wide footing:

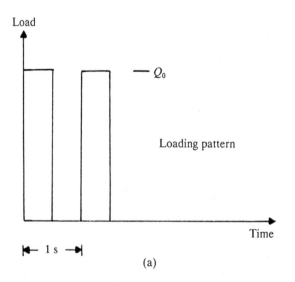

(a)

Figure 9.6   (Continued)

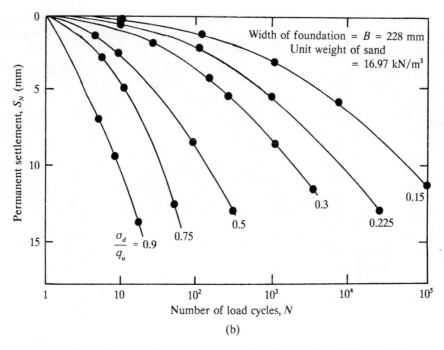

Figure 9.6 Plastic deformation due to repeated loading in plane strain case (redrawn after Raymond and Komos, 1978)

$$a = -0.0811 + 0.0115F \tag{9.2}$$

$$b = 0.12420 + 0.00127F \tag{9.3}$$

For 8.98-in.- (228-mm-) wide footing:

$$a = -0.1053 + 0.0421F \tag{9.4}$$

$$b = 0.0812 + 0.0031F \tag{9.5}$$

where $F = \sigma_d / q_u$ and S_N is measured in millimeters. Equations (9.1)–(9.5) are valid up to a load cycle of $N = 10^5$.

Figure 9.7 shows the experimental results of the variation of σ_d with $\log N$ for various values of S_N. For a given value of S_N, the plot of σ_d versus $\log N$ is approximately linear up to a value of $\sigma_d \approx \frac{1}{4} q_u$. For $\sigma_d < \frac{1}{4} q_u$, the slope of σ_d versus $\log N$ becomes smaller and the response tends toward elastic conditions.

From Eqs. (9.2)–(9.5), it may be seen that, *for a given soil,* the parameters a and b are functions of the width of the footing B. Thus, Eqs. (9.2) and (9.4) have been combined by Raymond and Komos to the form

$$a = -0.15125 + 0.0000693B^{1.18}(F + 6.09) \tag{9.6}$$

where B is the width of the footing. Similarly, Eqs. (9.3) and (9.5) can be combined as

$$b = 0.153579 + 0.0000363B^{0.821}(F - 23.1) \tag{9.7}$$

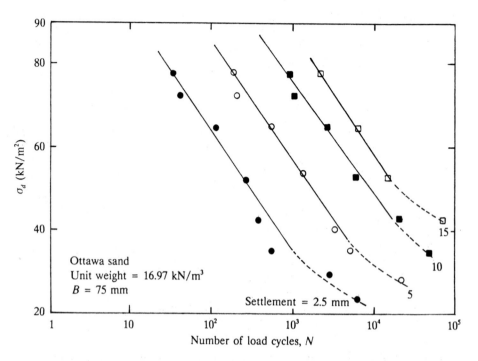

Figure 9.7 Variation of σ_d with log N for various values of permanent settlement (redrawn after Raymond and Komos, 1978)

Equations (9.6) and (9.7) are valid for only two different sizes of footing and for one soil. The general form of the equations for all sizes of footings and all soils can be written as

$$a = a_1 + a_2 B^{1.18} F + a_3 B^{1.18} \tag{9.8}$$

and

$$b = b_1 + b_2 B^{0.821} F - b_3 B^{0.821} \tag{9.9}$$

where $a_1, a_2, a_3, b_1, b_2, b_3$ are parameters for a given soil. However,

$$F = \frac{\sigma_d}{q_u}$$

and

$$q_u = \tfrac{1}{2}\gamma B N_\gamma \qquad \text{(for surface foundation)} \tag{9.10}$$

where N_γ is the static bearing capacity factor (Chapter 6) and γ is the unit weight of soil. Thus,

$$F = \frac{\sigma_d}{\tfrac{1}{2}\gamma B N_\gamma} \tag{9.11}$$

Substitution of Eq. (9.11) into Eqs. (9.8) and (9.9) yields

$$a = a_1 + a_4 \sigma_d B^{0.18} + a_3 B^{1.18} \tag{9.12}$$

and

$$b = b_1 + b_4 \sigma_d B^{-0.18} - b_3 B^{0.82} \tag{9.13}$$

where

$$a_4 = \frac{a_2}{\frac{1}{2}\gamma N_\gamma} \tag{9.14}$$

$$b_4 = \frac{b_2}{\frac{1}{2}\gamma N_\gamma} \tag{9.15}$$

and B is in millimeters.

If the values of $a_1, a_3, a_4, b_1, b_3, b_4$, which are the plastic properties of a given soil at a given density of compaction, can be determined by laboratory testing, the settlement of a given strip footing can be determined by combining Eqs. (9.1), (9.12), and (9.13). It needs to be pointed out that, for given values of σ_d and N, the value of S_N *decreases* with the increase of the width of the footing. This fact is demonstrated in Figure 9.8 for five different footings.

Analysis of this type may be used in the estimation of the settlement of railroad ties subjected to dynamic loads due to the movement of trains.

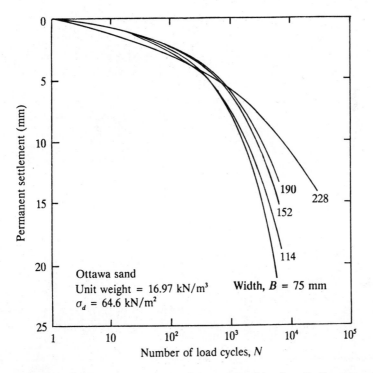

Figure 9.8 Plot of settlement versus number of load cycles for a constant value of σ_d (redrawn after Raymond and Komos, 1978)

9.4   Settlement of Machine Foundations on Granular Soils Subjected to Vertical Vibration

For machine foundations subjected to vertical vibrations, many investigators believe that the *peak acceleration* is the main controlling parameter for the settlement of the foundation. Depending on the relative density of granular soils, the solid particles come to an equilibrium condition under a given peak acceleration level. This *threshold acceleration* level must be exceeded before additional densification can take place.

The general nature of the settlement-time relationship for a foundation is shown in Figure 9.9. Note that in Figure 9.9, A_z is the amplitude of the foundation vibration and W is the weight of the foundation. The foundation settlement gradually increases with time and reaches a maximum value, beyond which it remains constant.

Brumund and Leonards (1972) have studied the settlement of circular foundations resting on sand subjected to vertical excitation by means of laboratory model tests. According to them, the *energy per cycle of vibration* imparted to the soil by the foundation can be used as the parameter for determination of settlement of foundations.

The model tests of Brumund and Leonards were conducted in a 2-ft³ (0.057-m³) container. They used 20–30 Ottawa sand, compacted to a relative density

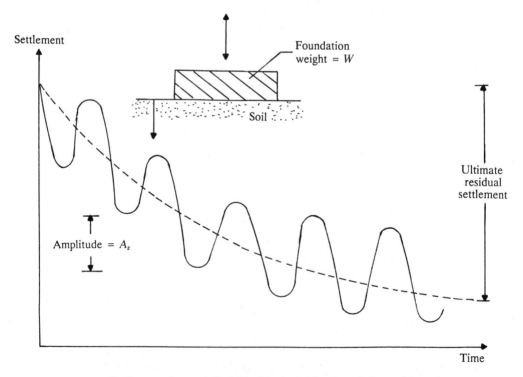

Figure 9.9 Settlement-time relationship for a machine foundation subjected to vertical vibration

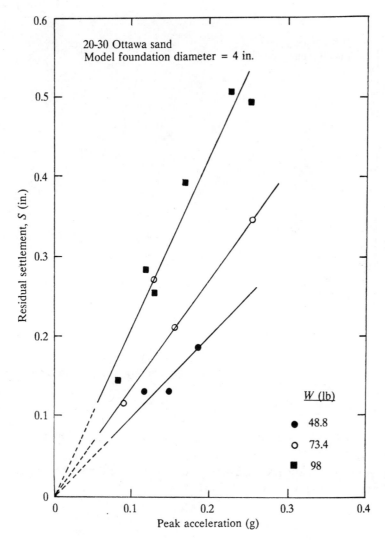

Figure 9.10 Plot of settlement versus peak acceleration for model foundation at a frequency of 20 Hz (after Brumund and Leonards, 1972)

of 70%. The model foundation used for the tests was 4 in. (101.6 mm) in diameter. The static ultimate bearing capacity was first experimentally determined before beginning the dynamic tests. The duration of vibration of the model foundation was chosen to be 20 min for all tests. Figure 9.10 shows the plot of the experimental results of settlement S against the peak acceleration for a *constant frequency of vibration*. For a given foundation weight W, the settlement increases linearly with the peak acceleration level. However, for a given frequency of vibration and peak acceleration level, the settlement increases with the increase of W. Figure 9.11 shows a plot of settlement S against the energy transmitted per cycle to the soil by the foundation. The data include the following:

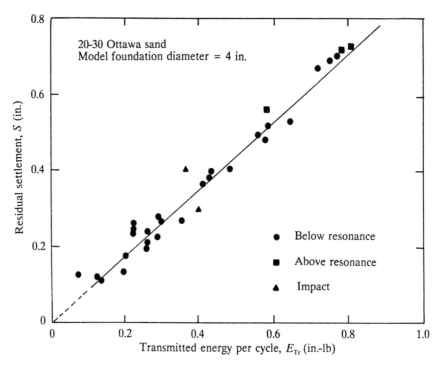

Figure 9.11 Plot of settlement versus transmitted energy per cycle (after Brumund and Leonards, 1972)

1. A frequency range of 14–59.3 cps (both above and below the resonant frequency)

2. A range in static pressure of 0.27–0.55 × static ultimate bearing capacity q_u. The static pressure q can be defined as

$$q = \frac{W}{A} \tag{9.16}$$

where A is the area of the foundation.

3. A range in maximum downward dynamic force of $0.3W-W$. The maximum downward dynamic force may theoretically be obtained from Eq. (2.90) as

$$F_{\text{dynam(max)}} = A_z \cdot \sqrt{k^2 + (c\omega)^2}$$

where A_z = amplitude of foundation vibration

$\quad k$ = spring constant = $\dfrac{4Gr_o}{1-\mu}$ (see Chapter 5)

$\quad G$ = shear modulus of soil

$\quad r_o$ = radius of foundation

$\quad \mu$ = Poisson's ratio of soil

$$c = \left(\frac{3.4}{1 - \mu}\right)r_o^2\sqrt{G\rho} \qquad \text{(see Chapter 5)}$$

ρ = density of soil

$\omega = 2\pi f \qquad (f = \text{frequency of vibration})$

The equation for determination of *energy transmitted* to the soil per cycle (E_{Tr}) may be obtained as follows:

$$E_{Tr} = \int F \, dz = F_{av} A_z \tag{9.17}$$

where F is the total contact force on the soil and F_{av} is the average contact force on the soil per cycle; however,

$$F_{av} = \tfrac{1}{2}(F_{max} + F_{min}) \tag{9.18}$$

$$F_{max} = W + F_{dynam(max)} \tag{9.19}$$

and

$$F_{min} = W - F_{dynam(max)} \tag{9.20}$$

Substituting Eqs. (9.19) and (9.20) into Eq. (9.18),

$$F_{av} = W \tag{9.21}$$

Thus, from Eqs. (9.17) and (9.21),

$$E_{Tr} = WA_z \tag{9.22}$$

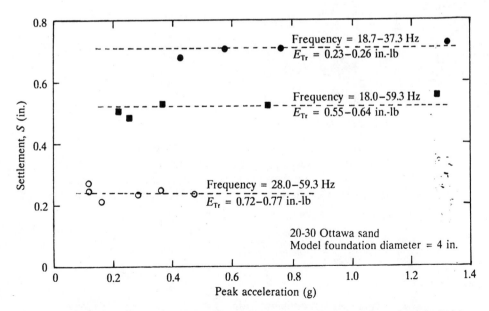

Figure 9.12   Settlement versus peak acceleration for three levels of transmitted energy (after Brumund and Leonards, 1972)

Figure 9.11 shows that the transmitted energy per cycle of oscillation E_{Tr} varies linearly with the settlement. A plot of the experimental results of settlement against peak acceleration for different ranges of E_{Tr} is plotted in Figure 9.12. This clearly demonstrates that, if the value of the transmitted energy is constant, the residual settlement remains constant irrespective of the level of peak acceleration.

The preceding concept is very important for the analysis of settlement of foundations of machineries subjected to vertical vibration. However, at this time, techniques of extrapolation of settlements of prototype foundations from laboratory model tests are not available. In any case, if the foundation soil is granular and loose, it is always advisable to take precautions to avoid possible problems in settlement. A specification of at least 70% relative density of compaction is often cited.

9.5   One-Dimensional Consolidation of Clay under Repeated Loading

One-dimensional consolidation of a clay layer under repeated loading has been studied by Wilson and Elgohary (1974). For the development of the mathematical relationships, the assumptions incorporated in Terzaghi's consolidation theory due to static load conditions (Terzaghi, 1943) were used. Those assumptions will not be repeated here as they are described in most standard soil mechanics texts (for example, see Das, 1983). However, an additional assumption has to be made; that is,

the coefficient of compressibility, a_v = the coefficient of expansion

The term a_v is defined as

$$a_v = -\frac{de}{d\bar{\sigma}}$$

where

e = void ratio

$\bar{\sigma}$ = effective stress on soil

Figure 9.13a shows a saturated clay layer of thickness H drained at the top only. Let a dynamic pressure $\sigma_{(t)}$ acting on the clay layer be defined by Figure 9.13b. This means that the dynamic load $\sigma_{(t)} = \sigma$ is on for a time T_1 and off for a time $T - T_1$. The basic differential equation of Terzaghi for consolidation may be written as

$$\frac{c_v \partial^2 u_{(z,t)}}{\partial z^2} = -\frac{\partial \bar{\sigma}_{(z,t)}}{\partial t} \tag{9.23}$$

where

c_v = coefficient of consolidation

$u_{(z,t)}$ = excess pore water pressure at depth z

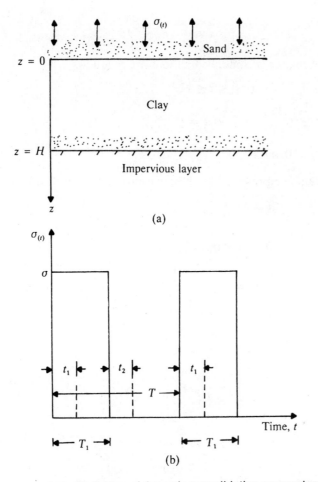

Figure 9.13   Definition of dynamic consolidation parameters

t = time

$\dfrac{\partial \bar{\sigma}_{(z,t)}}{dt}$ = rate of change of effective stress with time at a depth z

The excess pore water pressure at a depth z and time t may be written as

$$u_{(z,t)} = \sigma_t - \bar{\sigma}_{(z,t)} \tag{9.24}$$

where $\bar{\sigma}_{(z,t)}$ is that portion of the dynamic stress that has been transferred to the soil structure (i.e., effective stress). Since σ_t is a function of time, from Eq. (9.24) one can write

$$\frac{\partial^2 u_{(z,t)}}{\partial z^2} = -\frac{\partial^2 \bar{\sigma}_{(z,t)}}{\partial z^2} \tag{9.25}$$

Combining Eqs. (9.23) and (9.25),

$$c_v \frac{\partial^2 \overline{\sigma}_{(z,t)}}{\partial z^2} = \frac{\partial \overline{\sigma}_{(z,t)}}{\partial t} \tag{9.26}$$

For solving Eq. (9.26), the boundary conditions are as follows:

At time $t = 0$, $\overline{\sigma}_{(z,t)} = 0$ for $0 \leq z \leq H$

At any time $t \geq 0$, $\dfrac{\partial \overline{\sigma}_{(z,t)}}{\partial z} = 0$ for $z = H$

$\overline{\sigma}_{(z,t)} = \sigma_{(t)}$ at $z = 0$ and $t \geq 0$

The value of $\sigma_{(t)}$ can be represented by the equations

$$\sigma_{(t)} = \sigma \qquad \text{for } rT < t < rT + T_1 \tag{9.27}$$

and

$$\sigma_{(t)} = 0 \qquad \text{for } rT + T_1 < t < (r+1)T \tag{9.28}$$

where $r = 0, 1, 2, \ldots$.

The solution of Eq. (9.26) with the preceding boundary conditions gives the following results.

During the loading period:

$$u_{(z,t)} = \frac{4\sigma}{\pi} \sum_{m=0}^{\infty} \frac{1}{2m+1} \sin(MZ) \left[\frac{e^{\beta_m(T_1 - t_1)} - e^{\beta_m(T - t_1)}}{1 - e^{\beta_m T}} \right.$$

$$\left. - \frac{e^{-\beta_m(t - T_1)} - e^{-\beta_m(t - T)}}{1 - e^{\beta_m T}} + e^{-\beta_m t} \right] \tag{9.29}$$

where

$$M = (2m + 1)\tfrac{1}{2}\pi \tag{9.30}$$

$$Z = \frac{z}{H} \tag{9.31}$$

$$\beta_m = \frac{c_v M^2}{H^2} \tag{9.32}$$

t is the elapse of *total time* from the beginning of load application, and t_1 is the elapse of time from the beginning of each load cycle (see Figure 9.13).

The degree of consolidation at a depth z and time t can be given by

$$U_{(z,t)} = \frac{\sigma - u_{(z,t)}}{\sigma} \tag{9.33}$$

Substituting Eq. (9.29) into Eq. (9.33),

$$U_{(z,t)} = 1 - \frac{4}{\pi} \sum_{m=0}^{\infty} \frac{1}{2m+1} \sin(MZ) \left[\frac{e^{\beta_m(T_1 - t_1)} - e^{\beta_m(T - t_1)}}{1 - e^{\beta_m T}} \right.$$

$$\left. - \frac{e^{-\beta_m(t - T_1)} - e^{-\beta_m(t - T)}}{1 - e^{\beta_m T}} + e^{-\beta_m t} \right] \tag{9.34}$$

The average degree of consolidation at any time t ($U_{\text{av}(t)}$) can also be evaluated as

$$U_{av(t)} = \frac{\left[\sigma - \dfrac{1}{H} \displaystyle\int_0^H u_{(z,t)} \right]}{\sigma}$$

$$= 1 - \frac{8}{\pi^2} \sum_{m=0}^{\infty} \frac{1}{(2m+1)^2} \left[\frac{e^{\beta_m(T_1 - t_1)} - e^{\beta_m(T - t_1)}}{1 - e^{\beta_m T}} \right.$$

$$\left. - \frac{e^{-\beta_m(t - T_1)} - e^{-\beta_m(t - T)}}{1 - e^{\beta_m T}} + e^{-\beta_m t} \right] \tag{9.35}$$

where $U_{av(t)}$ = average degree of consolidation.
During the period when the load $\sigma_t = 0$

$$u_{(z,t)} = -\frac{4\sigma}{\pi} \sum_{m=0}^{\infty} \frac{1}{2m+1} \sin(MZ) \left[\frac{e^{\beta_m(T - T_1 - t_2)} - e^{\beta_m(T - t_2)}}{1 - e^{\beta_m T}} \right.$$

$$\left. + \frac{e^{-\beta_m(t - T_1)} - e^{-\beta_m(t - T)}}{1 - e^{\beta_m T}} - e^{-\beta_m t} \right] \tag{9.36}$$

$$U_{(z,t)} = \frac{4}{\pi} \sum_{m=0}^{\infty} \frac{1}{2m+1} \sin(MZ) \left[\frac{e^{\beta_m(T - T_1 - t_2)} - e^{\beta_m(T - t_2)}}{1 - e^{\beta_m T}} \right.$$

$$\left. + \frac{e^{-\beta_m(t - T_1)} - e^{-\beta_m(t - T)}}{1 - e^{\beta_m T}} - e^{-\beta_m t} \right] \tag{9.37}$$

For definition of time t_2, see Figure 9.13b.

At time $t = \infty$, the degree of consolidation $U_{(z,t)}$ for loading and unloading periods can be obtained from Eqs. (9.34) and (9.37) as follows:

For period during which the load is on:

$$U_{(z,t)} = 1 - \frac{4}{\pi} \sum_{m=0}^{\infty} \frac{1}{2m+1} \sin(MZ) \left[\frac{e^{\beta_m(T_1 - t_1)} - e^{\beta_m(T - t_1)}}{1 - e^{\beta_m T}} \right] \tag{9.38}$$

For period during which the load is off:

$$U_{(z,t)} = \frac{4}{\pi} \sum_{m=0}^{\infty} \frac{1}{2m+1} \sin(MZ) \left[\frac{e^{\beta_m(T - T_1 - t_2)} - e^{\beta_m(T - t_2)}}{1 - e^{\beta_m T}} \right] \tag{9.39}$$

From Eqs. (9.34) and (9.37), it can be seen that, for given values of T, T_1, H, and c_v, they represent two sets of curves: one for loading and the second for unloading. The nature of the plot of $U_{(z,t)}$ versus z/H at time $t = rT + \frac{1}{2}T_1$ (i.e., at the middle of a loading period) and at time $t = rT + \frac{1}{2}T_1 + \frac{1}{2}T$ (i.e., at the middle of the following unloading period) is shown in Figure 9.14.

The physical process of consolidation under this type of cyclic loading can be explained in the following manner. During the time the load is "on" the soil, the excess pore water pressure $u_{(z,t)}$ is positive; but it is negative during the period the load is "off." With the increase of time t, the *positive excess pore water pressure* during the time the load is on decreases and the negative pressure during the period of load removal increases. At very large t, an equilibrium condition is reached; i.e., the water flowing out of the soil due to positive excess

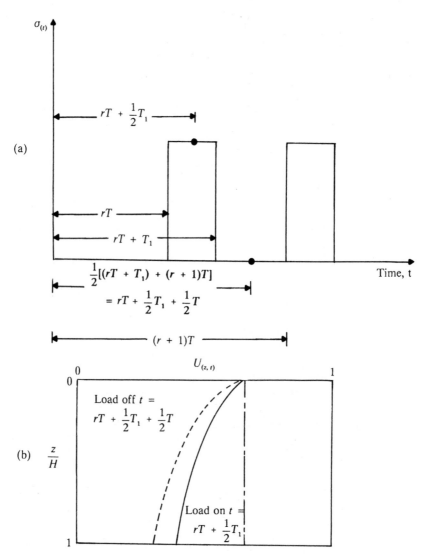

Figure 9.14 Nature of variation of the degree of consolidation with depth for dynamic loading

pore water pressure becomes equal to water flowing into the soil due to negative excess pore water pressure. It is not possible to reach 100% consolidation under cyclic loading.

9.6   Settlement of Sand Due to Cyclic Shear Strain

The experimental laboratory observations described in Section 9.2 have shown that when a sand layer is subjected to controlled vertical acceleration, considerable settlement does not occur up to a peak acceleration level of $a_{max} = g$. However, in several instances, the cyclic shear strains induced in the soil layers due to ground-shaking of seismic events have caused considerable damage

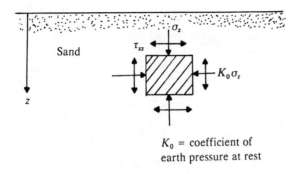

K_0 = coefficient of
earth pressure at rest

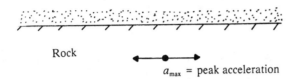

a_{max} = peak acceleration

Figure 9.15   Settlement of sand due to cyclic shear strain

(Figure 9.15). The controlling parameters for settlement in granular soils due to cyclic shear strain have been studied in detail by Silver and Seed (1971). Some of the results of this study are presented in this section.

The laboratory work of Silver and Seed was conducted on sand by using simple shear equipment developed by the Norwegian Geotechnical Institute. The frequency of the shear stress application to the sand specimens was 1 cps. Dry sand specimens were tested at various relative densities of compaction being subjected to varying normal stresses σ_z and amplitudes of shear strain γ'_{xz}. An example of the nature of variation of the vertical strain $\varepsilon_z = \Delta H/H$ (H = initial height of the specimen, ΔH = settlement) with number of cycles of shear strain application for a medium dense sand is shown in Figure 9.16. For these tests, the initial relative density (R_D) of compaction was 60%. Based on Figure 9.16, the following observations can be made.

a. For a given normal stress σ_z and amplitude of shear strain γ'_{xz}, the vertical strain increases with the number of strain cycles. However, a large portion of the vertical strain occurs in the first few cycles. For example, in Figure 9.16, the vertical strain occurring in the first 10 cycles is approximately equal to or more than that occurring in the next 40–50 cycles.

b. For a given value of the vertical stress and number of cycles N, the vertical strain increases with the increase of the shear strain amplitude.

However, one has to keep in mind that a small amount of compaction (i.e., increase in the relative density) could markedly reduce the settlement of a given soil. Silver and Seed (1971) also observed that at higher amplitudes of cyclic shear strain ($\gamma'_{xz} > 0.05\%$, for a given value of N), the vertical strain is not significantly affected by the magnitude of the vertical stress. This may not be true where the shear strain is less than 0.05%.

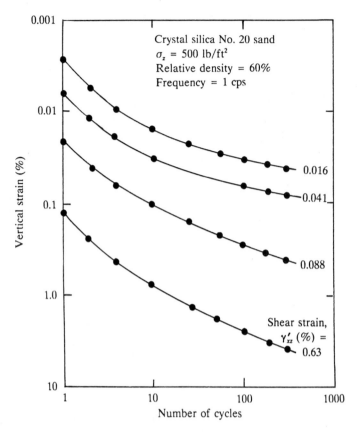

Figure 9.16 Variation of vertical strain with number of cycles (after Silver and Seed, 1971)

The basic understanding of the laboratory test results for the settlement due to cyclic shear strain application may now be used for calculation of settlement of sand layers due to seismic effect. This is presented in the following section.

9.7   Calculation of Settlement of Dry Sand Layers Subjected to Seismic Effect

Seed and Silver (1972) have suggested a procedure to calculate the settlement of a sand layer subjected to seismic effect. This procedure is outlined in a step-by-step manner.

1. Since the primary source of ground motion in a soil deposit during an earthquake is due to the upward propagation of motion from the underlying rock formation, adopt a representative history of horizontal acceleration for the base.

2. Divide the soil layer into n layers. They need not be of equal thickness.

3. Calculate the average value of the vertical effective stress $\bar{\sigma}_z$ for each layer (*Note:* In dry sand, total stress is equal to effective stress.)

4. Determine the representative relative densities for each layer.

5. Using the damping ratio and the shear moduli characteristics given in Section 4.19, calculate the history of shear strains at the middle of all n layers.

6. Convert the irregular strain histories obtained for each layer (Step 5) into average shear strains and equivalent number of uniform cycles (see Chapter 7).

7. Conduct laboratory tests with a simple shear equipment on representative soil specimens from each layer to obtain the vertical strains for the equivalent number of strain cycles calculated in Step 6. This has to be done for the average effective vertical stress levels $\bar{\sigma}_z$ calculated in Step 3 and the corresponding average shear strain levels calculated in Step 6.

8. Calculate the total settlement as

$$\Delta H = \varepsilon_{z(1)}H_1 + \varepsilon_{z(2)}H_2 + \cdots + \varepsilon_{z(n)}H_n \qquad (9.40)$$

where $\varepsilon_{z(1)}, \varepsilon_{z(2)}, \ldots$ are average vertical strains determined in Step 7 for layers 1, 2, ... and $H_1, H_2, \ldots$ are layer thicknesses.

The applicability of this procedure is explained in Example 9.1.

Example 9.1

A 20-m-thick sand layer is shown in Figure 9.17a. The unit weight of soil is 16.1 kN/m³. Using a design earthquake record, the average shear strain in the soil layer has been

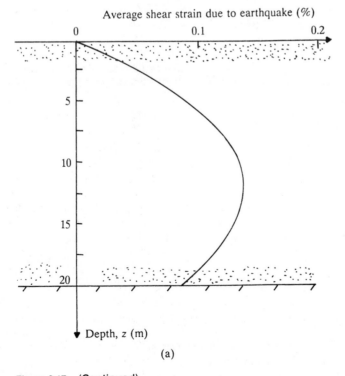

(a)

Figure 9.17 (Continued)

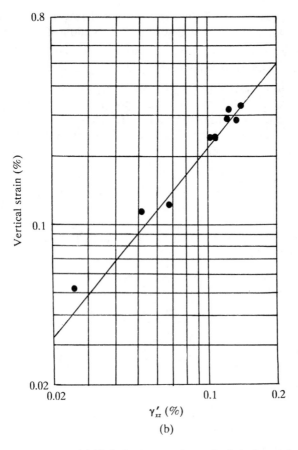

Figure 9.17 (a) Plot of average shear strain induced due to earthquake (sand unit weight = 16.1 kN/m³; relative density = 50%); (b) laboratory simple shear test results (number of cycles = 10)

evaluated and plotted in Figure 9.17a. (*Note:* It was assumed that $\gamma'_{av} \approx 0.65\gamma'_{max}$.) In this evaluation, the procedure outlined in Section 4.19 was followed with $G_{max} = 1000$ $K_{2(max)}\bar{\sigma}^{1/2}$ (lb/ft²) and damping = 20%. The number of equivalent cycles of shear strain application was estimated to be 10. Cyclic simple shear tests on representative specimens of this sand were conducted with their *corresponding vertical stresses* as in the field. The results of these tests are shown in Figure 9.17b.

Estimate the probable settlement of the sand layer.

Solution

From Figure 9.17b, it can be seen that, even though tests were conducted with different values of $\bar{\sigma}_z$, the results of ε_z versus γ'_{xz} are almost linear in a log-log plot. This shows that the magnitude of the effective overburden pressure has practically no influence on the vertical strain. Thus, for this calculation, the average line of the experimental results is used. For calculation of settlement, the following table can be prepared.

Depth (m)	H (m)	Shear strain at the middle of layer, γ'_{xz} (%)	ε_z (%)	$H\varepsilon_z \times 10^{-2}$ (m)
0 – 2.5	2.5	0.025	0.043	0.1075
2.5– 5	2.5	0.065	0.13	0.325
2 – 7.5	2.5	0.100	0.22	0.55
7.5–10	2.5	0.125	0.28	0.700
10 –12.5	2.5	0.140	0.31	0.775
12.5–15	2.5	0.135	0.3	0.750
15 –17.5	2.5	0.125	0.28	0.700
17.5–20	2.5	0.105	0.23	0.575

$$\Delta H = \sum H\varepsilon_z$$

$$= 4 \cdot 4825 \times 10^{-2} \text{ m}$$

$$\approx 44 \cdot 8 \text{ mm}$$

∎

9.8   Settlement of a Dry Sand Layer Due to Multidirectional Shaking

Pyke, Seed, and Chan (1975) have made studies to calculate the settlement of a dry sand layer subjected to multidirectional shaking; i.e., shaking with accelera-

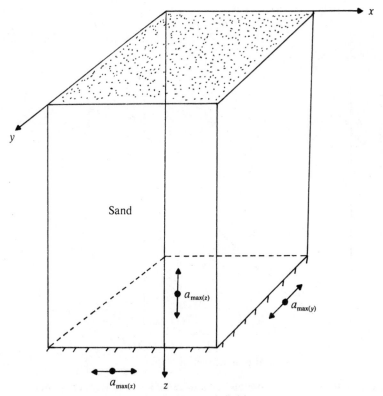

Figure 9.18 Multidirectional shaking—definition

tions in the x, y, and z directions as shown in Figure 9.18. The conclusions of this study show that the settlements caused by combined *horizontal motions* are approximately equal to the sum of the settlements caused by the components acting separately. The effect of the vertical acceleration is again to increase the settlement. Figure 9.19 shows the effect of the vertical acceleration on settlement on Monterey No. 0 sand with an initial relative density of 60%. As an example, let us consider the problem of settlement given in Example 9.1. If the same sand layer is subjected to similar base accelerations in the x and y directions, and if the *average* vertical acceleration in the layer is about 0.2 g, the total settlement can be estimated as follows:

Settlement due to the component in the x direction = 44.8 mm

Settlement due to the component in the y direction = 44.8 mm

Total settlement due to horizontal motions = 89.6 mm

For $a_{z(max)} = 0.2$ g, from Figure 9.19 the ratio of settlement is about 1.3. Thus, the total settlement due to all three components is equal to 1.3(89.6) = 116.48 mm.

It needs to be pointed out that vertical acceleration acting alone without

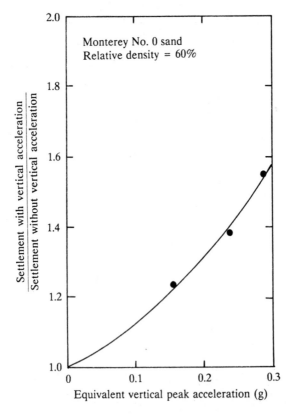

Figure 9.19 Effect of vertical motion superimposed on horizontal motion (after Seed and Chan, 1975)

horizontal motion has practically no effect of settlement up to about 1 g (Section 9.2). However, when it acts in combination with the horizontal motions, it produces a marked increase of total settlement.

PROBLEMS

9.1 The results of a set of laboratory simple shear tests on a dry sand are given below (vertical stress $\bar{\sigma}_z = 400$ lb/ft^2; number of cycles $= 12$; frequency $= 1$ cps; initial relative density of specimens $= 65\%$).

Peak shear strain γ'_{xz} (%)	Vertical strain (%)	Peak shear strain γ'_{xz} (%)	Vertical strain (%)
0.02	0.035	0.10	0.095
0.04	0.06	0.15	0.20
0.06	0.075	0.2	0.28
0.08	0.09		

Plot the results on log-log graph paper. Approximate the results in the form of an equation

$$\gamma'_{xz} = m\varepsilon_z^n$$

9.2 A dry sand deposit is 40 ft thick and its relative density is 65%. This layer of sand may be subjected to an earthquake. The number of equivalent cycles of shear stress application due to an earthquake is estimated to be 12. Following is the variation of the *average* expected shear strain with depth.

Depth (ft)	Average shear strain (%)	Depth (ft)	Average shear strain (%)
0	0	25	0.186
5	0.08	30	0.170
10	0.135	35	0.160
15	0.155	40	0.140
20	0.175		

Estimate the probable settlement of the sand layer using the laboratory test results given in Problem 9.1.

9.3 Repeat Problem 9.2 for the following (depth of sand layer $= 10$ m):

Depth (m)	Average shear strain (%)
0	0
2.5	0.1
5	0.14
7.5	0.135
10	0.117

REFERENCES

Brumund, W. F., and Leonards, G. A. (1972). "Subsidence of Sand Due to Surface Vibration," *Journal of the Soil Mechanics and Foundations Division*, ASCE, Vol. 98, No. SM1, pp. 27–42.

D'Appolonia, D. J., and D'Appolonia, E. (1967). "Determination of the Maximum Density of Cohesionless Soils," *Proceedings*, 3rd Asian Regional Conference on Soil Mechanics and Foundation Engineering, Haifa, Israel, Vol. 1, p. 266.

D'Appolonia, E. (1970). "Dynamic Loadings," *Journal of the Soil Mechanics and Foundations Division*, ASCE, Vol. 96, No. SM1, pp. 49–72.

Das, B. M. (1983). *Advanced Soil Mechanics*. McGraw-Hill, New York.

Krizek, R. J., and Fernandez, J. I. (1971). "Vibratory Densification of Damp Clayey Sands," *Journal of the Soil Mechanics and Foundations Division*, ASCE, Vol. 97, No. SM8, pp. 1069–1079.

Ortigosa, P., and Whitman, R. V. (1968). "Densification of Sand by Vertical Vibrations with Almost Constant Stress," *Publication No. 206*, Department of Civil Engineering, Massachusetts Institute of Technology, Cambridge.

Pyke, R., Seed, H. B., and Chan, C. K. (1975). "Settlement of Sands Under Multidirectional Shaking," *Journal of the Geotechnical Engineering Division*, ASCE, Vol. 101, No. GT4, pp. 379–398.

Raymond, G. P., and Komos, F. E. (1978). "Repeated Load Testing of a Model Plane Strain Footing," *Canadian Geotechnical Journal*, Vol. 15, No. 2, pp. 190–201.

Seed, H. B., and Silver, M. L. (1972). "Settlement of Dry Sands During Earthquakes," *Journal of the Soil Mechanics and Foundations Division*, ASCE, Vol. 98, No. SM4, pp. 381–397.

Silver, M. L., and Seed, H. B. (1971). "Volume Changes in Sands During Cyclic Loading," *Journal of the Soil Mechanics and Foundations Division*, ASCE, Vol. 97, No. SM9, pp. 1171–1182.

Terzaghi, K. (1943). *Theoretical Soil Mechanics*, Wiley, New York.

Wilson, N. E., and Elgohary, M. M. (1974). "Consolidation of Soils Under Cyclic Loading," *Canadian Geotechnical Journal*, Vol. 11, No. 3, pp. 420–423.

Chapter 10

LIQUEFACTION OF SOIL

10.1   Introduction

During earthquakes, major destruction of various types of structures occurs due to the creation of fissures, abnormal and/or unequal movement, and loss of strength or stiffness of the ground. The loss of strength or stiffness of the ground results in the settlement of buildings, failure of earth dams, landslides and other hazards. The process by which loss of strength occurs in soil is called *soil liquefaction*. The phenomenon of soil liquefaction is primarily associated with medium- to fine-grained *saturated cohesionless soils*. Examples of soil liquefaction–related damage are the June 16, 1964, earthquake at Niigata, Japan, and also the 1964 Alaskan earthquake.

One of the first attempts to explain the liquefaction phenomenon in sandy soils was made by Casagrande (1936) and is based on the concept of *critical void ratio*. Dense sand, when subjected to shear, tends to dilate; loose sand, under similar conditions, tends to decrease in volume. The void ratio at which sand does not change in volume when subjected to shear is referred to as the *critical void ratio*. Casagrande explained that deposits of sand that have a void ratio larger than the critical void ratio tend to decrease in volume when subjected to vibration by a *seismic effect*. If drainage is unable to occur, the pore water pressure increases. Based on the effective stress principles, at any depth of a soil deposit

$$\sigma' = \sigma - u \tag{10.1}$$

where

σ' = effective stress

σ = total stress

u = pore water pressure

If the magnitude of σ remains practically constant, and the pore water pressure u gradually increases, a time may come when σ will be equal to u. At that time, σ' will be equal to zero. Under this condition, the sand does not possess any shear strength, and it develops into a liquified state. However, one must keep in mind the following facts, which show that the critical void ratio concept may not be sufficient for a quantitative evaluation of soil liquefaction potential of sand deposits:

a. Critical void ratio is not a constant value, but changes with confining pressure.

b. Volume changes due to dynamic loading conditions are different than the one-directional static load conditions realized in the laboratory by direct shear and triaxial shear tests.

For that reason, since the mid-1960s intensive investigations have been carried out around the world to determine the soil parameters that control liquefaction. In this chapter, the findings of some of these studies are discussed.

10.2   Fundamental Concept of Liquefaction

Figure 10.1 shows the gradual densification of a sand by repeated back-and-forth straining in a simple shear test. For this case drainage from the soil occurs

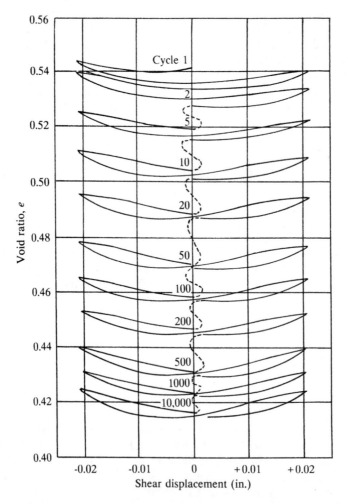

Figure 10.1   Void ratio versus cyclic shear displacement for densification of a sand with successive cycles of shear (after Youd, 1972)

freely. Each cycle of straining reduces the void ratio of the soil by a certain amount, although at a decreasing rate. It is important to note that there exists a *threshold shear strain*, below which no soil densification can take place, irrespective of the number of cycles.

Decrease in volume of the sand, as shown in Figure 10.1, can take place only if drainage occurs freely. However, under earthquake conditions, due to rapid cyclic straining this will not be the condition. Thus, during straining gravity loading is transferred from soil solids to the pore water. The result will be an increase of pore water pressure with a reduction in the capacity of the soil to resist loading. This is schematically shown in Figure 10.2. In this figure, let A be the point on the compression curve that represents the void ratio (e_0) and effective state of stress (σ_A') at a certain depth in a saturated sand deposit. Due to a certain number of earthquake-related cyclic straining, let $AB = \Delta e$ be the equivalent change of void ratio of the soil at that depth *if full drainage is allowed*. However, if drainage is prevented, the void ratio will remain as e_0 and the effective stress will be reduced to the level of σ_C', with an increase of pore water pressure of magnitude Δu. So the state of the soil can be represented by point C. If the number of cyclic straining is large enough, the magnitude of Δu may become equal to σ_A', and the soil will liquefy.

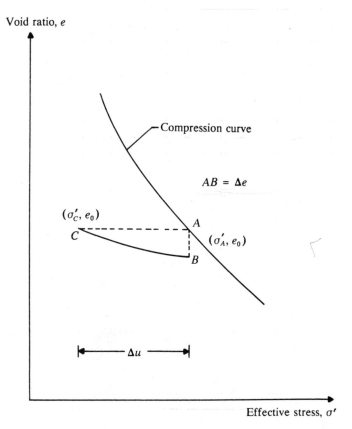

Figure 10.2 Mechanism of pore water pressure generation due to cyclic loading in undrained conditions

10.3   Laboratory Studies to Simulate Field Conditions for Soil Liquefaction

If one considers a soil element in the field, as shown in Figure 10.3a, when earthquake effects are not present, the vertical effective stress on the element is equal to σ', which is equal to σ_v, and the horizontal effective stress on the element equals $K_0\sigma_v$, where K_0 is the at-rest earth pressure coefficient.

Due to ground-shaking during an earthquake, a cyclic shear stress τ_h will be imposed on the soil element. This is shown in Figure 10.3b. Hence, any laboratory test to study the liquefaction problem must be designed in a manner so as to simulate the condition of a constant normal stress and a cyclic shear stress on a plane of the soil specimen. Various types of laboratory test procedure have been adopted in the past, such as the dynamic triaxial test (Seed and Lee, 1966; Lee and Seed, 1967), cyclic simple shear test (Peacock and Seed, 1968; Finn, Bransby, and Pickering, 1970; Seed and Peacock, 1971), cyclic torsional shear

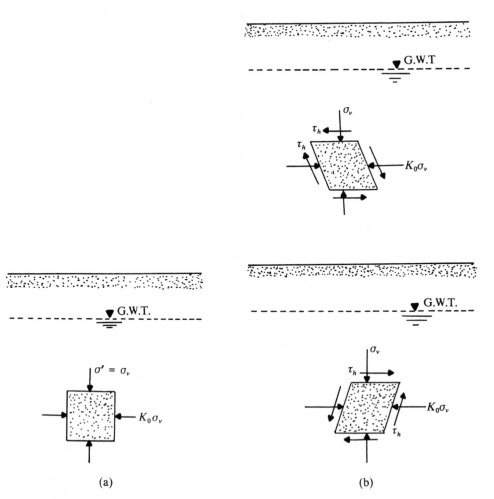

(a)   (b)

Figure 10.3 Application of cyclic shear stress on a soil element due to an earthquake

test (Yoshimi and Oh-oka, 1973; Ishibashi and Sherif, 1974), and shaking table test (Prakash and Mathur, 1965). However, the most commonly used laboratory test procedures are the dynamic triaxial tests and the simple shear tests. These are discussed in detail in the following sections.

DYNAMIC TRIAXIAL TEST

10.4   General Concepts and Test Procedures

Consider a saturated soil specimen in a triaxial test, as shown in Figure 10.4a, which is consolidated under an all-around pressure of σ_3. The corresponding Mohr's circle is shown in Figure 10.4b. If the stresses on the specimen are

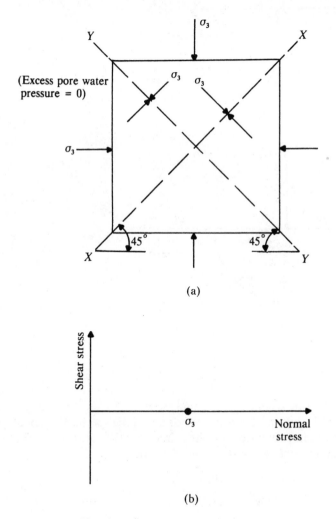

(a)

(b)

Figure 10.4   (Continued)

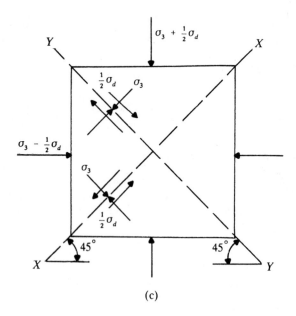

(c)

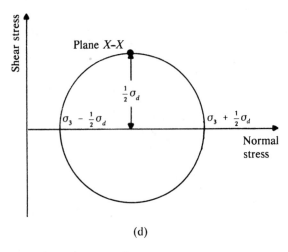

(d)

Figure 10.4   (Continued)

changed such that the axial stress is equal to $\sigma_3 + \frac{1}{2}\sigma_d$ and the radial stress is $\sigma_3 - \frac{1}{2}\sigma_d$ (Figure 10.4c), and drainage into or out of the specimen is not allowed, then the corresponding total stress Mohr's circle is of the nature shown in Figure 10.4d. Note that the stresses on the plane $X-X$ are

Total normal stress $= \sigma_3$,   shear stress $= +\frac{1}{2}\sigma_d$

and the stresses on the plane $Y-Y$ are

Total normal stress $= \sigma_3$,   shear stress $= -\frac{1}{2}\sigma_d$

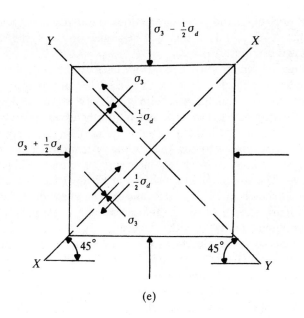

(e)

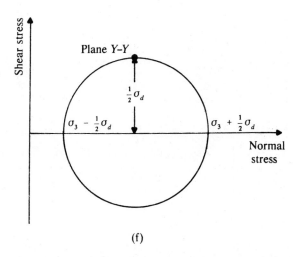

(f)

Figure 10.4   Simulation of cyclic shear stress on a plane for a triaxial test specimen

Similarly, if the specimen is subjected to a stress condition as shown in Figure 10.4e, the corresponding total stress Mohr's circle will be as shown in Figure 10.4f. The stresses on the plane $X-X$ are

Total normal stress $= \sigma_3$,   shear stress $= -\frac{1}{2}\sigma_d$

The stresses on the plane $Y-Y$ are

Total normal stress $= \sigma_3$,   shear stress $= +\frac{1}{2}\sigma_d$

It can be seen that, if cyclic normal stresses of magnitude $\frac{1}{2}\sigma_d$ are applied simultaneously in the horizontal and vertical directions, one can achieve a stress condition along planes $X-X$ and $Y-Y$ that will be similar to the cyclic shear stress application shown in Figure 10.3b.

However, for saturated sands, actual laboratory tests can be conducted by applying an all-around consolidation pressure of σ_3 and then applying a cyclic load having an amplitude of σ_d in the axial direction only without allowing drainage as shown in Figure 10.5a. The axial strain and the excess pore water pressure can be measured along with the number of cycles of load (σ_d) application. The question may now arise as to how the loading system shown in Figure 10.5a would produce stress conditions shown in Figure 10.4c and e. This can be explained as follows. The stress condition shown in Figure 10.5b is the sum of the stress conditions shown in Figure 10.5c and d. The effect of the stress condition shown in Figure 10.5d is to reduce the excess pore water pressure of the specimen by an amount equal to $\frac{1}{2}\sigma_d$ without causing any change in the axial strain. Thus, the effect of the stress conditions shown in Figure 10.5b (which is the same as Figure 10.4c) can be achieved by only subtracting a pore water pressure $u = \frac{1}{2}\sigma_d$ from that observed from the loading condition shown in Figure 10.5c. Similarly, the loading condition shown in Figure 10.5e is the loading condition in Figure 10.5f plus the loading condition in Figure 10.5g. The effect

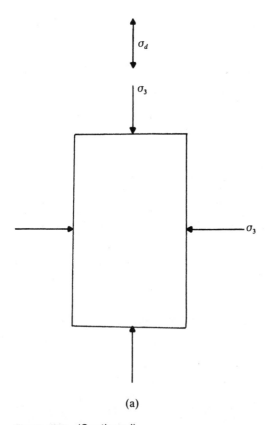

(a)

Figure 10.5   (Continued)

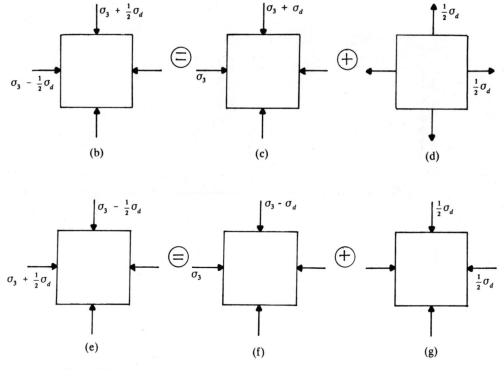

Figure 10.5

of the stress condition shown in Figure 10.5g is only to increase the pore water pressure by an amount $\frac{1}{2}\sigma_d$. Thus the effect of the stress conditions shown in Figure 10.5e (which is the same as in Figure 10.4e) can be achieved by only adding $\frac{1}{2}\sigma_d$ to the pore water pressure observed from the loading condition in Figure 10.5f.

10.5   Typical Results from Cyclic Triaxial Test

Several cyclic undrained triaxial tests on saturated soil specimens have been conducted by Seed and Lee (1966) on Sacramento River sand retained between No. 50 and No. 100 U.S. sieves. The results of a typical test in *loose sand* (void ratio, $e = 0.87$) is shown in Figure 10.6. For this test, the initial all-around pressure and the initial pore water pressure were 28.4 lb/in.^{2} (196.2 kN/m^2) and 14.2 lb/in.^{2} (98.2 kN/m^2), respectively. Thus, the all-around consolidation pressure σ_3 is equal to 14.2 lb/in.^{2} (98.1 kN/m^2). The cyclic deviator stress σ_d was applied with a frequency of 2 cps. Figure 10.7 is a plot of the axial strain, change in pore water pressure u, and the change in pore water pressure corrected to mean extreme principal stress conditions (i.e., subtracting or adding $\frac{1}{2}\sigma_d$ from or to the observed pore water pressure) against the number of cycles of load application. Figure 10.7c shows that the change in pore water pressure becomes equal to σ_3 during the ninth cycle, indicating that the effective confining pressure is equal to zero.

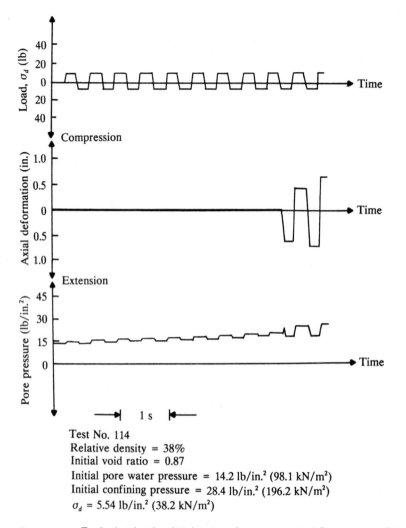

Figure 10.6 Typical pulsating load test on loose saturated Sacramento River sand (redrawn after Seed and Lee, 1966)

During the tenth cycle, the axial strain exceeded 20% and the soil liquefied. The relationship between the magnitude of σ_d against the number of cycles of pulsating stress applications for the liquefaction of the same loose sand [$e = 0.87$, $\sigma_3 = 14.2$ lb/in.^{2} (98.1 kN/m^2)] is shown in Figure 10.8. Note that the number of cycles of pulsating stress application increases with the decrease of the value of σ_d.

The nature of variation of the axial strain and the *corrected* pore water pressure for a pulsating load test in a dense Sacramento River sand is shown in Figure 10.9. After about 13 cycles, the change in pore water pressure becomes equal to σ_3; however, the axial strain amplitude did not exceed 10% after even 30 cycles of load application. This is a condition with a peak cyclic pore pressure ratio of 100%, with limited strain potential due to the remaining resistance of the

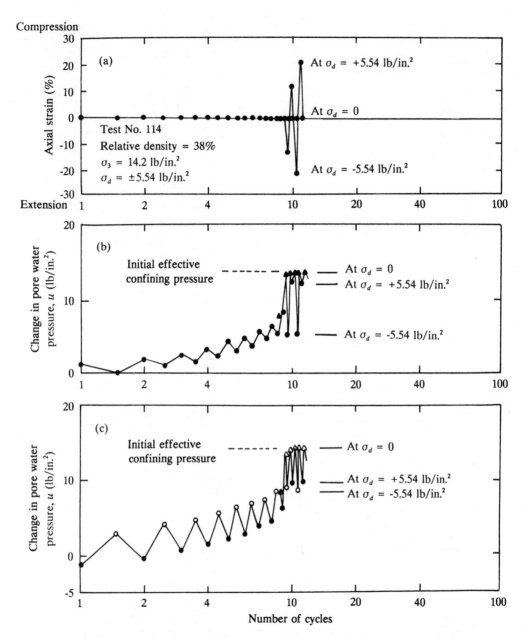

Figure 10.7 Typical pulsating load test on loose Sacramento River sand: (a) plot of axial strain versus number of cycles of load application; (b) observed change in pore water pressure versus number of cycles of load application; (c) change in pore water pressure (corrected to mean principal stress condition) versus number of cycles of load application (after Seed and Lee, 1966)

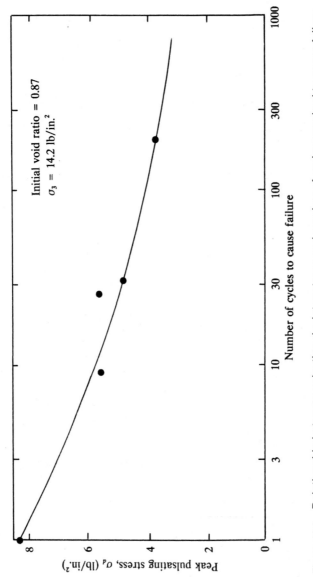

Figure 10.8 Relationship between pulsating deviator stress and number of cycles required to cause failure in Sacramento River sand (redrawn after Seed and Lee, 1966)

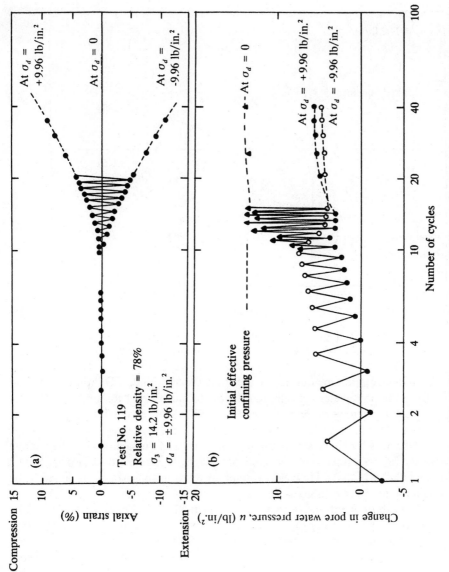

Figure 10.9   Typical pulsating load test on dense Sacramento River sand: (a) plot of axial strain versus number of cycles of load application; (b) corrected change of pore water pressure versus number of cycles of load application (after Seed and Lee, 1966).

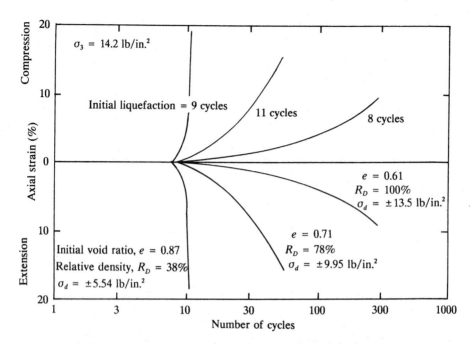

Figure 10.10   Axial strain after initial liquefaction for pulsating load tests at three densities for Sacramento River sand (after Seed and Lee, 1966)

soil to deformation, or due to the fact that the soil dilates. Dilation of the soil reduces the pore water pressure and helps stabilization of soil under load. This may be referred to as *cyclic mobility* (Seed, 1979). More discussion on this subject is given in Section 10.11.

A summary of axial strain, number of cycles for liquefaction, and the relative density for Sacramento River sand are given in Figure 10.10 [for $\sigma_3 = 14.2$ lb/in.^{2} (98.1 kN/m^2)]. However, a different relationship may be obtained if the confining pressure σ_3 is changed.

It has been mentioned earlier that the critical void ratio of a sand cannot be used as a unique criterion for a quantitative evaluation of the liquefaction potential. This can now be seen from Figure 10.11, which shows the critical void ratio line for Sacramento River sand. Based on the initial concept of critical void ratio, one would assume that a soil specimen represented by a point to the left of the critical void ratio line would *not* be susceptible to liquefaction; likewise, a specimen that plots to the right of the critical void ratio line *would* be vulnerable to liquefaction. In order to test this concept, the cyclic load test results on two specimens are shown as *A* and *B* in Figure 10.11. Under a similar pulsating stress $\sigma_d = \pm 17.06$ lb/in.^{2} (117.72 kN/m^2), specimen *A* liquefied in 57 cycles, whereas specimen *B* did not fail even in 10,000 cycles. This is contrary to the aforementioned assumptions. Thus, the liquefaction potential depends on five important factors:

1. Relative density R_D
2. Confining pressure σ_3

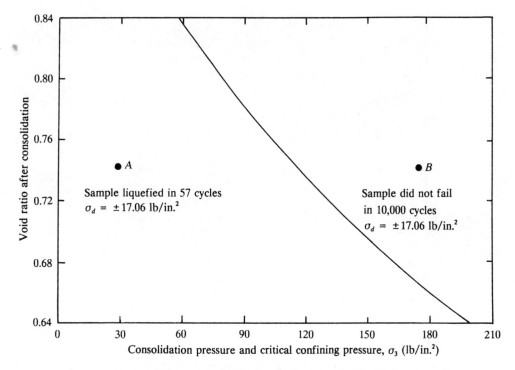

Figure 10.11   Critical confining pressure-void ratio relationship for Sacramento River sand (redrawn after Seed and Lee, 1966)

3. Peak pulsating stress σ_d

4. Number of cycles of pulsating stress application

5. Overconsolidation ratio

The importance of the first four factors is discussed in the following section. The overconsolidation ratio is discussed in Section 10.9. Soil grain characteristics are also known to have some effects on the liquefaction potential.

10.6   Influence of Various Parameters on Soil Liquefaction Potential

Influence of the Initial Relative Density

The effect of the initial relative density of a soil on liquefaction is shown in Figure 10.12. All tests shown in Figure 10.12 are for $\sigma_3 = 14.2$ lb/in.^{2} (98.1 kN/m^2). The *initial liquefaction* corresponds to the condition when the pore water pressure becomes equal to the confining pressure σ_3. In most cases, *20% double amplitude* strain is considered as failure. It may be seen that, for a given value of σ_d, the initial liquefaction and the failure occur simultaneously for loose sand (Figure 10.12a). However, as the relative density increases, the difference between the

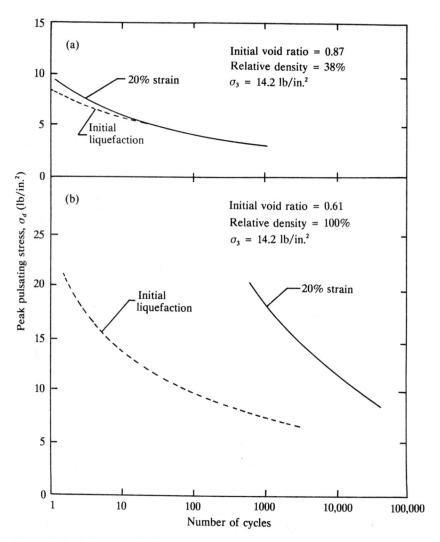

Figure 10.12 Influence of Initial relative density on liquefaction for Sacramento River sand (redrawn after Lee and Seed, 1967)

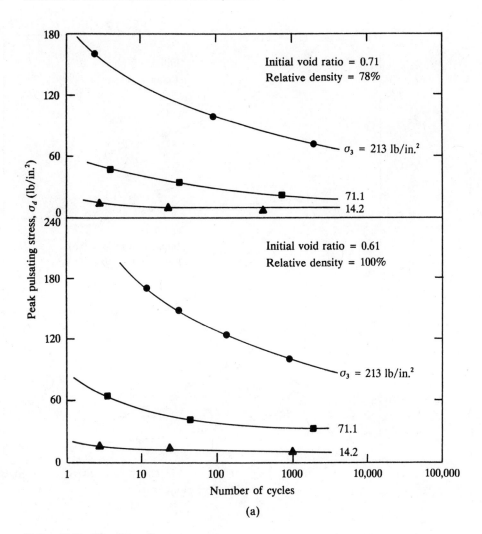

(a)

Figure 10.13   (Continued)

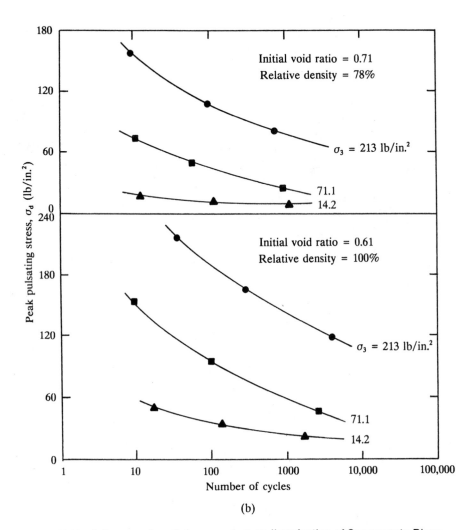

Figure 10.13   Influence of confining pressure on liquefaction of Sacramento River sand: (a) initial liquefaction; (b) 20% strain (redrawn after Lee and Seed, 1967)

number of cycles to cause 20% double amplitude strain and to cause initial liquefaction increases.

Influence of Confining Pressure

The influence of the confining pressure σ_3 on initial liquefaction and 20% double amplitude strain condition is shown in Figure 10.13. For a given initial relative density and peak pulsating stress, the number of cycles to cause initial liquefaction or 20% strain increases with the increase of the confining pressure. This is true for all relative densities of compaction.

Influence of the Peak Pulsating Stress

Figure 10.14 shows the variation of the peak pulsating stress σ_d with the confining pressure for initial liquefaction in 100 cycles (Figure 10.14a) and for 20% axial strain in 100 cycles (Figure 10.14b). Note that for a given initial void ratio (i.e., relative density R_D) and number of cycles of load application, the variation of σ_d for initial liquefaction with σ_3 is practically linear. A similar relation also exists for loose sand with a 20% axial strain condition.

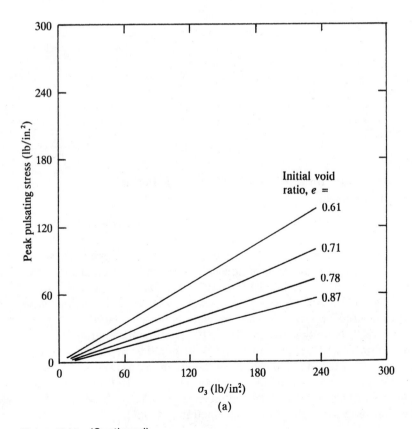

Figure 10.14 (Continued)

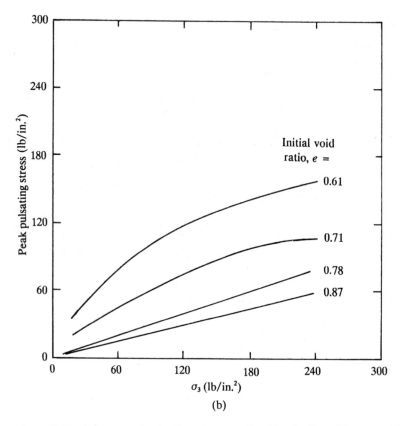

Figure 10.14   Influence of pulsating stress on the liquefaction of Sacramento River sand (a) initial liquefaction in 100 cycles; (b) 20% strain in 100 cycles (redrawn after Lee and Seed, 1967)

10.7   **Development of Standard Curves for Initial Liquefaction**

By compilation of the results of liquefaction tests conducted by several investigators on various types of sand, average standard curves for initial liquefaction for a given number of load cycle application can be developed. These curves can then be used for evaluation of liquefaction potential in the field. Some of these plots developed by Seed and Idriss (1971) are given in Figure 10.15.

Figure 10.15a is a plot of $\frac{1}{2}\sigma_d/\sigma_3$ versus D_{50} to cause initial liquefaction in 10 cycles of stress application. The plot is for an initial relative density of compaction of 50%. Note that D_{50} in Figure 10.15a is the mean grain size, i.e., the size through which 50% of the soil will pass. It should be kept in mind that $\frac{1}{2}\sigma_d$ is the magnitude of the maximum cyclic shear stress imposed on a soil specimen (see planes $X-X$ and $Y-Y$ of Figure 10.4d, f). Figure 10.15b is a similar plot for initial liquefaction in 30 cycles of stress application. These curves are used in Section 10.15 for evaluation of liquefaction potential.

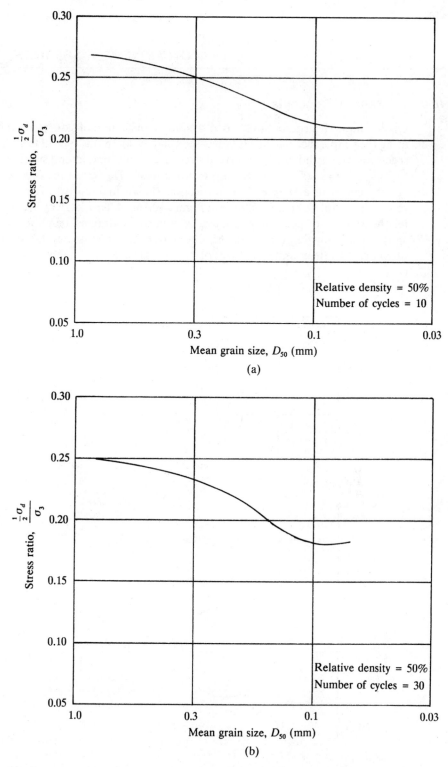

Figure 10.15 Stress ratio causing liquefaction of sands in 10 and 30 cycles (after Seed and Idriss, 1971)

CYCLIC SIMPLE SHEAR TEST

10.8   General Concepts

Cyclic simple shear tests can be used to study liquefaction of saturated sands by using the simple shear apparatus (also see Chapter 4). In this type of test, the soil specimen is consolidated by a vertical stress σ_v. At this time, lateral stress is equal to $K_0\sigma_v$ (K_0 = coefficient of earth pressure at rest). The initial stress conditions of a specimen in a simple shear device are shown in Figure 10.16a; the corresponding Mohr's circle is shown in Figure 10.16b. After that, a cyclic horizontal shear stress of peak magnitude τ_h is applied (undrained condition) to the specimen as shown in Figure 10.16c. The pore water pressure and the strain are observed with the number of cycles of horizontal shear stress application.

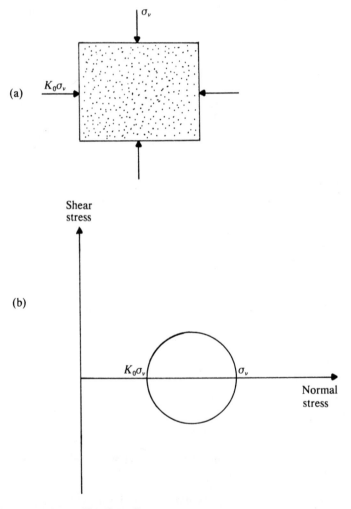

Figure 10.16 (Continued)

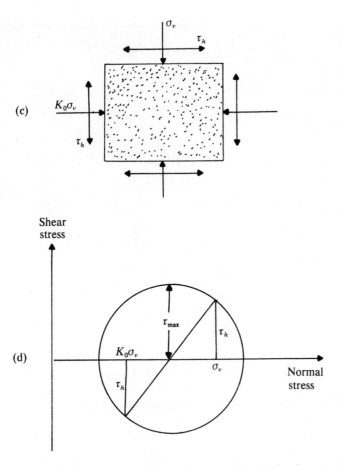

Figure 10.16   Maximum shear stress for cyclic simple shear test

Using the stress conditions on the soil specimen at a certain time during the cyclic shear test, a Mohr's circle is plotted in Figure 10.16d. Note that the maximum shear stress on the specimen in simple shear is not τ_h, but

$$\tau_{max} = \sqrt{\tau_h^2 + [\tfrac{1}{2}\sigma_v(1 - K_0)]^2} \tag{10.2}$$

10.9   Typical Test Results

Typical results of some soil liquefaction tests on Monterey sand using simple shear apparatus are shown in Figure 10.17. Note that these are for the initial liquefaction condition. From the figure the following facts may be observed:

1. For a given value of σ_v and relative density R_D, a decrease of τ_h requires an increase of the number of cycles to cause liquefaction.

2. For a given value of R_D and number of cycles of stress application, a decrease of σ_v requires a decrease of the peak value of τ_h for causing liquefaction.

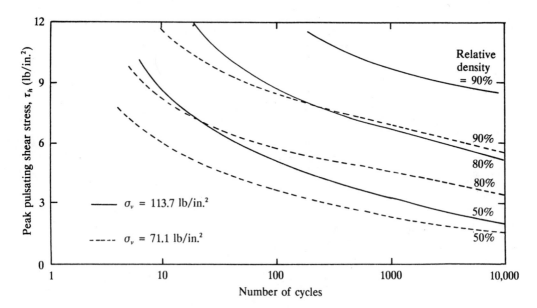

Figure 10.17   Initial liquefaction in cyclic simple shear test on Monterey sand
(redrawn after Peacock and Seed, 1968)

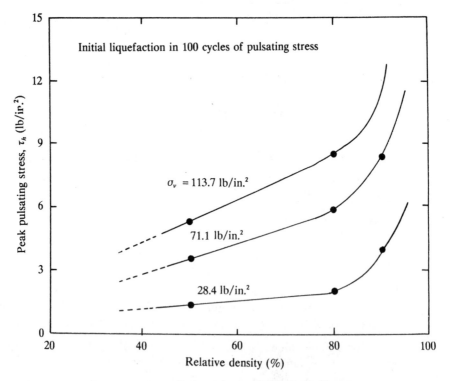

Figure 10.18   Effect of relative density on cyclic shear stress causing initial
liquefaction of Monterey sand (redrawn after Peacock and Seed, 1968)

3. For a given value of σ_v and number of cycles of stress application, τ_h for causing liquefaction increases with the increase of the relative density.

Another important factor—the variation of the peak value of τ_h for causing initial liquefaction with the initial relative density of compaction (for a given value of σ_v and number of stress cycle applications)—is shown in Figure 10.18. For a relative density up to about 80%, the peak value of τ_h for initial liquefaction increases linearly with R_D.

Influence of Test Condition

In simple shear test equipment, there is always some nonuniformity of stress conditions. This causes specimens to develop liquefaction under lower applied horizontal cyclic stresses as compared to that in the field. This happens even though care is taken to improve the preparation of the specimens and rough platens are used at the top and bottom of the specimens to be tested. For that reason, for a given value of σ_v, R_D, and number of cyclic shear stress applications, the peak value of τ_h in the field is about 15%–50% higher than that obtained from the cyclic simple shear test. This fact has been demonstrated by Seed and Peacock (1971) for a uniform medium sand ($R_D \approx 50\%$) in which the field values are about 20% higher than the laboratory values.

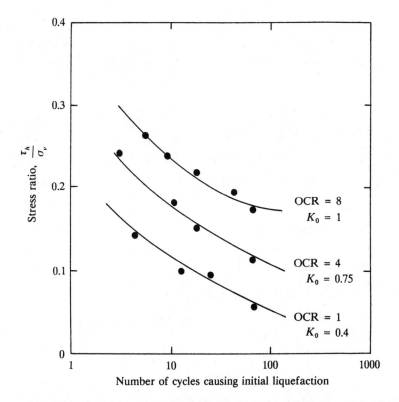

Figure 10.19   Influence of overconsolidation ratio on stresses causing liquefaction in simple shear tests (redrawn after Seed and Peacock, 1971)

Influence of Overconsolidation Ratio on the Peak Value of τ_h Causing Liquefaction

For the cyclic simple shear test, the value of τ_h is highly dependent on the value of the initial lateral earth pressure coefficient at rest (K_0). The value of K_0 is, in turn, dependent on the overconsolidation ratio (OCR). The variation of τ_h/σ_v for initial liquefaction with the overconsolidation ratio as determined by the cyclic simple shear test is shown in Figure 10.19 on p. 421. <u>For a given relative density and number of cycles causing initial liquefaction, the value of τ_h/σ_v decreases with the decrease of K_0</u>. It needs to be mentioned at this point that *all* the *cyclic triaxial* studies for liquefaction are conducted for the initial value of $K_0 = 1$.

10.10   Rate of Excess Pore Water Pressure Increase

Seed and Booker (1977) and DeAlba, Chan, and Seed (1975) measured the rate of excess pore water pressure increase in saturated sands during liquefaction using cyclic simple shear tests. The range of the variation of pore water pressure generation u_g during cyclic loading is shown in Figure 10.20. The average value of the variation of u_g can be expressed in a nondimensional form as (Seed, Martin, and Lysmer, 1975)

$$\boxed{\frac{u_g}{\sigma_v} = \left(\frac{2}{\pi}\right) \arcsin\left(\frac{N}{N_i}\right)^{1/2\alpha}} \tag{10.3}$$

where

     u_g = excess pore water pressure generated

     σ_v = initial consolidation pressure

     N = number of cycles of shear stress application

     N_i = number of cycles of shear stress needed for initial liquefaction

     α = constant (≈ 0.7)

Hence, the rate of change of u_g with N can be given as

$$\boxed{\frac{\partial u_g}{\partial N} = \left(\frac{2\sigma_v}{\alpha \pi N_i}\right)\left[\frac{1}{\sin^{2\alpha-1}\left(\frac{\pi}{2}r_u\right)\cos\left(\frac{\pi}{2}r_u\right)}\right]} \tag{10.4}$$

where

$$r_u = \frac{u_g}{\sigma_v} \tag{10.5}$$

The preceding relationship is very useful in the study of the stabilization of potentially liquefiable sand deposits.

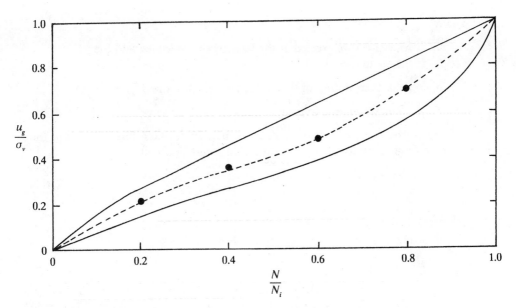

Figure 10.20   Rate of pore water pressure buildup in cyclic simple shear test (after Seed and Booker, 1977)

10.11   Large-Scale Simple Shear Tests

In the study of soil liquefaction of granular soils, certain aspects of the test procedures have remained a matter for concern. Some of those concerns are as follows:

a. Stress concentration in small-scale simple shear tests leads to some inaccuracy in the results (Castro, 1969).

b. Stress concentration at the base and cap of cyclic triaxial test specimens and the possibility of necking leads to nonuniformity of strain and redistribution of water content (Castro, 1975).

c. Attempts to study liquefaction by using shaking table tests (e.g., Emery, Finn, and Lee, 1972; Finn, 1972; Finn, Emery, and Gupta, 1970; O-Hara, 1972; Ortigosa, 1972; Tanimoto, 1967; Whitman, 1970; Yoshimi, 1967) have also raised some questions, since the results, in some cases, have been influenced by the confining effects of the sides of the box.

For that reason, DeAlba, Seed, and Chan (1976) conducted large-scale simple shear tests with one-directional cyclic stress application. The specimens of sand used for testing had dimensions of 90 in. × 42 in. × 4 in. (depth) (2300 mm × 1100 mm × 100 mm). Each specimen was constructed over a shaking table. A rubber membrane was placed over the sand to prevent drainage. An inertia mass was also placed on top of the sand. Movement of the shaking table produced cyclic stress conditions in the sand. The cyclic shear stress was determined as

$$\tau_h = \frac{W}{g} a_m \tag{10.6}$$

where

W = total pressure exerted at the base by the specimen and the inertia mass

a_m = peak acceleration of the uniform cyclic motion

g = acceleration due to gravity

From the measured displacement of the inertia mass during shaking, the average single-amplitude cyclic shear strain could be obtained as

$$\gamma' = \pm\frac{\Delta}{2h} \tag{10.7}$$

where

γ' = average single amplitude cyclic shear strain

h = specimen height

Figure 10.21 shows the variation of τ_h/σ_v against the number of cycles for initial liquefaction ($N = N_i$) for various values of the relative density of sand (R_D). Note that this has been corrected for the compliance effects of the specimens and the pore water pressure–measuring system and the effects of membrane penetration. The nature of these plots is similar to those shown in Figure 10.17. Figure 10.22 shows a comparison of the variation of τ_h/σ_v versus N_i (for $R_D = 50\%$) obtained from the reported results of Ortigosa (1972), O-Hara (1972), Finn, Emery, and Gupta (1971), and the large-scale simple shear test results of DeAlba, Chan, and Seed (1976). The differences between the results are primarily due to (1) the effect of membrane penetration and compliance effects, (2) the length-to-height ratio of the specimens and hence the boundary conditions, and (3) the nature of sample preparation. It is thus evident from Figure 10.22 that care should be taken to provide proper boundary conditions if meaningful data are to be obtained from shaking table tests.

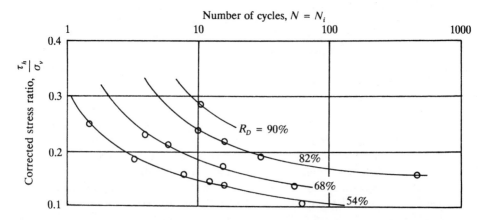

Figure 10.21   Corrected τ_h/σ_v versus N_i for initial liquefaction from large-scale simple shear tests (after DeAlba, Seed, and Chan, 1976)

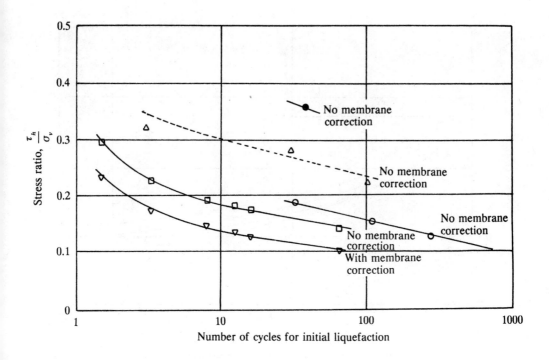

LEGEND

	Author	Length/height ratio	Material	Sample preparation
●	Ortigosa (1972)	2.3:1	Medium sand	Poured dry, compacted
△	O-Hara (1972)	3.4:1	Fine sand	
○	Finn et al. (1971) (extrapolated)	10.3:1	Medium sand	Pluviated through water
□	DeAlba et al. (1976)	22.5:1	Medium sand	Pluviated through air
▽	DeAlba et al. (1976)	22.5:1	Medium sand	Pluviated through air

Figure 10.22   Comparison of shaking table test results—$R_D = 50\%$ (after DeAlba, Seed, and Chan, 1976)

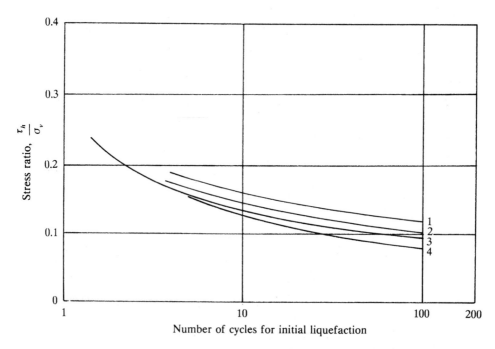

Figure 10.23 shows the comparison of τ_h/σ_v versus number of cycles for initial liquefaction

Curve No.	Reference	Type of Test
1	Yoshimi and Oh-oka (1973)	Ring torsion
2	Finn (1972)	Simple shear
3	DeAlba et al. (1976)	Shaking table (corrected)
4	Seed and Peacock (1971)	Simple shear

Figure 10.23 Comparison of shaking table and simple shear liquefaction test results—$R_D = 50\%$ (after DeAlba, Seed, and Chan, 1976)

Figure 10.23 shows the comparison of τ_h/σ_v versus number of cycles for initial liquefaction of saturated sand at $R_D = 50\%$ obtained from various studies using small-scale and large-scale simple shear devices. The sample preparation techniques in all the studies were similar. Based on Figure 10.23, it can be concluded that the results are in good agreement and the errors due to stress concentration in small-scale simple shear tests are not very large.

The variation of single-amplitude cyclic shear strain [Eq. (10.7)] with N for *dense sands* obtained from large-scale simple shear tests is shown in Figure 10.24. Note that the magnitude of γ' increased gradually with N after initial liquefaction up to a maximum limiting value and remained constant thereafter. Figure 10.25 shows the relationships between cyclic stress ratio and number of stress cycles producing average shear strains of 5%, 10%, 15%, and 20% calculated from displacements measured from the large-scale simple shear tests. The results of Figure 10.25 have been replotted as values of cyclic stress ratio causing initial liquefaction, or different levels of shear strain (for $N = 10$), versus relative

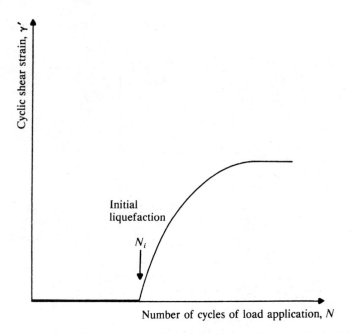

Figure 10.24 Nature of variation of γ' with number of cycles of load application

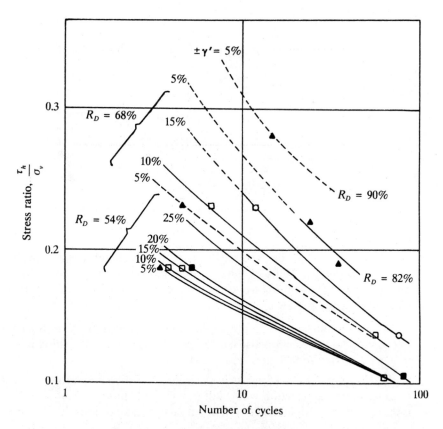

Figure 10.25 Relationship between τ_h/σ_v and number of cycles causing different levels of strain (after DeAlba, Seed, and Chan, 1976)

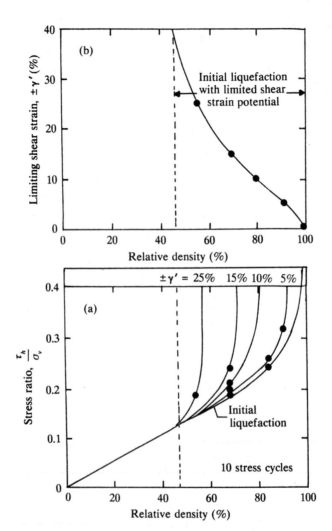

Figure 10.26 Limiting shear strains—10 cycles of stress (after DeAlba, Seed, and Chan, 1976)

densities in Figure 10.26a. The results show that each curve is asymptotic to a certain value of R_D. Hence a curve of limiting shear strain versus R_D can be obtained as shown in Figure 10.26b. Based on Figure 10.26, the following conclusions can be drawn.

1. For initial $R_D \leq 45\%$, the application of cyclic stress ratio high enough to cause initial liquefaction also causes unlimited shear strain. This corresponds to a condition of liquefaction.

2. For initial $R_D > 45\%$, the application of cyclic stress ratio high enough to cause initial liquefaction will result in a limited amount of shear strain. This is the case of soil with limited strain potential or the condition of cyclic mobility.

3. The limiting strain potential decreases with the increase of the initial relative density of soil.

DEVELOPMENT OF A PROCEDURE FOR DETERMINATION
OF FIELD LIQUEFACTION

10.12   Correlation of the Liquefaction Results from Simple Shear and Triaxial Tests

The conditions for determination of field liquefaction problems are related to the ratio of τ_h/σ_v; this is also true for the case of cyclic simple shear tests. However, in the case of triaxial tests, the results are related to the ratio of $\frac{1}{2}\sigma_d/\sigma_3$. It appears that a correlation between τ_h/σ_v and $\frac{1}{2}\sigma_d/\sigma_3$ needs to be developed (for a given number of cyclic stress application to cause liquefaction). Seed and Peacock (1971) considered the following alternative criteria for correlation for the onset of soil liquefaction.

1. The *maximum ratio of the shear stress* developed during cyclic loading *to the normal stress* during consolidation on any plane of the specimen can be a controlling factor. For triaxial specimens, this is equal to $\frac{1}{2}\sigma_d/\sigma_3$, and for simple shear specimens it is about $\tau_h/(K_0\sigma_v)$. Thus,

$$\frac{\tau_h}{K_0\sigma_v} = \frac{\frac{1}{2}\sigma_d}{\sigma_3}$$

or

$$\left[\frac{\tau_h}{\sigma_v}\right]_{\text{simple shear}} = K_0\left[\frac{\frac{1}{2}\sigma_d}{\sigma_3}\right] \tag{10.8}$$

2. Another possible condition for the onset of liquefaction can be the *maximum ratio of change in shear stress* during cyclic loading *to the normal stress* during consolidation on any plane. For simple shear specimens this is about $\tau_h/(K_0\sigma_v)$, and for triaxial specimens it is $\frac{1}{2}\sigma_d/\sigma_3$. This leads to the same equation as Eq. (10.8).

3. The third possible alternative can be given by the *ratio of the maximum shear stress* induced in a specimen during cyclic loading *to the mean principal stress* on the specimen during consolidation. For simple shear specimens:

Maximum shear stress
during cyclic loading $= \sqrt{\tau_h^2 + [\frac{1}{2}\sigma_v(1 - K_0)]^2}$ (10.2)

Mean principal stress during
consolidation (Figure 10.16a) $= \frac{1}{3}(\sigma_v + K_0\sigma_v + K_0\sigma_v) = \frac{1}{3}\sigma_v(1 + 2K_0)$

(10.9)

For triaxial specimens, maximum shear stress during cyclic loading $= \frac{1}{2}\sigma_d$ and mean principal stress during consolidation $= \sigma_3$; so

$$\frac{\sqrt{\tau_h^2 + [\frac{1}{2}\sigma_v(1 - K_0)]^2}}{\frac{1}{3}\sigma_v(1 + 2K_0)} = \frac{\frac{1}{2}\sigma_d}{\sigma_3}$$

$$\frac{\sqrt{(\tau_h/\sigma_v)^2 + [\frac{1}{2}(1 - K_0)]^2}}{\frac{1}{3}(1 + 2K_0)} = \frac{\frac{1}{2}\sigma_d}{\sigma_3}$$

$$\left[\frac{\tau_h}{\sigma_v}\right]_{\text{simple shear}} = \sqrt{\left(\frac{\frac{1}{2}\sigma_d}{\sigma_3}\right)^2 [\frac{1}{3}(1+2K_0)]^2 - [\frac{1}{2}(1-K_0)]^2}$$

or

$$\left[\frac{\tau_h}{\sigma_v}\right] = \left(\frac{\frac{1}{2}\sigma_d}{\sigma_3}\right)\sqrt{\frac{1}{9}(1+2K_0)^2 - \frac{\frac{1}{4}(1-K_0)^2}{(\frac{1}{2}\sigma_d/\sigma_3)^2}} \tag{10.10}$$

4. The fourth possible alternative may be the *ratio of the maximum change in shear stress* on any plane during cyclic loading *to the mean principal stress* during consolidation. Thus, for simple shear specimens, it is equal to $3\tau_h/[\sigma_v(1+2K_0)]$, and for triaxial specimens it is $\frac{1}{2}\sigma_d/\sigma_3$; so

$$\left[\frac{\tau_h}{\sigma_v}\right]_{\text{simple shear}} = \frac{1}{3}(1+2K_0)\left(\frac{\frac{1}{2}\sigma_d}{\sigma_3}\right) \tag{10.11}$$

Thus, in general, it can be written as

$$\left(\frac{\tau_h}{\sigma_v}\right)_{\text{simple shear}} = \alpha'\left(\frac{\frac{1}{2}\sigma_d}{\sigma_3}\right)_{\text{triax}} \tag{10.12}$$

where $\alpha' = K_0$ for Cases 1 and 2,

$$\alpha' = \sqrt{\frac{1}{9}(1+2K_0)^2 - \frac{\frac{1}{4}(1-K_0)^2}{(\frac{1}{2}\sigma_d/\sigma_3)^2}} \qquad \text{for Case 3}$$

and

$$\alpha' = \frac{1}{3}(1+2K_0) \qquad \qquad \text{for Case 4}$$

The values of α' for the four cases considered here are given in Table 10.1.

From Table 10.1 it may be seen that for normally consolidated sands, the value of α' is generally in the range of 45%–50%, with an average of about 47%.

Finn, Emery, and Gupta (1971) have shown that, for initial liquefaction of normally consolidated sands, α' is equal to $\frac{1}{2}(1+K_0)$. The value of K_0 can be given by the relation (Jaky, 1944)

$$K_0 = 1 - \sin \phi \tag{10.13}$$

Castro (1975) has proposed that the initial liquefaction may be controlled

Table 10.1   Values of α' [Eq. (10.12)][a]

K_0	Case 1	Case 2	Case 3	Case 4
0.4	0.4	0.4	—	0.6
0.5	0.5	0.5	0.25[b]	0.67
0.6	0.6	0.6	0.54[b]	0.73
0.7	0.7	0.7		0.80
0.8	0.8	0.8	0.83[b]	0.87
0.9	0.9	0.9		0.93
1.0	1.0	1.0	1.0	1.0

[a] After Seed and Peacock (1971).

by the criteria of the *ratio of the octahedral shear stress* during cycle loading *to the effective octahedral normal stress during consolidation*. The effective octahedral normal stress during consolidation σ'_{oct} is given by the relation

$$\sigma'_{oct} = \tfrac{1}{3}(\sigma'_1 + \sigma'_2 + \sigma'_3) \tag{10.14}$$

where σ'_1, σ'_2, σ'_3 are, respectively, the major, intermediate, and minor *effective* principal stresses.

The octahedral shear stress τ_{oct} during cyclic loading is

$$\tau_{oct} = \tfrac{1}{3}[(\sigma_1 - \sigma_3)^2 + (\sigma_1 - \sigma_2)^2 + (\sigma_2 - \sigma_3)^2]^{1/2} \tag{10.15}$$

where σ_1, σ_2, σ_3 are, respectively, the major, intermediate, and minor principal stresses during cyclic loading. For cyclic triaxial tests,

$$\left(\frac{\tau_{oct}}{\sigma'_{oct}}\right)_{triax} = \frac{2}{3}\sqrt{2}\left(\frac{\tfrac{1}{2}\sigma_d}{\sigma_3}\right)_{triax} \tag{10.16}$$

For cyclic simple shear tests,

$$\left(\frac{\tau_{oct}}{\sigma'_{oct}}\right)_{simple\ shear} = \left(\frac{\sqrt{6}}{1 + 2K_0}\right)\left(\frac{\tau_h}{\sigma_v}\right)_{simple\ shear} \tag{10.17}$$

Thus,

$$\left(\frac{\tau_{oct}}{\sigma'_{oct}}\right)_{simple\ shear} = \left(\frac{\tau_{oct}}{\sigma'_{oct}}\right)_{triax}$$

or

$$\left(\frac{\sqrt{6}}{1 + 2K_0}\right)\left(\frac{\tau_h}{\sigma_v}\right)_{simple\ shear} = \frac{2}{3}\sqrt{2}\left(\frac{\tfrac{1}{2}\sigma_d}{\sigma_3}\right)_{triax}$$

or

$$\left(\frac{\tau_h}{\sigma_v}\right)_{simple\ shear} = \frac{2}{3}\sqrt{2}\left(\frac{1 + 2K_0}{\sqrt{6}}\right)\left(\frac{\tfrac{1}{2}\sigma_d}{\sigma_3}\right)$$

$$= \frac{\tfrac{2}{3}(1 + 2K_0)}{\sqrt{3}}\left(\frac{\tfrac{1}{2}\sigma_d}{\sigma_3}\right)_{triax} \tag{10.18}$$

Comparing Eqs. (10.12) and (10.18),

$$\alpha' = \frac{\tfrac{2}{3}(1 + 2K_0)}{\sqrt{3}} \tag{10.19}$$

10.13   Correlation of the Liquefaction Results from Triaxial Tests to Field Conditions

Section 10.9 explained that the field value of τ_h/σ_v for initial liquefaction is about 15%–50% higher than that obtained from simple shear tests. Thus,

$$\left(\frac{\tau_h}{\sigma_v}\right)_{field} = \beta\left(\frac{\tau_h}{\sigma_v}\right)_{simple\ shear} \tag{10.20}$$

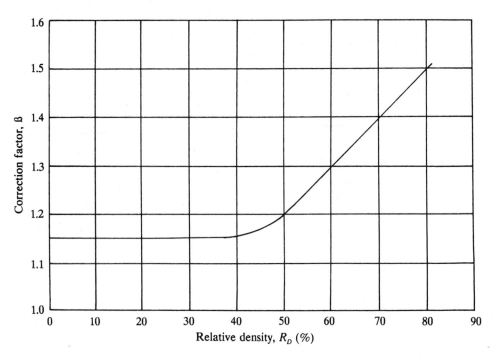

Figure 10.27 Variation of correction factor β with relative density [Eq. (10.20)]

The approximate variation of β with relative density of sand is given in Figure 10.27. Combining Eqs. (10.12) and (10.20), one obtains

$$\left(\frac{\tau_h}{\sigma_v}\right)_{\text{field}} = \beta\left(\frac{\tau_h}{\sigma_v}\right)_{\text{simple shear}} = \alpha'\beta\left(\frac{\frac{1}{2}\sigma_d}{\sigma_3}\right)_{\text{triax}} = C_r\left(\frac{\frac{1}{2}\sigma_d}{\sigma_3}\right)_{\text{triax}} \tag{10.21}$$

where $C_r = \alpha'\beta$.

Using an average value of $\alpha' = 0.47$ and the values of β given in Figure 10.27, the variation of C_r with relative density can be obtained. This is shown in Figure 10.28.

Eq. (10.21) presents the correlations for initial liquefaction between the stress ratios in the field, cyclic simple shear tests, and cyclic triaxial tests for a given sand at the *same relative density*. However, when laboratory tests are conducted at relative density, say, $R_{D(1)}$, whereas the field conditions show the sand deposit to be at a relative density of $R_{D(2)}$, one has to convert the laboratory test results to correspond to a relative density of $R_{D(2)}$. It has been shown in Figure 10.18 that τ_h for initial liquefaction in the laboratory in a given number of cycles is approximately proportional to the relative density (for $R_D \leq 80\%$). Thus,

$$\tau_{h[R_{D(2)}]} = \tau_{h[R_{D(1)}]}\left[\frac{R_{D(2)}}{R_{D(1)}}\right] \tag{10.22}$$

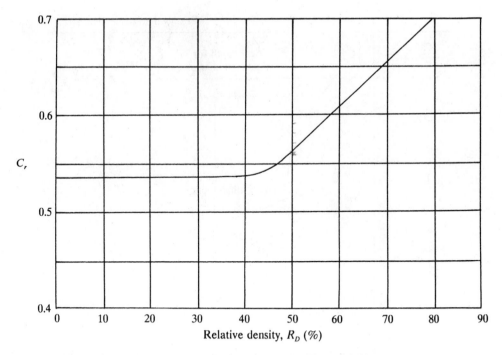

Figure 10.28   Variation of C_r with relative density [Eq. (10.21)]

where $\tau_{h[R_{D(1)}]}$ is the cyclic peak shear stress required to cause initial liquefaction in the laboratory for a given value of σ_v and number of cycles, by simple shear test; and $\tau_{h[R_{D(2)}]}$ is the cyclic peak shear stress required to cause initial liquefaction in the field for the same value of σ_v *and number of cycles*, by simple shear test. Combining Eqs. (10.21) and (10.22),

$$\left(\frac{\tau_h}{\sigma_v}\right)_{\text{field}[R_{D(2)}]} = C_r \left(\frac{\frac{1}{2}\sigma_d}{\sigma_3}\right)_{\text{triax}[R_{D(1)}]} \cdot \frac{R_{D(2)}}{R_{D(1)}} \qquad (10.23)$$

$R_{D(1)} = \text{LAB}$
$R_{D(2)} = \text{FIELD}$

10.14   Zone of Initial Liquefaction in the Field

There are five general steps for determining the zone in the field where soil liquefaction due to an earthquake can be initiated:

1. Establish a design earthquake.

2. Determine the time history of shear stresses induced by the earthquake at various depths of sand layer.

3. Convert the shear stress–time histories into N number of equivalent stress cycles (see Section 7.8). These can be plotted against depth, as shown in Figure 10.29.

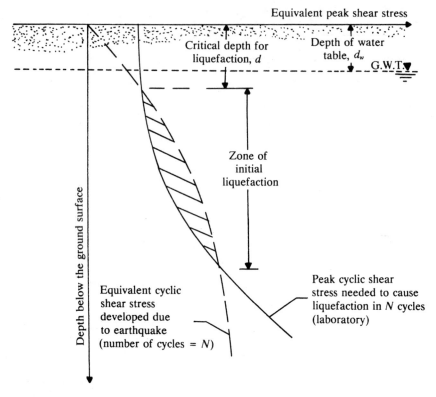

Figure 10.29 Zone of initial liquefaction in the field

4. Using the laboratory test results, determine the magnitude of the cyclic stresses required to cause initial liquefaction in the field in N cycles (determined from Step 3) at various depths. Note that the cyclic shear stress levels change with depth due to change of σ_v. These can be plotted with depth as shown in Figure 10.29.

5. The zone in which the cyclic shear stress levels required to cause initial liquefaction (Step 4) are equal to or less than the equivalent cyclic shear stresses induced by an earthquake is the zone of possible liquefaction. This is shown in Figure 10.29.

10.15 Relation Between Maximum Ground Acceleration and the Relative Density of Sand for Soil Liquefaction

This section discusses a simplified procedure developed by Seed and Idriss (1971) to determine the relation between the maximum ground acceleration due to an earthquake and the relative density of a sand deposit in the field for the initial liquefaction condition. Figure 10.30a shows a layer of sand deposit in which we consider a column of soil of height h and unit area of cross section. Assuming the soil column to behave as a *rigid body*, the maximum shear stress at a depth

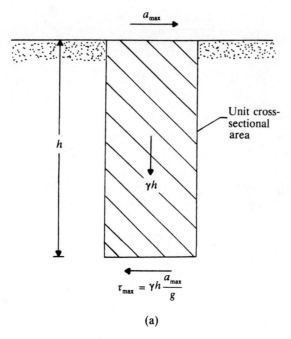

$$\tau_{max} = \gamma h \frac{a_{max}}{g}$$

(a)

Figure 10.30   (Continued)

h due to a maximum ground surface acceleration of a_{max} can be given by

$$\tau_{max} = \left(\frac{\gamma h}{g}\right) a_{max} \tag{10.24}$$

where τ_{max} is the maximum shear stress, γ is the unit weight of soil, and g is acceleration due to gravity.

However, the soil column is not a rigid body. For the deformable nature of the soil, the maximum shear stress at a depth h, determined by Eq. (10.24), needs to be modified as

$$\tau_{max(modif)} = C_D \left[\left(\frac{\gamma h}{g}\right) a_{max}\right] \tag{10.25}$$

where C_D is a stress reduction factor. The range of C_D for different soil profiles is shown in Figure 10.30b, along with the average value up to a depth of 40 ft (12.2 m).

It has been shown that the maximum shear stress determined from the shear stress–time history during an earthquake can be converted into an equivalent number of significant stress cycles. According to Seed and Idriss, one can take

$$\tau_{av} = 0.65\tau_{max(modif)} = 0.65C_D\left[\left(\frac{\gamma h}{g}\right) a_{max}\right] \tag{10.26}$$

The corresponding number of significant cycles N for τ_{av} is given in Table 10.2.

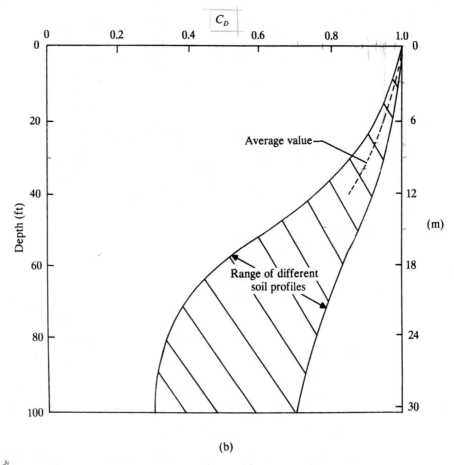

(b)

Figure 10.30 (a) Maximum shear stress at a depth for a rigid soil column; (b) range of the shear stress reduction factor C_D for the deformable nature of soil (after Seed and Idriss, 1971)

Table 10.2 Significant Number of Stress Cycles N Corresponding to τ_{av}

Earthquake magnitude	N
7	10
7.5	20
8	30

Note that although the values of N given in the table are somewhat different from those given in Figure 7.15, it does not make a considerable difference in the calculations.

One can now combine Eq. (10.23), which gives the correlation of laboratory results of cyclic triaxial tests to the field conditions, and Eq. (10.26) to determine the relationships between a_{max} and R_D. This can be better shown with the aid of a numerical example.

In general, the critical depth of liquefaction (see Figure 10.29) occurs at a depth of about 20 ft (6.1 m) when the depth of water table d_w is 0–10 ft (0–3.05 m); similarly, the critical depth is about 30 ft (9.15 m) when the depth of the water table is about 15 ft.

Liquefaction occurs in sands having a mean grain size D_{50} of 0.075–0.2 mm. Consider a case where

$$D_{50} = 0.075 \text{ mm}$$

$$d_w = 15 \text{ ft } (\approx 4.58 \text{ m})$$

$\gamma =$ unit weight above the ground water table (GWT) $= 115 \text{ lb/ft}^3$
(18.08 kN/m³)

$\gamma_{sat} =$ unit weight of soil below GWT $= 122.4 \text{ lb/ft}^3$ (19.24 kN/m³)

$\gamma' =$ effective unit weight of soil below GWT

$$= (122.4 - 62.4) = 60. \text{ lb/ft}^3 \text{ (9.43 kN/m}^3)$$

$$\text{significant number of stress cycles} = 10$$

$$\text{(earthquake magnitude} = 7)$$

The critical depth of liquefaction d is about 30 ft (9.15 m). At that depth the *total normal stress* is equal to

$$15(\gamma) + 15\gamma_{sat} = 15(115) + 15(122.4) = 3561 \text{ lb/ft}^2 \qquad (170.63 \text{ kN/m}^2)$$

From Eq. (10.26)

$$\tau_{av} = 0.65 C_D \left[\left(\frac{\gamma h}{g} \right) a_{max} \right]$$

The value of C_D for $d = 30$ ft is 0.925 (Fig. 10.30). Thus,

$$\tau_{av} = \frac{(0.65)(0.925)(3561)a_{max}}{g} = \frac{2141 a_{max}}{g} \tag{10.27}$$

Again, from Eq. (10.23),

$$\tau_{h(field)[R_{D(2)}]} = \sigma_v C_r \left(\frac{\frac{1}{2}\sigma_d}{\sigma_3} \right)_{triaxial[R_{D(1)}]} \frac{R_{D(2)}}{R_{D(1)}}$$

At a depth 30 ft below the ground surface, the *initial effective stress* σ_v is equal to $15(\gamma) + 15(\gamma') = 15(115) + 15(60) = 2625 \text{ lb/ft}^2$ (125.78 kN/m²). From Figure 10.15, for $D_{50} = 0.075$ mm, $(\frac{1}{2}\sigma_d/\sigma_3)_{triax, R_D=50\%} \approx 0.215$. Hence

$$\tau_{h(\text{field})(R_{D(2)})} = \frac{2625[C_r(0.215)]R_{D(2)}}{50} = 11.29C_rR_{D(2)} \qquad (10.28)$$

For liquefaction, τ_{av} of Eq. (10.27) should be equal to $\tau_{h(\text{field})[R_{D(2)}]}$. Hence,

$$\frac{2141a_{max}}{g} = 11.29C_rR_{D(2)}$$

or

$$\frac{a_{max}}{g} = 0.0053C_rR_{D(2)} \qquad (10.29)$$

It is now possible to prepare Table 10.3 to determine the variation of a_{max}/g with $R_{D(2)}$. Note that $R_{D(2)}$ is the relative density in the field.

Figure 10.31 shows a plot of a_{max}/g versus the relative density as determined from Table 10.3. For this given soil (i.e., given D_{50}, d_w, and number of significant stress cycles N), if the relative density in the field and a_{max}/g are such that they plot as point A in Figure 10.31 (i.e., above the curve showing the relationship of Eq. (10.29)], then liquefaction would occur. On the other hand, if the relative density and a_{max}/g plot as point B [i.e., below the curve showing the relationship of Eq. (10.29)], then liquefaction would not occur.

Diagrams of the type shown in Figure 10.31 could be prepared for various

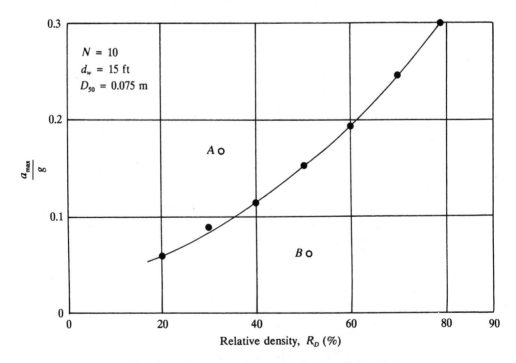

Figure 10.31 Plot of a_{max}/g versus relative density from Table 10.3

Table 10.3   Relation Between
a_{max}/g versus $R_{D(2)}$ Eq. (10.29)]

$R_{D(2)}$ (%)	Ratio[a] C_r	$\dfrac{a_{max}}{g}$
20	0.54	0.0572
30	0.54	0.0856
40	0.54	0.1144
50	0.565	0.1497
60	0.61	0.1938
70	0.66	0.2449
80	0.705	0.2989

[a] From Figure 10.28.

combinations of D_{50}, d_w, and N. Since, in the field, for liquefaction the range of D_{50} is 0.075–0.2 mm and the range of N is about 10–20, one can take the critical combinations (i.e., $D_{50} = 0.075$ mm, $N = 20$; $D_{50} = 0.2$ mm, $N = 10$) and plot

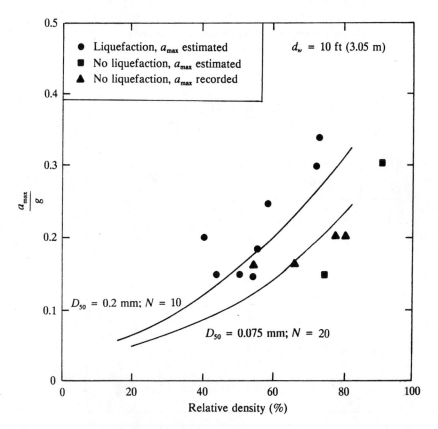

Figure 10.32 Evaluation of liquefaction potential for sand below the ground surface (redrawn after Seed and Idriss, 1971)

graphs as shown in Figure 10.32. These graphs provide a useful guide in the evaluation of liquefaction potential in the field.

10.16   Liquefaction Analysis from Standard Penetration Resistance

Another way of evaluating the soil liquefaction potential is to prepare correlation charts with the standard penetration resistance. After the occurrence of the Niigata earthquake of 1964, Kishida (1966), Kuizumi (1966), and Ohasaki (1966) studied the areas in Niigata where liquefaction had and had not occurred. They developed criteria, based primarily on standard penetration resistance of sand deposits, to differentiate between liquefiable and nonliquefiable conditions. Subsequently, a more detailed collection of field data for liquefaction potential was made by Seed and Peacock (1971). These results and some others were presented by Seed, Mori, and Chan (1977) in a graphical form, which is a plot of τ_h/σ_v

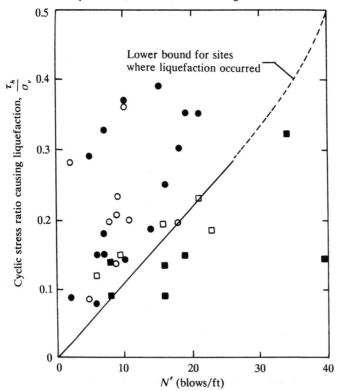

● Liquefaction, stress ratio based on estimated acceleration
○ Liquefaction, stress ratio based on good acceleration data
■ No liquefaction, stress ratio based on estimated acceleration
□ No liquefaction, stress ratio based on good acceleration data

Figure 10.33   Correlation between τ_h/σ_v and N' (after Seed, 1979)

versus N'. This is shown in Figure 10.33. In this figure note that N' is the corrected standard penetration resistance for an effective overburden pressure of 1 ton/ft^2. Figure 10.33 shows the lower bounds of the correlation curve causing liquefaction in the field. However, correlation charts such as this cannot be used with confidence in the field, primarily because they do not take into consideration the magnitude of the earthquake and the duration of shaking.

In order to develop a better correlation chart, Seed (1979) considered the results of the large-scale simple shear test conducted by DeAlba, Chan, and Seed (1976), which were discussed in Section 10.11. These results were corrected to take into account the significant factors that affect the field condition, and they are shown in Table 10.4. It is important to realize that the $(\tau_h/\sigma_v)_{test}$ values listed in Table 10.4 are those required for a peak cyclic pore pressure ratio of 100% and cyclic shear strain of $\pm 5\%$. Also, the correlation between R_D and N' shown in columns 1 and 2 are via the relationship established by Bieganousky and Marcuson (1977).

Excellent agreement is observed when the values of N' and the corresponding $(\tau_h/\sigma_v)_{field}$ values (columns 2 and 6) shown in Table 10.4 are superimposed on the lower-bound correlation curve shown in Figure 10.33. Hence the lower-bound curve of Figure 10.33 is for an earthquake magnitude $M = 7.5$. Proceeding in a similar manner and utilizing the results shown in Table 10.4, lower-bound curves for $M = 6, 7.5$, and 8.25 can be obtained as shown in Figure 10.34. Also shown in this figure is the variation of the limited strain potential in percent (for effective overburden pressure of 1 ton/ft^2). Figure 10.34 can be used for determination of the liquefaction potential in the field. In doing so, it is important to remember that

$$N' = C_N N_F \tag{10.30}$$

where

Table 10.4   Data from Large-scale Simple Shear Tests on Freshly Deposited Sand[a]

Relative density, R_D	N' (blows/ft)	$M = 5-6$ 5 cycles		$M = 7-7.5$ 15 cycles		$M = 8-8.25$ 25 cycles	
		$\left(\dfrac{\tau_h}{\sigma_v}\right)_{test}$	$\left(\dfrac{\tau_h}{\sigma_v}\right)_{field}$	$\left(\dfrac{\tau_h}{\sigma_v}\right)_{test}$	$\left(\dfrac{\tau_h}{\sigma_v}\right)_{field}$	$\left(\dfrac{\tau_h}{\sigma_v}\right)_{test}$	$\left(\dfrac{\tau_h}{\sigma_v}\right)_{field}$
(1)	(2)	(3)	(4)	(5)	(6)	(7)	(8)
54	13.5	0.22	0.25	0.17	0.19	0.155	0.175
68	23	0.30	0.335	0.24	0.27	0.21	0.235
82	33	0.44	0.49	0.32	0.37	0.28	0.315
90	39	0.59	0.66	0.41	0.46	0.36	0.405

Note: N' = standard penetration resistance corrected to an effective overburden pressure of 1 ton/ft^2; M = magnitude of earthquake.

[a] After Seed (1979).

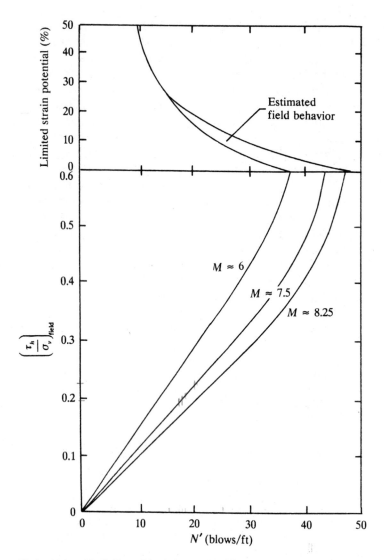

Figure 10.34   Variation of $(\tau_h/\sigma_v)_{\text{field}}$ with N' and M (after Seed, 1979)

N_F = field standard penetration test values

C_N = correction factor to convert to an effective overburden pressure
(σ_v') of 1 ton/ft^2

The correction factor can be expressed as (Liao and Whitman, 1986)

$$C_N = \sqrt{\dfrac{1}{\sigma_v'}}$$

(10.31)

where σ_v' is in tons per square foot or kilograms per square centimeter.

A slight variation of Figure 10.34 is given by Seed, Idriss, and Arango (1983) and Seed and Idriss (1982).

Discussion regarding soil liquefaction has so far been limited to the case of clean sand; however liquefaction can, and has been, observed in silty sands. Information regarding the liquefaction of silty sand is somewhat limited. Seed et al. (1984) presented limited correlations between $(\tau_h/\sigma_v)_{field}$, N', and percent fines (F) for an earthquake magnitude $M = 7.5$, which can be summarized as follows:

Percent of fines, F	N'	Lower bound of $(\tau_h/\sigma_v)_{field}$ for which liquefaction is likely $(M = 7.5)$
≤ 5	5	0.055
	10	0.115
	15	0.17
	20	0.22
	25	0.295
	30	0.5
10	5	0.098
	10	0.16
	15	0.225
	20	0.295
	25	0.5
35	5	0.13
	10	0.185
	15	0.26
	20	0.4

Example 10.1

Following are the field standard penetration numbers in a deposit of sand. Groundwater table is encountered at a depth of 10 ft measured from the ground surface. Given for the sand:

Dry unit weight $= 110 \, \text{lb/ft}^3$

Saturated unit weight $= 122.4 \, \text{lb/ft}^3$

Determine, for an earthquake magnitude of 7.5, if liquefaction will occur at the site. Assume that the maximum intensity of ground acceleration is $a_{max} = 0.15$ g.

Depth (ft)	N_F (blows/ft)
5	6
10	8
15	10
20	14
25	16
30	20
35	20

Solution

Step 1. The following table can now be prepared.

Depth (ft)	Vertical effective stress (tons/ft^2)	C_N [Eq. (10.31)]	N'^a (blows/ft)	$\left(\dfrac{\tau_h}{\sigma_v}\right)^b_{\text{field}}$	τ_h (tons/ft^2)
5	0.275	1.91	11	0.128	0.035
10	0.55	1.35	11	0.128	0.07
15	0.70	1.20	12	0.14	0.098
20	0.85	1.08	15	0.168	0.143
25	1.0	1.00	16	0.184	0.184
30	1.15	0.93	19	0.21	0.242
35	1.30	0.88	18	0.195	0.254

^{a} $N' = C_N N_F$ (rounded off).
^{b} From Figure 10.34.

Step 2. Calculation of τ_{av} using Equation (10.26) can be done by the following table.

Depth (ft)	Total vertical stress (tons/ft^2)	$\dfrac{a_{max}}{g}$	C_D^a	τ_{av}^b (tons/ft^2)
10	0.55	0.15	0.98	0.053
15	0.856	0.15	0.97	0.125
20	1.163	0.15	0.96	0.167
25	1.469	0.15	0.95	0.209
30	1.775	0.15	0.94	0.250
35	2.081	0.15	0.90	0.281

^{a} Figure 10.30(b),
^{b} $\tau_{av} = 0.65 C_D[(\gamma h/g)a_{max}]$

Step 3. Check to see if $\tau_{av} \geq \tau_h$. In that case, liquefaction would occur. From the preceding tables, it can be seen that between depths of 10 ft and 35 ft, τ_{av} is greater than τ_h, so liquefaction occurs. ∎

10.17   Other Correlations for Field Liquefaction Analysis

Correlation with Cone Penetration Resistance

In many cases during field exploration, the variation of the cone penetration resistance is measured with depth. Similar to the standard penetration number N_F, the field cone penetration resistance needs to be corrected to a standard effective overburden pressure. Thus, for clean sand (Ishihara, 1985)

$$q_c' = C_N q_c \tag{10.32}$$

where

q_c = field cone penetration resistance (kg/cm^2)

C_N = correction factor

q_c' = corrected cone penetration number (kg/cm^2)

If the value of q_c' for $\sigma_v' = 1$ kg/cm^2 (or 1 ton/ft^2) is needed, then Eq. (10.31) may be used. It has been noted from several field tests that

$$q_{c(\sigma_v'=1 \text{ kg/cm}^2)}' = AN' \tag{10.33}$$

where $A = 4$ to 5 for clean sands.

Assuming the value of A to be about 4,

$$q_{c(\sigma_v'=1 \text{ kg/cm}^2)}' \approx 4N'$$

Thus,

$$\boxed{N' \approx \frac{q_{c(\sigma_v'=1 \text{ kg/cm}^2)}'}{4}} \tag{10.34}$$

Once the estimated values for N' are known, Figure 10.34 can be used to check the possibility for liquefaction in the field.

Use of Threshold Strain

It was discussed in Section 10.2 that for densification of sand under drained condition, a threshold shear strain level must be exceeded. Similarly, under undrained conditions, a threshold cyclic shear strain level needs to be exceeded to cause buildup of excess pore water pressure and thus possible liquefaction. So if it can be shown that a cyclic shear strain in soil as a result of an earthquake does not exceed a certain threshold level, liquefaction cannot occur. This would provide a conservative evaluation due to the fact that liquefaction may not always occur even if the strains do exceed the threshold level (Committee on Earthquake Engineering, Commission on Engineering and Technical Systems, 1985).

The peak shear strain caused by an earthquake ground motion can be estimated from Eq. (10.25) as

$$\gamma' = \frac{\tau_{\text{max(modif)}}}{G} = \frac{C_D[(\gamma h/g)a_{\text{max}}]}{G} \tag{10.35}$$

where

γ' = peak shear strain

G = shear modulus

or

$$\boxed{\gamma' = \frac{C_D \rho h a_{\text{max}}}{G} = \frac{C_D h a_{\text{max}}}{(G/\rho)(G_{\text{max}}/G_{\text{max}})} = \frac{C_D h a_{\text{max}}}{(G/G_{\text{max}})v_s^2}} \tag{10.36}$$

where

v_s = shear wave velocity in soil

G_{max} = maximum shear modulus (see Chapter 4)

The magnitude of G/G_{max} can be assumed to be about 0.8. Substituting into Eq. (10.36) and combining with an average value of C_D,

$$\gamma' = \frac{1.2a_{max}h}{v_s^2} \tag{10.37}$$

By measuring v_s with depth h, the variation of γ' can be calculated. The typical value of the threshold strain is about 0.01% (Dobry et al., 1981). If the magnitude of the calculated γ' does not exceed this threshold limit, then there is safety against liquefaction.

Correlation with Overlying Liquefaction-Resistant Stratum

The earthquake of magnitude 7.7 that occurred on May 26, 1983, in the northern part of Japan has provided enough data to study the effect of an overlying liquefaction-resistant stratum on the liquefaction potential of sand *with standard penetration resistance $N \leq 10$*. Figure 10.35 defines the terms H_1 and H_2 which

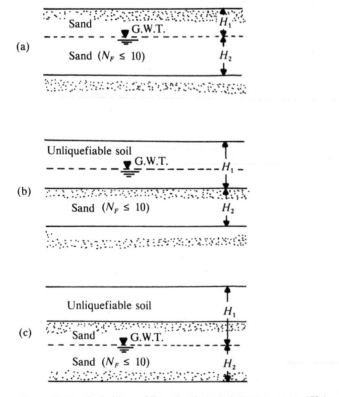

Figure 10.35 Definition of liquefaction-resistant stratum (H_1) and liquefiable stratum (H_2)

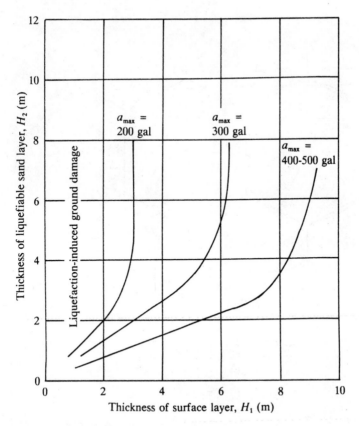

Figure 10.36 Ishihara's proposed boundary curves for site identification of liquefaction-induced damage

are, respectively, the liquefaction resistant stratum and the liquefiable stratum. Based on field observations, Ishihara (1985) developed a correlation chart between H_1, H_2, and maximum acceleration a_{max}. This correlation chart is shown in Figure 10.36.

10.18   Remedial Action to Mitigate Liquefaction

In order to ensure the functionality and safety of engineering projects that are likely to be subjected to damage due to possible liquefaction of the subsoil, several actions can be taken:

1. Removal or replacement of undesirable soil. If liquefaction of a soil layer under a structure is a possibility, then it may be excavated and recompacted with or without additives. Otherwise the potentially liquefiable soil may be replaced with nonliquefiable soil.

2. Densification of the *in situ* material. This can be achieved by using several techniques such as vibroflotation, dynamic compaction, and compaction piles.

3. *In situ* soil improvement by grouting and chemical stabilization.

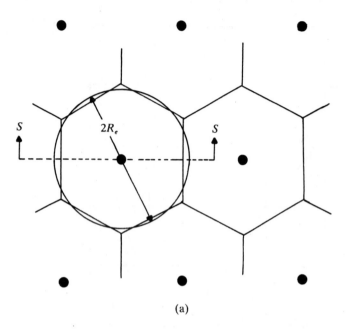

(a)

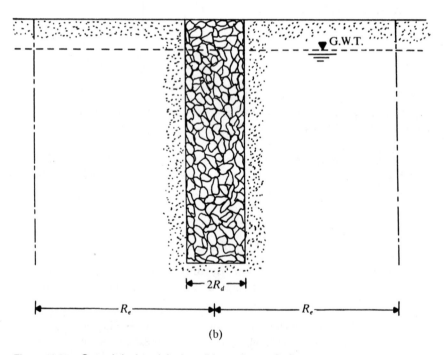

(b)

Figure 10.37 Gravel drains: (a) plan; (b) section at S–S

4. Use of relief wells such as *gravel* or *rock drains* for the control of undesirable pore water pressure. Figure 10.37 is a schematic diagram of gravel or rock drains. The purpose of the installation of gravel or rock drains is to dissipate the excess pore water pressure almost as fast as it is generated in the sand deposit due to cyclic loading. The design principles of gravel and rock drains have been developed by Seed and Booker (1977) and are described here. Assuming that Darcy's law is valid, the continuity of flow equation in the sand layer may be written as

$$\frac{\partial}{\partial x}\left(\frac{k_h}{\gamma_w}\frac{\partial u}{\partial x}\right) + \frac{\partial}{\partial y}\left(\frac{k_h}{\gamma_w}\frac{\partial u}{\partial y}\right) + \frac{\partial}{\partial z}\left(\frac{k_v}{\gamma_w}\frac{\partial u}{\partial z}\right) = \frac{\partial \varepsilon}{\partial t} \tag{10.38}$$

where

k_h = coefficient of permeability of the sand in the horizontal direction

k_v = coefficient of permeability of the sand in the vertical direction

u = excess pore water pressure

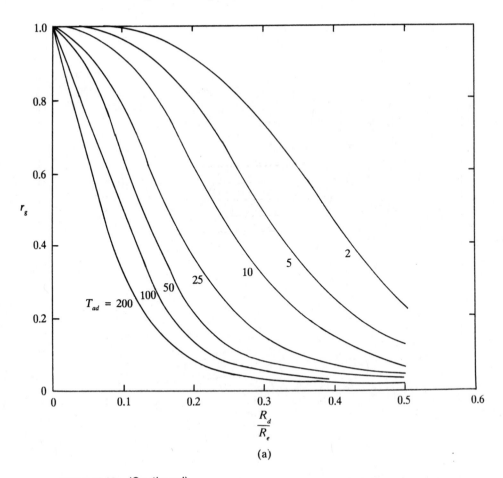

(a)

Figure 10.38   (Continued)

γ_w = unit weight of water

ε = volumetric strain (compression positive)

During a time interval dt, the pore water pressure in a soil element changes by du. However, if a cyclic shear stress is applied on a soil element, there is an increase of pore water pressure. In a time dt, there are dN number of cyclic shear stresses; the corresponding increase of pore water pressure is $(\partial u_g / \partial N)\,dN$ (where u_g is the excess pore water pressure generated by cyclic shear stress—see also Section 10.10). Thus, the net change in pore water pressure in time dt is equal to $[du - (\partial u_g / \partial N)\,dN)$, and

$$\partial \varepsilon = m_{v_3}[\partial u - (\partial u_g / \partial N)\,dN]$$

or

$$\frac{\partial \varepsilon}{\partial t} = m_{v_3}\left(\frac{\partial u}{\partial t} - \frac{\partial u_g}{\partial N}\frac{\partial N}{\partial t}\right) \tag{10.39}$$

where m_{v_3} = coefficient of volume compressibility.

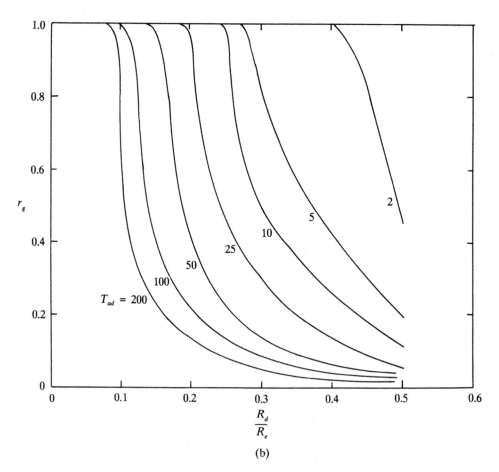

Figure 10.38 (Continued)

Combining Eqs. (10.38) and (10.39),

$$\frac{\partial}{\partial x}\left(\frac{k_h}{\gamma_w}\frac{\partial u}{\partial x}\right) + \frac{\partial}{\partial y}\left(\frac{k_h}{\gamma_w}\frac{\partial u}{\partial y}\right) + \frac{\partial}{\partial z}\left(\frac{k_h}{\gamma_w}\frac{\partial u}{\partial z}\right) = m_{v_3}\left(\frac{\partial u}{\partial t} - \frac{\partial u_g}{\partial N}\frac{\partial N}{\partial t}\right) \qquad (10.40)$$

If m_{v_3} is a constant and *radial symmetry* exists, then Eq. (10.40) can be written in cylindrical coordinates as

$$\frac{k_h}{\gamma_w m_{v_3}}\left(\frac{\partial^2 u}{\partial^2 r} + \frac{1}{r}\frac{\partial u}{\partial r}\right) + \frac{k_v}{\gamma_w m_{v_3}}\frac{\partial^2 u}{\partial z^2} = \frac{\partial u}{\partial t} - \frac{\partial u_g}{\partial N}\frac{\partial N}{\partial t} \qquad (10.41)$$

For the condition of purely radial flow, Eq. (10.41) takes the form

$$\frac{k_h}{\gamma_w m_{v_3}}\left(\frac{\partial^2 u}{\partial^2 r} + \frac{1}{r}\frac{\partial u}{\partial r}\right) = \frac{\partial u}{\partial t} - \frac{\partial u_g}{\partial N}\frac{\partial N}{\partial t} \qquad (10.42)$$

In order to solve Eq. (10.42), it is necessary to evaluate the terms k_h, m_{v_3}, $\partial N/\partial t$, and $\partial u_g/\partial N$. The value of k_h can be easily determined from field pumping tests. The coefficient of volume compressibility can be determined from cyclic

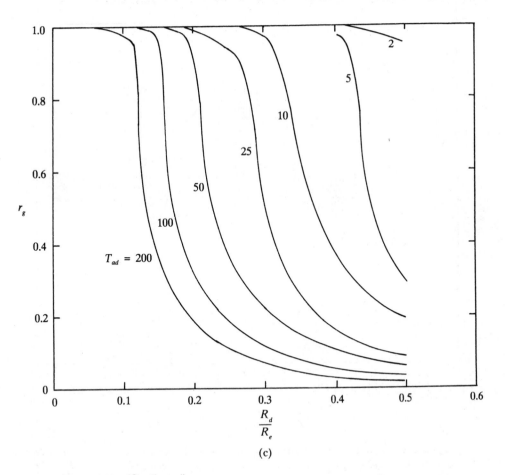

(c)

Figure 10.38   (Continued)

triaxial tests (Lee and Albaisa, 1974). The term $\partial N/\partial t$ can be expressed as

$$\frac{\partial N}{\partial t} = \frac{N_s}{t_d} \tag{10.43}$$

where N_s = significant number of uniform stress cycles due to an earthquake and t_d = duration of an earthquake.

The rate of excess pore water pressure buildup, $\partial u_g/\partial N$, in a saturated undrained cyclic simple shear test is given by Eq. (10.4) (Section 10.10). For radial flow conditions, the relation given by Eq. (10.42) has been solved by Seed and Booker (1977). It has been shown that the ratio u/σ_v is a function of the following parameters:

$$\frac{R_d}{R_e} = \frac{\text{radius of rock or gravel drains}}{\text{effective radius of the rock or gravel drains}} \tag{10.44}$$

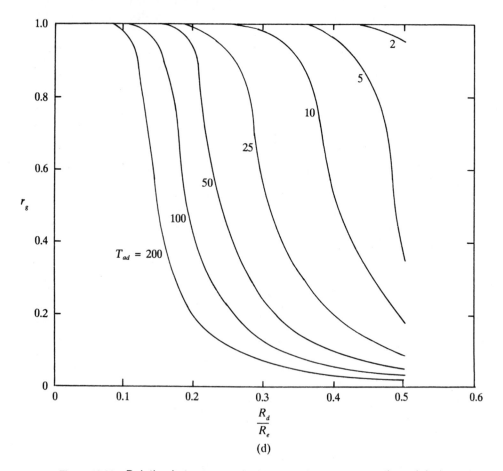

(d)

Figure 10.38 Relation between greatest pore water pressure ratio and drain system parameters: (a) $N_S/N_i = 1$; (b) $N_S/N_i = 2$; (c) $N_S/N_i = 3$; (d) $N_S/N_i = 4$ (after Seed and Booker, 1977)

N_s/N_i, and

$$T_{ad} = \frac{k_h}{\gamma_w}\left(\frac{t_d}{m_{v_3}R_d^2}\right) \tag{10.45}$$

Using these parameters, the solution to Eq. (10.42) is given in a nondimensional form in Figure 10.38 for design of rock or gravel drains. In Figure 10.38, the term r_g is defined as

$$r_g = \frac{\text{greatest limiting value of } u_g \text{ chosen for design}}{\sigma_v} \tag{10.46}$$

In obtaining the solutions given in Figure 10.38, it was assumed that the coefficient of permeability of the material used in the gravel or rock drains is infinity. However, in practical cases, it would be sufficient to have a value of

$$\frac{k_{h(\text{rock or gravel})}}{k_{(\text{sand})}} \approx 200$$

Example 10.2

For a sand deposit, it is given that

$$m_{v_3} = 2.8 \times 10^{-5} \text{ m}^2/\text{kN}$$

$$k_h = 0.02 \text{ mm/s} = 2 \times 10^{-5} \text{ m/s}$$

For a design earthquake, the equivalent number of uniform stress cycles (for uniform stress $= \tau_{av}$) was determined to be 30. The duration of the earthquake is about 65 s.

From laboratory tests, it was determined that 12 cycles of cyclic stress application (the peak magnitude of the cyclic stress is equal to τ_{av}) would be enough to cause initial liquefaction in the sand.

Assuming that the radius of the gravel drains to be used is 0.25 m, and $r_g = 0.6$, determine the spacing of gravel drains.

Solution

From Eq. (10.45),

$$T_{ad} = \frac{k_h}{\gamma_w}\left(\frac{t_d}{m_{v_3}R_d^2}\right) = \frac{2 \times 10^{-5}}{9.81}\left[\frac{65}{2.8 \times 10^{-5}(0.25)^2}\right] = 75.72$$

$$\frac{N_s}{N_i} = 30/12 = 2.5, \qquad r_g = 0.6$$

Referring to Figure 10.38b, for $T_{ad} = 75.72$, $N_s/N_i = 2$, $r_g = 0.6$,

$$\frac{R_d}{R_e} \approx 0.17$$

From Figure 10.38c, for $T_{ad} = 75.72$, $N_S/N_1 = 3$, $r_g = 0.6$,

$$\frac{R_d}{R_e} \approx 0.2$$

Thus, for $N_S/N_i = 2.5$, $R_d/R_e \approx \frac{1}{2}(0.17 + 0.2) = 0.185$. Hence,

$$R_e = \frac{R_d}{0.185} = \frac{0.25}{0.185} = \underline{1.35\ m}$$ ∎

PROBLEMS

10.1 Explain the terms *initial liquefaction* and *cyclic mobility*.

10.2 For a sand deposit the following is given:

$$\text{Mean grain size } (D_{50}) = 0.2 \text{ mm}$$

$$\text{Depth of water table} = 10 \text{ ft}$$

$$\text{Unit weight of soil above G.W.T.} = 105 \text{ lb/ft}^3$$

$$\text{Unit weight of soil below G.W.T.} = 120 \text{ lb/ft}^3$$

$$\text{Expected earthquake magnitude} = 7.5$$

Make all calculations and prepare a graph showing the variation of a_{max}/g and the relative density in the field for liquefaction to occur.

10.3 Repeat Problem 10.2 for a mean grain size of 0.075 mm.

10.4 Repeat Problem 10.2 for the following conditions:

$$\text{Mean grain size } (D_{50}) = 0.075 \text{ mm}$$

$$\text{Depth of water table} = 4.6 \text{ m}$$

$$\text{Unit weight of soil above G.W.T.} = 15 \text{ kN/m}^3$$

$$\text{Unit weight of soil below G.W.T.} = 18 \text{ kN/m}^3$$

$$\text{Expected earthquake magnitude} = 8$$

10.5 Repeat Problem 10.4 for a mean grain size of 0.2 mm.

10.6 Consider the soil and the groundwater table conditions given in Problem 10.2. Assume that the relative density in the field is 60%. The maximum expected intensity of ground shaking (a_{max}/g) is 0.2 and the magnitude of earthquake is 7.5.
   a. Calculate and plot the variation of the shear stress τ_{av} induced in the sand deposit with depth 0–70 ft. Use Eq. (10.26).
   b. Calculate the variation of the shear stress required to cause liquefaction with depth. Plot the shear stress determined in the same graph as used in (a). Use Eq. (10.23).
   c. From the plotted graph, determine the depths at which liquefaction is initiated.

10.7 Repeat Problem 10.6(a)–(c) for the data given in Problem 10.4. Assume the relative density of the sand to be 60% and the maximum expected intensity of ground shaking to be 0.15g.

10.8 The standard penetration test results of a sand deposit at a certain site are given here.

Depth (ft)	N_F (blows/ft)
5	8
10	7
15	12
20	15
25	17
30	17

The groundwater table is located at a depth of 7 ft below the ground surface. The dry and saturated unit weights of sand are 105 lb/ft^3 and 118 lb/ft^3, respectively. For an expected earthquake magnitude $M = 6$ and maximum acceleration $a_{max} = 0.1g$, will liquefaction occur?

10.9 In a sand deposit, the groundwater table is located at a depth of 2 m measured from the ground surface. Following are the shear wave velocities in the sand deposit.

Depth (m)	Shear wave velocity, v_s (m/s)
2–4	450
4–6	600
6–10	675

For a maximum ground acceleration $a_{max} = 0.16g$, determine whether liquefaction is likely to occur.

10.10 Solve the gravel drain problem given Example 10.2 for $r_g = 0.7$.

10.11 Repeat Example 10.2 for the gravel drain with the following data:

$$m_{v_3} = 3.5 \times 10^{-5} \text{ m}^2/\text{KN}$$

$$k_h = 1.4 \times 10^{-5} \text{ m/s}$$

$$\frac{\text{Equivalent number of uniform}}{\text{stress cycles due to earthquake}} = 20$$

$$\text{Duration of earthquake} = 50 \text{ s}$$

$$\frac{\text{Number of uniform stress}}{\text{cycles for liquefaction}} = 12$$

$$\text{Radius of gravel drains} = 0.3 \text{ m}$$

$$r_g = 0.7$$

REFERENCES

Bieganousky, W. A., and Marcuson, W. F., III (1977). "Liquefaction Potential of Dams and Foundations, Report 2. Laboratory Standard Penetration Test on Platte River Sand and Standard Concrete Sand," *WES Report No. 76-2*, U.S. Army Waterways Experiment Station, Vicksburg, Mississippi.

Casagrande, A. (1936). Characteristics of Cohesionless Soils Affecting the Stability of Slopes and Earthfills," *Journal of the Boston Society of Civil Engineers*, January, Vol. 23, pp. 257–276.

Castro, G. (1969). "Liquefaction of Sands," *Harvard Soil Mechanics Series No. 81*, Cambridge, Massachusetts.

Castro, G. (1975). "Liquefaction and Cyclic Mobility of Saturated Sands," *Journal of the Geotechnical Engineering Division*, ASCE, Vol. 101, No. GT6, pp. 551–569.

Commission on Engineering and Technical Systems—Committee on Earthquake Engineering (1985). "Liquefaction of Soils During Earthquakes," National Academy Press, Washington, D.C.

DeAlba, P., Chan, C. K., and Seed, H. B. (1975). "Determination of Soil Liquefaction Characteristics by Large Scale Laboratory Tests," *Report No. EERC 75-14*, Earthquake Engineering Research Center, University of California, Berkeley, California.

DeAlba, P., Seed, H. B., and Chan, C. K. (1976). "Sand Liquefaction in Large-Scale Simple Shear Tests," *Journal of the Geotechnical Engineering Division*, ASCE, Vol. 102, No. GT9, pp. 909–927.

Dobry, R., Stoke, K. H., Land, R. S., and Youd, T. L. (1981). "Liquefaction for *S*-wave Velocity," *Preprint 81-544*, ASCE National Convention, St. Louis, Missouri.

Emery, J. J., Finn, W. D. L., and Lee, K. W. (1972). "Uniformity of Saturated Sand Samples," *Soil Mechanics Series*, University of British Columbia, Vancouver, British Columbia, Canada.

Finn, W. D. L. (1972). "Soil Dynamics—Liquefaction of Sands," *Proceedings*, International Conference on Microzonation for Safer Construction Research and Application, Seattle, Washington, Vol. 1.

Finn, W. D. L., Bransby, P. L., and Pickering D. J. (1970). "Effect of Strain History on Liquefaction of Sands," *Journal of the Soil Mechanics and Foundations Division*, ASCE, Vol. 96, No. SM6, pp. 1917–1934.

Finn, W. D. L., Emery, J. J., and Gupta, Y. P. (1970). "A Shaking Table Study of the Liquefaction of Saturated Sands during Earthquake Engineering," *Proceedings*, 3rd European Symposium on Earthquake Engineering, Sofia, Bulgaria.

Finn, W. D. L., Emery, J. J., and Gupta, Y. P. (1971). "Soil Liquefaction Studies Using a Shaking Table," *Closed Loop Magazine*, Fall/Winter, MTS Systems Corporation, Minneapolis, Minnesota.

Finn, W. D. L., Pickering, D. J., and Bransby, P. L. (1971). "Sand Liquefaction in Triaxial and Simple Shear Tests," *Journal of the Soil Mechanics and Foundations Division*, ASCE, Vol. 97, No. SM4, pp. 639–659.

Ishibashi, I., and Sherif, M. A. (1974). "Soil Liquefaction by Torsional Simple Shear Device," *Journal of the Geotechnical Engineering Division*, ASCE, Vol. 100, No. GT8, pp. 871–888.

Ishihara, K. (1985). "Stability of Natural Deposits During Earthquakes," *Proceedings*, 11th International Conference on Soil Mechanics and Foundation Engineering, Vol. 1, pp. 321–376.

Jaky, J. (1944). "The Coefficient of Earth Pressure at Rest," *Journal of the Society of the Hungarian Architectural Engineers*, Vol. 21, pp. 355–358.

Kishida, H. (1966). "Damage to Reinforced Concrete Buildings in Niigata City with Special Reference to Foundation Engineering," *Soils and Foundations*, Tokyo, Japan, Vol. 7, No. 1, pp. 75–92.

Kuizumi, Y. (1966). "Changes in Density of Sand Subsoil Caused by the Niigata Earthquake," *Soils and Foundations*, Tokyo, Japan, Vol. 8, No. 2, pp. 38–44.

Lee, K. L., and Albaisa, A. (1974). "Earthquake Induced Settlements in Saturated Sands," *Journal of the Geotechnical Engineering Division*, ASCE, Vol. 100, No. GT4, pp. 387–404.

Lee, K. L., and Seed, H. B. (1967). "Cyclic Stress Conditions Causing Liquefaction of Sand," *Journal of the Soil Mechanics and Foundations Division*, ASCE, Vol. 93, No. SM1, pp. 47–70.

Liao, S. and Whitman, R. V. (1986). "Overburden Correction Factors for SPT in Sand," *Journal of Geotechnical Engineering*, ASCE, Vol. 112, No. GT3, pp. 373–377.

O-Hara, S. (1972). "The Results of Experiment on the Liquefaction of Saturated Sands with a Shaking Box: Comparison with Other Methods," *Technology Reports of the Yamaguchi University*, Vol. 1, No. 1, Yamaguchi, Japan.

Ohasaki, Y. (1966). "Niigata Earthquake 1964, Building Damage and Soil Conditions," *Soils and Foundations*, Tokyo, Japan, Vol. 6, No. 2, pp. 14–37.

Ortigosa, P. (1972). "Licuacion de Arenas Sometidas a Vibraciones Horizontales," *Revista del Instituto de Investigaciones de Ensoyes de Materials*, Vol. II, No. 3.

Peacock, W. H., and Seed, H. B. (1968). "Sand Liquefaction Under Cyclic Loading Simple Shear Conditions," *Journal of the Soil Mechanics and Foundations Division*, ASCE, Vol. 94, No. SM3, pp. 689–708.

Prakash, S., and Mathur, J. N. (1965). "Liquefaction of Fine Sand Under Dynamic Loading," *Proceedings*, 5th Symposium of the Civil and Hydraulic Engineering Departments, Indian Institute of Science, Bangalore, India.

Seed, H. B. (1979). "Soil Liquefaction and Cyclic Mobility Evaluation for Level Ground During Earthquakes," *Journal of the Geotechnical Engineering Division*, ASCE, Vol. 105, No. GT2, pp. 201–255.

Seed, H. B., and Booker, J. R. (1977). "Stabilization of Potential Liquefiable Sand Deposits Using Gravel Drains," *Journal of the Geotechnical Engineering Division*, ASCE, Vo. 103, No. GT7, pp. 757–768.

Seed, H. B., and Idriss, I. M. (1971). "Simplified Procedure for Evaluating Soil Liquefaction Potential," *Journal of the Soil Mechanics and Foundations Division*, ASCE, Vol. 97, No. SM9, pp. 1249–1273.

Seed, H. B., and Idriss, I. M. (1982). "Ground Motion and Soil Liquefaction During Earthquakes," *Monograph Series*, Earthquake Engineering Research Institute, University of California, Berkeley, California.

Seed, H. B., Idriss, I. M., and Arango, I. (1983). "Evaluation of Liquefaction Potential Using Field Performance Data," *Journal of Geotechnical Engineering*, ASCE, Vol. 109, No. GT3, pp. 458–482.

Seed, H. B., and Lee, K. L. (1966). "Liquefaction of Saturated Sands During Cyclic Loading," *Journal of the Soil Mechanics and Foundations Division*, ASCE, Vol. 92, No. SM6, pp. 105–134.

Seed, H. B., Martin, P. O., and Lysmer, J. (1975). "The Generation and Dissipation of Pore Water Pressure During Soil Liquefaction," *Report No. EERC 75-26*, Earthquake Engineering Research Institute, University of California, Berkeley, California.

Seed, H. B., Mori, K., and Chan, C. K. (1977). "Influence of Seismic History on Liquefaction of Sands," *Journal of the Geotechnical Engineering Division*, ASCE, Vol. 103, No. GT4, pp. 246–270.

Seed, H. B., and Peacock, W. H. (1971). "The Procedure for Measuring Soil Liquefaction Characteristics," *Journal of the Soil Mechanics and Foundations Division*, ASCE, Vol. 97, No. SM8, pp. 1099–1119.

Seed, H. B., Tokimatsu, K., Harder, L. F., and Chung, R. M. (1984). "The Influence of SPT Procedures in Soil Liquefaction Resistance Evaluations," *Report No. EERC-84/15*, Earthquake Engineering Research Institute, University of California, Berkeley, California.

Tanimoto, K. (1967). "Liquefaction of a Sand Layer Subjected to Shock and Vibratory Loads," *Proceedings*, 3rd Asian Regional Conference on Soil Mechanics and Foundation Engineering, Haifa, Israel, Vol. 1.

Whitman, R. V. (1970). "Summary of Results from Shaking Table Tests at University of Chile Using a Medium Sand," *Progress Report No. 9*, Effect of Local Soil Conditions upon Earthquake Damage, *Research Report R70-25, Soils Publication No. 258*, Massachusetts Institute of Technology, Cambridge, Massachusetts.

Yoshimi, Y. (1967). "An Experimental Study of Liquefaction of Saturated Sands," *Soils and Foundations*, Tokyo, Japan, Vol. 7, No. 2.

Yoshimi, Y. (1970). "Liquefaction of Saturated Sand During Vibration Under Quasi-Plane-Strain Conditions," *Proceedings*, 3rd Japan Earthquake Engineering Symposium, Tokyo, Japan.

Yoshimi, Y., and Oh-oka, H. (1973). "A Ring Torsion Apparatus for Simple Shear Tests," *Proceedings*, 8th International Conference on Soil Mechanics and Foundation Engineering, Vol. 1.2, Moscow, USSR.

Youd, T. L. (1972). "Compaction of Sand by Repeated Straining," *Journal of the Soil Mechanics and Foundations Division*, ASCE, Vol. 98, No. SM7, pp. 709–725.

$$\tan \theta = \frac{opp}{adj}$$

$$\frac{T}{R} = \tan \frac{\Delta}{2}$$

$$\sec = \frac{hyp}{adj}$$

$$opp$$

$$\sin \frac{\Delta}{2} = \frac{T}{Q}$$

$$Q = \frac{T}{\sin \frac{\Delta}{2}} = R$$

$$\frac{R}{Q}$$

$$T \csc - R$$

$$\cos \frac{\Delta}{2} = R \sec$$

$$\cos \frac{\Delta}{2} = \frac{adj}{hyp}$$

$$\cos \frac{\Delta}{2} = \frac{x}{R}$$

$$R \cos \frac{\Delta}{2}$$

Chapter 11

MACHINE FOUNDATIONS ON PILES

11.1   Introduction

It was mentioned in Section 5.4 that for low-speed machineries subjected to vertical vibration, the natural frequency of the foundation-soil system should be at least twice the operating frequency. In the design of these types of foundations, if changes in size and mass of the foundation do not lead to a satisfactory design, a pile foundation may be considered. It is also possible that the subsoil conditions are such that the vibration of a shallow machine foundation may lead to undesirable settlement. In many circumstances the load-bearing capacity of the soil may be low compared to the static and dynamic load imposed by the machine and the shallow foundation. In that case the design will then dictate consideration of the use of piles. In this chapter, the fundamental concepts of pile foundations of vibrating machines will be considered. It should also be kept in mind that the piles supporting machine foundations are for cases of *low amplitudes of vibration* in contrast to those encountered under earthquake-type loading. For that reason, when encountered with the selection of proper parameters for soil such as the shear modulus G, the value(s) that correspond to low amplitudes of strain should be used.

PILES SUBJECTED TO VERTICAL VIBRATION

In general, piles can be grouped into two broad categories:

1. *End-bearing piles.* These piles penetrate through soft soil layers up to a hard stratum or rock. The hard stratum or rock can be considered as rigid.

2. *Friction piles, or floating piles.* The tips of these piles do not rest on hard stratum. The piles resist the applied load by means of frictional resistance developed at the soil-pile interface.

11.2   End-Bearing Piles

Figure 11.1 shows a pile driven up to a rock layer. The length of the pile is equal to L, and the load on the pile coming from the foundation is W. This problem can be approximately treated as a vertical rod *fixed at the base* (that is, at the rock layer) and free on top. For determining the natural frequency of the piles, three possible cases may arise.

Case 1. If W is very small (≈ 0), the natural frequency of vibration can be given by following Eq. (3.57) as

$$\boxed{f_n = \frac{\omega_n}{2\pi} = \frac{1}{4L}\sqrt{\frac{E_P}{\rho_P}}} \tag{11.1}$$

where

f_n = natural frequency of vibration

ω_n = natural circular frequency

E_P = modulus of elasticity of the pile material

ρ_P = density of the pile material

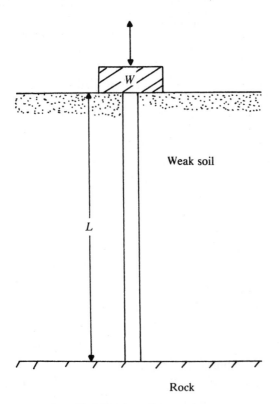

Figure 11.1 End-bearing pile

Case 2. If W is of the same order of magnitude as the weight of the pile, the natural frequency of vibration can be given by Eq. (4.20). (Note similar end conditions between Figure 4.13 and Figure 11.1.) Thus,

$$\frac{AL\gamma_P}{W} = \left[\frac{\omega_n L}{v_{c(P)}}\right] \tan\left[\frac{\omega_n L}{v_{c(P)}}\right] \tag{11.2}$$

or

$$\boxed{\frac{L\gamma_P}{\sigma_0} = \left[\frac{\omega_n L}{v_{c(P)}}\right] \tan\left[\frac{\omega_n L}{v_{c(P)}}\right]} \tag{11.3}$$

where

$\quad A$ = area of the cross section of the pile

$\quad \gamma_P$ = unit weight of the pile material

$\quad \omega_n$ = natural circular frequency

$\quad v_{c(P)}$ = longitudinal wave propagation velocity in the pile

$$\sigma_0 = \frac{W}{A}$$

Figure 11.2 shows a plot of $\omega_n L / v_{c(P)}$ against $L\gamma_P/\sigma_0$ that can be used to determine ω_n and f_n. Note that

$$f_n = \frac{\omega_n}{2\pi} \tag{11.4}$$

Case 3. If W is larger and the weight of the pile is negligible in comparison, then from Equation (11.2)

$$\frac{AL\gamma_P}{W} \approx \left[\frac{\omega_n L}{v_{c(P)}}\right]^2$$

However,

$$v_{c(P)} = \sqrt{\frac{E_P}{\rho_P}} = \sqrt{\frac{E_P g}{\gamma_P}}$$

where g = acceleration due to gravity. So,

$$\omega_n = \sqrt{\frac{AE_P g}{LW}}$$

or

$$\boxed{f_n = \frac{1}{2\pi}\sqrt{\frac{E_P g}{\sigma_0 L}}} \tag{11.5}$$

where σ_0 = axial stress = W/A.

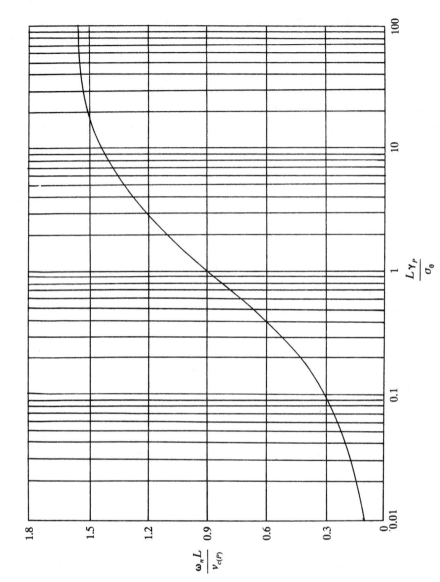

Figure 11.2   Plot of Eq. (11.3)

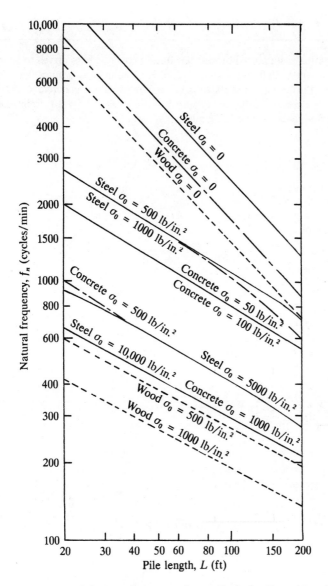

Figure 11.3   Resonant frequency for vertical vibration of a point bearing pile (after Richart, 1962)

Richart (1962) prepared a graph for f_n with various values of pile length L and σ_0, and this is shown is Figure 11.3. In preparing Figure 11.3, the following material properties have been used.

Material	E_P (lb/in.^{2})	γ_P (lb/ft^3)
Steel	29.4×10^6	480
Concrete	3.0×10^6	150
Wood	1.2×10^6	40

Example 11.1

A machine foundation is supported by four prestressed concrete piles driven to bedrock. The length of each pile is 80 ft, and they are 12 in. × 12 in. in cross section. The weight of the machine and the foundation is 300 kips. Given: unit weight of concrete = 150 lb/ft^3 and the modulus of elasticity of the concrete used for the piles = 3.5 × 10^6 lb/in^2. Determine the natural frequency of the pile-foundation system.

Solution

There are four piles. The weight carried by each pile is

$$W = \frac{300 \times 10^3 \text{ lb}}{4} = 75 \times 10^3 \text{ lb}$$

The weight of each pile is

$$AL\gamma_P = \underbrace{\left(\frac{12 \times 12}{144}\right)}_{\substack{\uparrow \\ \text{Area of} \\ \text{cross section}}} \times \underbrace{(80)}_{\substack{\uparrow \\ \text{Length}}} \times (150) = 12 \times 10^3 \text{ lb}$$

$$v_{c(P)} = \sqrt{\frac{E_P}{\rho_P}} = \sqrt{\frac{(3.5 \times 10^6)(144)}{(150/32.2)}} = 10,401.5 \text{ ft/s}$$

From Eq. (11.2),

$$\frac{AL\gamma_P}{W} = \left[\frac{\omega_n L}{v_{c(P)}}\right] \tan\left[\frac{\omega_n L}{v_{c(P)}}\right]$$

So

$$\frac{75 \times 10^3}{12 \times 10^3} = \left[\frac{(\omega_n)(80)}{(10,401.5)}\right] \tan\left[\frac{(\omega_n)(80)}{(10,401.5)}\right]$$

or

$$6.25 = [(\omega_n)(0.00769)] \tan[(\omega_n)(0.00769)]$$

$$0.00769\omega_n \approx 1.36$$

$$\omega_n = 176.85 \text{ rad/s}$$

So

$$\underline{f_n = 1688.8 \text{ cycles/min}}$$

11.3   Friction Piles

Figure 11.4a shows a pile having a length of embedment equal to L and a radius of R. The pile is subjected to a dynamic load

$$Q = Q_0^{i\omega t} \tag{11.6}$$

It is possible to idealize the pile to a mass-spring-dashpot system, as shown in Figure 11.4b. The mass m shown in Figure 11.4b can be assumed to be the mass

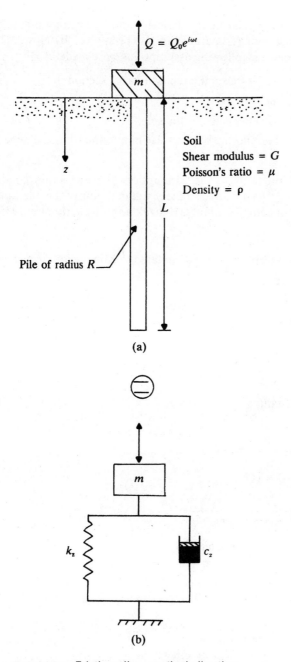

(a)

(b)

Figure 11.4   Friction pile—vertical vibration

of the cap and machinery. The mathematical formulation for obtaining the stiffness (k_z) and the damping (c_z) parameters has been given by Novak (1977). In developing the theory, the following assumptions were made:

1. The pile is vertical, elastic, and circular in cross section.

2. The pile is floating.

3. The pile is perfectly connected to the soil.

4. The soil above the pile tip behaves as infinitesimal, thin, independent linearly elastic layers.

The last assumption leads to the assumption of plane strain condition. Referring to Figure 11.5, the dynamic stiffness and damping of the pile can then be described in terms of complex stiffness (Novak and El-Sharnouby, 1983) as

$$K = K_1 + iK_2 \tag{11.7}$$

The applied force Q and displacement z are related to K in the following manner:

$$Q = Kz = (K_1 + iK_2)z \tag{11.8}$$

where

$$i = \sqrt{-1}$$

K_1 = real part of K = Re K

K_2 = imaginary part of K = Im K

Hence, the spring constant is

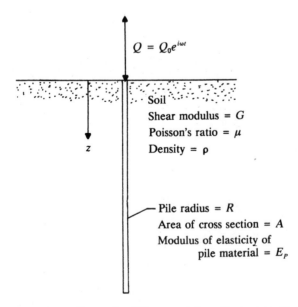

Figure 11.5 Dynamic stiffness and damping constant for a single pile—vertical mode of vibration

$$k_z = K_1 = \text{Re}\, K \tag{11.9}$$

and the equivalent viscous damping is

$$c_z = \frac{K_2}{\omega} = \frac{\text{Im}\, K}{\omega} \tag{11.10}$$

Combining Equations (11.7), (11.9), and (11.10),

$$K = k_z + i\omega c_z \tag{11.11}$$

So, the force-displacement relation can be expressed as

$$Q = (k_z + i\omega c_z)z$$

or

$$Q = k_z z + c_z \dot{z} \tag{11.12}$$

where $\dot{z} = dz/dt$.

The relationships for k_z and c_z have been given by Novak and El-Sharnouby (1983) as

$$\boxed{k_z = \left(\frac{E_P A}{R}\right)f_{z1}} \tag{11.13}$$

and

$$\boxed{c_z = \left(\frac{E_P A}{\sqrt{G/\rho}}\right)f_{z2}} \tag{11.14}$$

where

$$E_P = \text{modulus of elasticity of the pile material}$$

$$A = \text{area of pile cross section}$$

$$G = \text{shear modulus of soil}$$

$$\rho = \text{density of soil}$$

$$f_{z1}, f_{z2} = \text{nondimensional parameters}$$

The variations of f_{z1} and f_{z2} for end-bearing piles are shown in Figures 11.6 and 11.7. Similarly, the f_{z1} and f_{z2} variations for floating piles are shown in Figures 11.8 and 11.9.

Pile foundations are generally constructed in groups. The stiffness and damping constants of a pile group are not simple summations of the stiffness and damping constants of individual piles. Novak (1977) suggested that when piles are closely spaced, the displacement of one pile is increased due to the displacement of all other piles and, conversely, the stiffness and damping of the group are reduced. Hence, the stiffness of the pile group can be obtained as

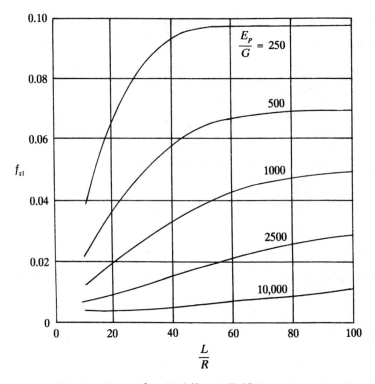

Figure 11.6 Variation of f_{z1} with L/R and E_P/G for end-bearing piles (after Novak and El-Sharnouby, 1983)

$$k_{z(g)} = \frac{\sum\limits_{1}^{n} k_z}{\sum\limits_{r=1}^{n} \alpha_r} \qquad\qquad (11.15)$$

and

$$c_{z(g)} = \frac{\sum\limits_{1}^{n} c_z}{\sum\limits_{r=1}^{n} \alpha_r} \qquad\qquad (11.16)$$

where

$k_{z(g)}$ = spring constant for the pile group

$c_{z(g)}$ = dashpot constant for the pile group

n = number of piles in the group

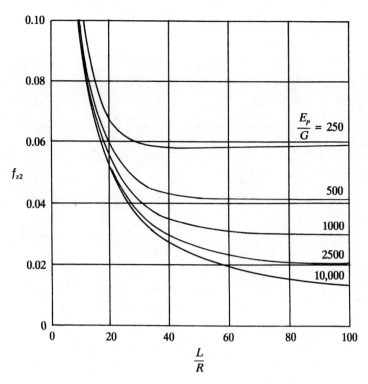

Figure 11.7 Variation of f_{z2} with L/R and E_P/G for end-bearing piles (after Novak and El-Sharnouby, 1983)

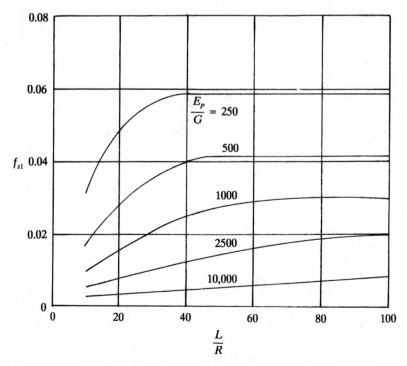

Figure 11.8 Variation of f_{z1} with L/R and E_P/G for floating piles (after Novak and El-Sharnouby, 1983)

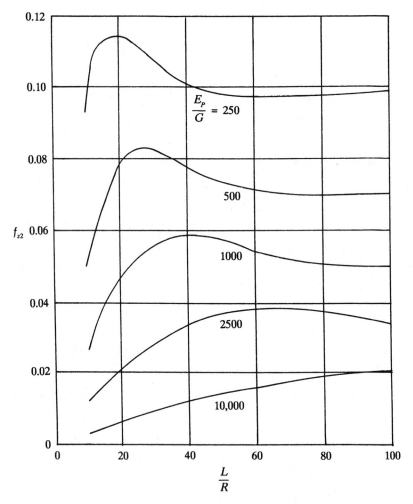

Figure 11.9 Variation of f_{z2} with L/R and E_P/G for floating piles (after Novak and El-Sharnouby, 1983)

$\alpha_r =$ the interaction factor describing the contribution of the rth pile to the displacement of the reference pile (that is, $\alpha_1 = 1$)

Since no analytical solutions for the dynamic interaction of piles are available at the present time, an estimate of α_r can be obtained from the static solution of Poulos (1968). This is shown in Figure 11.10.

For group piles, a cap is constructed over the piles (Figure 11.11). In the estimation of the stiffness and damping constants, the contribution of the pile cap should be taken into account. The relationships describing the stiffness and geometric damping of embedded foundations are given in Chapter 5 as

$$k_{z(\text{cap})} = Gr_0\left[\bar{C}_1 + \frac{G_s}{G}\frac{D_f}{r_0}\bar{S}_1\right]$$
(5.120)

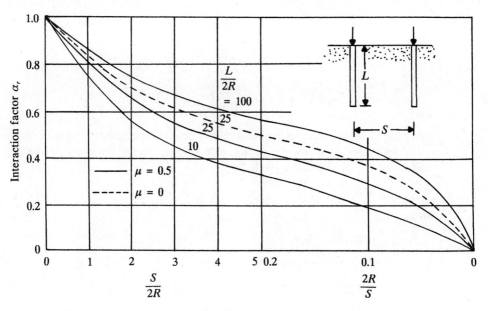

Figure 11.10 Variation of interaction factor α_r (after Poulos, 1968)

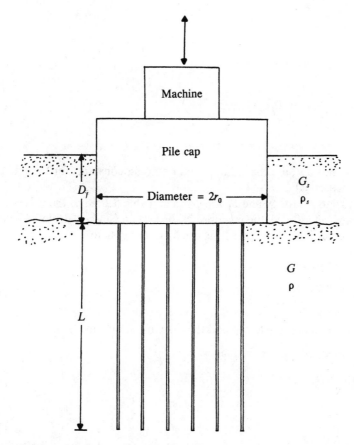

Figure 11.11 Group pile with pile cap

and

$$c_{z(cap)} = r_0^2 \sqrt{\rho G} \left[\bar{C}_2 + \bar{S}_2 \frac{D_f}{r_0} \sqrt{\frac{G_s \rho_s}{G \rho}} \right] \tag{5.121}$$

Since the soil located below the pile cap may be of poor quality and it may shrink away with time, it would be on the safe side to ignore the effect of the cap base—that is, $\bar{C}_1 = 0$ and $\bar{C}_2 = 0$. So

$$k_{z(cap)} = G_s D_f \bar{S}_1 \tag{11.17}$$

and

$$k_{z(cap)} = D_f r_0 \bar{S}_2 \sqrt{G_s \rho_s} \tag{11.18}$$

Thus, for the group pile and cap,

$$k_{z(T)} = \frac{\sum\limits_{1}^{n} k_z}{\sum\limits_{r=1}^{n} \alpha_r} + G_s D_f \bar{S}_1 \tag{11.19}$$

and

$$c_{z(T)} = \frac{\sum\limits_{1}^{n} c_z}{\sum\limits_{r=1}^{n} \alpha_r} + D_f r_0 \bar{S}_2 \sqrt{G_s \rho_s} \tag{11.20}$$

where $k_{z(T)}$ and $c_{z(T)}$ are the stiffness and damping constants for the pile group and cap, respectively.

The variations for $\bar{S}_1$ and $\bar{S}_2$ are given in Table 5.2. Once the values of $k_{z(T)}$ and $c_{z(T)}$ are determined, the response of the system can be calculated using the principles described in Chapter 2, as briefly outlined here.

a. Damping ratio:

$$D_z = \frac{c_{z(T)}}{2\sqrt{k_{z(T)} m}} \tag{11.21}$$

where m = mass of the pile cap and the machine supported by it.

b. Undamped natural frequency:

$$\omega_n = \sqrt{\frac{k_{z(T)}}{m}} \tag{11.22}$$

$$f_n = \frac{1}{2\pi} \sqrt{\frac{k_{z(T)}}{m}} \tag{11.23}$$

c. Damped natural frequency:

$$f_m = f_n\sqrt{1 - 2D_z^2} \qquad \text{(for constant force excitation)} \qquad (11.24)$$

$$f_m = \frac{f_n}{\sqrt{1 - 2D_z^2}} \qquad \text{(for rotating mass–type excitation)} \qquad (11.25)$$

d. Amplitude of vibration at resonance:

$$A_z = \frac{Q_0}{k_{z(T)}} \frac{1}{2D_z\sqrt{1 - D_z^2}} \qquad \text{(for constant force excitation)} \qquad (11.26)$$

$$A_z = \frac{m_1 e}{m} \frac{1}{2D_z\sqrt{1 - D_z^2}} \qquad \text{(for rotating mass–type excitation)} \qquad (11.27)$$

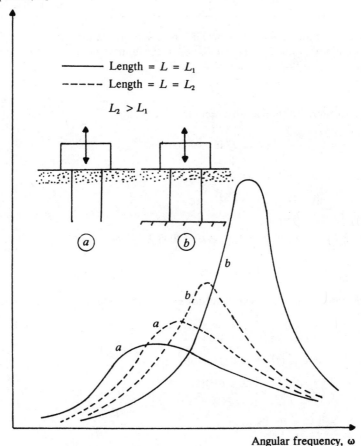

Figure 11.12 Nature of variation of A_z with ω for floating and point-bearing piles

e. Amplitude of vibration at frequency other than resonance:

$$A_z = \frac{\dfrac{Q_0}{k_{z(T)}}}{\sqrt{\left(1 - \dfrac{\omega^2}{\omega_n^2}\right)^2 + 4D_z^2 \dfrac{\omega^2}{\omega_n^2}}} \qquad \text{(for constant force excitation)} \qquad (11.28)$$

$$A_z = \frac{\dfrac{m_1 e}{m}\left(\dfrac{\omega}{\omega_n}\right)^2}{\sqrt{\left(1 - \dfrac{\omega^2}{\omega_n^2}\right)^2 + 4D_z^2 \dfrac{\omega^2}{\omega_n^2}}} \qquad \text{(for rotating mass–type excitation)} \quad (11.29)$$

The nature of variation of A_z with ω for floating piles and point-bearing piles is shown in Figure 11.12. From this figure, it can be seen that the relaxation of the pile tips reduces both the resonant frequency and amplitude of vibration.

Example 11.2

A group of four piles is supporting a machine foundation, as shown in Figure 11.13. Determine $k_{z(T)}$ and $c_{z(T)}$. Given: $E_P = 3 \times 10^6$ lb/in^2, $G = G_s = 4000$ lb/in^2, $\gamma = \gamma_s = 118$ lb/ft^3. Assume Poisson's ratio of soil, $\mu = 0.5$.

Solution

Equivalent radius of pile cross section:

$$R = \left(\frac{12 \times 12}{\pi}\right)^{1/2} = 6.77 \text{ in.}$$

Length of piles $= L = 40$ ft:

$$\frac{L}{R} = \frac{40 \times 12}{6.77} = 70.9$$

Given $E_P = 3 \times 10^6$ lb/in.^{2}; $G_s = G = 4000$ lb/in.^{2}. So

$$\frac{E_P}{G} = \frac{3 \times 10^6}{4000} = 750$$

Referring to Figures 11.8 and 11.9, for $E_P/G = 750$ and $L/R = 70.9$, the magnitudes of f_{z1} and f_{z2} are

$$f_{z1} \approx 0.034 \quad \text{and} \quad f_{z2} \approx 0.06$$

Hence, from Eqs. (11.13) and (11.14)

$$k_z = \left(\frac{E_P A}{R}\right)f_{z1} = \left[\frac{(3 \times 10^6)(12 \times 12)}{6.77}\right](0.034)$$

$$= 2.17 \times 10^6 \text{ lb/in.}$$

$$c_z = \left(\frac{E_P A}{\sqrt{G/\rho}}\right)f_{z2}$$

However,

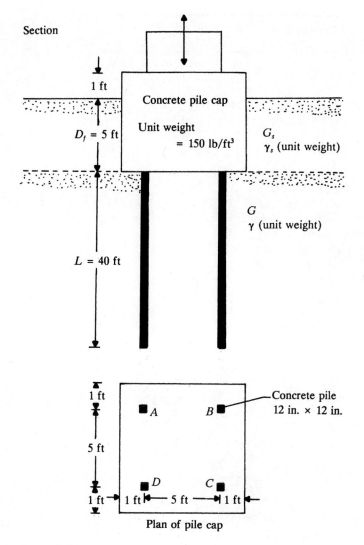

Section

1 ft

Concrete pile cap

$D_f = 5$ ft

Unit weight
= 150 lb/ft^3

G_s
γ_s (unit weight)

G
γ (unit weight)

$L = 40$ ft

1 ft

Concrete pile
12 in. × 12 in.

■ A B ■

5 ft

■ D C ■

1 ft   1 ft   5 ft   1 ft

Plan of pile cap

Figure 11.13

$$\sqrt{\frac{G}{\rho}} = \sqrt{\frac{4000}{(118/12^3)/(32.2 \times 12)}} = 4757.5 \text{ in./s}$$

So

$$c_z = \left[\frac{(3 \times 10^6)(12 \times 12)}{4757.5} \right](0.06)$$

$$= 0.545 \times 10^4 \text{ lb-s/in.}$$

In order to determine the stiffness and damping constants for group piles, Eqs. (11.15) and (11.16) can be used. However $\sum_{r=1}^{n} \alpha_r$ needs to be considered first. This can be done by using Figure 11.10 and preparing the following table.

Reference pile →	A^a	B	C	D
Interacting pile				
A	1.00	0.54	0.48	0.54
B	0.54	1.00	0.54	0.48
C	0.48	0.54	1.00	0.54
D	0.54	0.48	0.54	1.00
	2.56	2.56	2.56	2.56

^a *Note:* Reference pile A.

For interaction between piles A and A, $S = 0$ and $S/2R = 0$. So $\alpha_r = 1$. For interaction between piles A and B, $S = 5$ ft, $2R = (2)(6.77) = 13.54$ in., and $S/2R = (5 \times 12)/13.54 = 4.43$. So $\alpha_r \approx 0.54$. Similarly, for interaction between piles A and D, $\alpha_r \approx 0.54$. Between piles A and C,

$$\frac{S}{2R} = \frac{[\sqrt{(5)^2 + (5)^2}]12}{13.54} = 6.27$$

or

$$\frac{2R}{S} = 0.16 \qquad \alpha_r \approx 0.48$$

The average value of $\sum_{r=1}^{n} \alpha_r = 2.56$. Hence, using Eq. (11.15),

$$k_{z(g)} = \frac{(4)(2.17 \times 10^6)}{2.56} = 3.39 \times 10^6 \text{ lb/in.}$$

$$c_{z(g)} = \frac{(4)(0.545 \times 10^6)}{2.56} = 0.852 \times 10^4 \text{ lb-s/in.}$$

Again, for the contributions of the pile cap, $k_{z(cap)}$ [Eq. (11.17)] and $c_{z(cap)}$ [Eq. (11.18)] need to be determined. Given:

$D_f = 5$ ft; $\qquad\qquad\qquad G_s = 4000$ lb/in.^{2}

$\bar{S}_1 = 2.7$ (Table 5.2); $\qquad \bar{S}_2 = 6.7$ (Table 5.2)

$$r_0 = \sqrt{\frac{7 \times 7}{\pi}} = 3.949 \text{ ft} \approx 47.39 \text{ in.}$$

From Eq. (11.17),

$$k_{z(cap)} = G_s D_f \bar{S}_1 = (4000)(5 \times 12)(2.7)$$

$$= 0.648 \times 10^6 \text{ lb/in.}$$

Similarly, from Eq. (11.18),

$$c_{z(cap)} = D_f r_0 \bar{S}_2 \sqrt{G_s \rho_s}$$

$$= (5 \times 12)(47.39)(6.7) \left[\sqrt{\frac{(4000)(118/12^3)}{32.2 \times 12}} \right]$$

$$= 1.6 \times 10^4 \text{ lb-s/in.}$$

So

$$k_{z(T)} = k_{z(g)} + k_{z(cap)} = 3.39 \times 10^6 + 0.648 \times 10^6$$

$$= 4.038 \times 10^6 \text{ lb/in.}$$

$$c_{z(T)} = c_{z(g)} + c_{z(cap)} = 0.852 \times 10^4 + 1.6 \times 10^4$$

$$= 2.452 \times 10^4 \text{ lb-s/in.}$$

Example 11.3

Refer to Example 11.2. If the weight of the machine being supported is 15,000 lb, determine the damping ratio.

Solution

Weight of the pile cap:

$$(7)(7)(6)(150) = 44,100 \text{ lb}$$

Total weight of pile cap and machine:

$$44,100 + 10,000 = 54,100 \text{ lb}$$

From Eq. (11.21),

$$D_z = \frac{c_{z(T)}}{2\sqrt{k_{z(T)}m}} = \frac{2.452 \times 10^4}{2\sqrt{(4.038 \times 10^6)[54,100/(32.2 \times 12)]}} = \underline{0.516}$$

SLIDING, ROCKING, AND TORSIONAL VIBRATION

11.4   Sliding and Rocking Vibration

Novak (1974) and Novak and El-Sharnouby (1983) derived the stiffness and damping constants for a single pile in a similar manner as described for the case of vertical vibration in Section 11.3. Following are the relationships for the spring and dashpot coefficients for single piles.

Sliding Vibration of Single Pile

$$k_x = \frac{E_P I_P}{R^3} f_{x1} \qquad\qquad (11.30)$$

$$c_x = \frac{E_P I_P}{R^2 v_s} f_{x2} \qquad\qquad (11.31)$$

Table 11.1   The Stiffness and Dampness Parameters
for Sliding Vibration ($L/R > 25$)

Poisson's ratio of soil, μ	$\dfrac{E_P}{G}$	f_{x1}	f_{x2}
0.25	10,000	0.0042	0.0107
	2,500	0.0119	0.0297
	1,000	0.0236	0.0579
	500	0.0395	0.0953
	250	0.0659	0.1556
0.4	10,000	0.0047	0.0119
	2,500	0.0132	0.0329
	1,000	0.0261	0.0641
	500	0.0436	0.1054
	250	0.0726	0.1717

Note: G = shear modulus of soil.

where

E_P = modulus of elasticity of the pile material

I_P = moment of inertia of the pile cross section

v_s = shear wave velocity in soil

R = radius of the pile

The variations of f_{x1} and f_{x2} are given in Table 11.1, which is based on the analysis of Novak (1974) and Novak and El-Sharnouby (1983).

When piles are installed in groups and subjected to sliding vibration, the spring constant and the damping coefficient of the group can be given as

$$k_{x(g)} = \frac{\sum\limits_{1}^{n} k_x}{\sum\limits_{r=1}^{n} \alpha_{L(r)}} \tag{11.32}$$

and

$$c_{x(g)} = \frac{\sum\limits_{1}^{n} c_x}{\sum\limits_{r=1}^{n} \alpha_{L(r)}} \tag{11.33}$$

where

$\alpha_{L(r)}$ = interaction factor (Poulos, 1971)

$k_{x(g)}$ = spring constant for the pile group

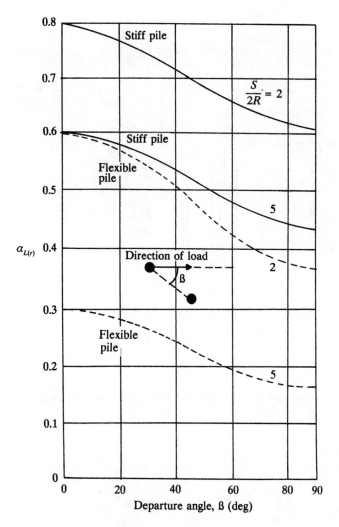

Figure 11.14   Variation of $\alpha_{L(r)}$ (after Poulos, 1971)

$c_{x(g)}$ = damping coefficient for the pile group

   n = number of piles in the group.

The variation of $\alpha_{L(r)}$ is given in Figure 11.14.

   As in the case of vertical vibration, the effect of the pile cap (Figure 11.15) needs to be taken into account in the determination of total stiffness and damping constant. In Eqs. (5.123) and (5.124), the relationships for k_x and c_x for embedded foundations have been described as

$$k_x = Gr_0 \left[\bar{C}_{x1} + \frac{G_s}{G} \frac{D_f}{r_0} \bar{S}_{x1} \right] \tag{5.123}$$

$$c_x = r_0^2 \sqrt{\rho G} \left[\bar{C}_{x2} + \bar{S}_{x2} \frac{D_f}{r_0} \sqrt{\frac{G_s \rho_s}{G \rho}} \right] \tag{5.124}$$

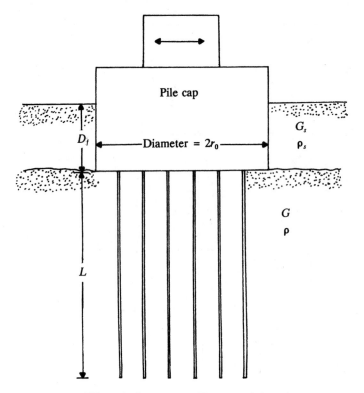

Figure 11.15 Effect of pile cap on stiffness and damping constants—sliding vibration

Assuming $\bar{C}_{x1} = 0$ and $\bar{C}_{x2} = 0$

$$k_{x(\text{cap})} = G_s D_f \bar{S}_{x1} \tag{11.34}$$

and

$$c_{x(\text{cap})} = D_f r_0 \bar{S}_{x2} \sqrt{G_s \rho_s} \tag{11.35}$$

Hence, for the group pile and cap,

$$k_{x(T)} = \frac{\displaystyle\sum_1^n k_x}{\displaystyle\sum_{r=1}^n \alpha_{L(r)}} + G_s D_f \bar{S}_{x1} \tag{11.36}$$

and

$$c_{x(T)} = \frac{\displaystyle\sum_1^n c_x}{\displaystyle\sum_{r=1}^n \alpha_{L(r)}} + D_f r_0 \bar{S}_{x2} \sqrt{G_s \rho_s} \tag{11.37}$$

The damping ratio D_x for the system can then be determined as

$$D_x = \frac{c_{x(T)}}{2\sqrt{k_{x(T)}m}}$$ (11.38)

where m = mass of the pile cap and the machine supported. The damped natural frequency f_m is given as

$$f_m = \frac{1}{2\pi}\left[\sqrt{\frac{k_{z(T)}}{m}}\right][\sqrt{1 - 2D_x^2}]$$ (for constant force excitation)   (11.39)

and

$$f_n = \frac{1}{2\pi}\frac{\sqrt{k_{x(T)}/m}}{\sqrt{1 - 2D_x^2}}$$ (11.40)

The amplitudes of vibration can be calculated using Eqs. (5.58), (5.59), (5.60), and (5.61). While using these equations, k_x needs to be replaced by $k_{x(T)}$.

Rocking Vibration for Single Pile

$$k_\theta = \frac{E_P I_P}{R} f_{\theta 1}$$ (11.41)

$$c_\theta = \frac{E_P I_P}{v_s} f_{\theta 2}$$ (11.42)

The terms E_P, I_P, v_s, and R have been defined in relation to Eqs. (11.30) and (11.31). The numerical values of $f_{\theta 1}$ and $f_{\theta 2}$ obtained by Novak (1974) and Novak and El-Sharnouby (1987) are given in Table 11.2.

For coupling between horizontal translation and rocking, the cross stiffness and damping constants are as follows:

Table 11.2   The Stiffness and Damping Parameters for Rocking Vibration ($L/R > 25$)

Poisson's ratio of soil, μ	$\dfrac{E_P}{G}$	$f_{\theta 1}$	$f_{\theta 2}$
0.25	10,000	0.2135	0.1577
	2,500	0.2998	0.2152
	1,000	0.3741	0.2598
	500	0.4411	0.2953
	250	0.5186	0.3299
0.4	10,000	0.2207	0.1634
	2,500	0.3097	0.2224
	1,000	0.3860	0.2677
	500	0.4547	0.3034
	250	0.5336	0.3377

Note: G = shear modulus of soil.

Table 11.3 Values of $f_{x\theta 1}$ and $f_{x\theta 2}$

Poisson's ratio of soil, μ	$\dfrac{E_P}{G}$	$f_{x\theta 1}$	$f_{x\theta 2}$
0.25	10,000	-0.0217	-0.0333
	2,500	-0.0429	-0.0646
	1,000	-0.0668	-0.0985
	500	-0.0929	-0.1337
	250	-0.1281	-0.1786
0.4	10,000	-0.0232	-0.0358
	2,500	-0.0459	-0.0692
	1,000	-0.0714	-0.1052
	500	-0.0991	-0.1425
	250	-0.1365	-0.1896

Note: G = shear modulus of soil.

$$k_{x\theta} = \frac{E_P I_P}{R^2} f_{x\theta 1} \tag{11.43}$$

$$c_{x\theta} = \frac{E_P I_P}{R v_s} f_{x\theta 2} \tag{11.44}$$

The numerical values for $f_{x\theta 1}$ and $f_{x\theta 2}$ are given in Table 11.3, which is based on the works of Novak (1974) and Novak and El-Sharnouby (1983).

For group piles the stiffness $[k_{\theta(g)}]$ and damping $[c_{\theta(g)}]$ constants can be written as

$$k_{\theta(g)} = \sum_{1}^{n} [k_\theta + k_z x_r^2 + k_x Z_c^2 - 2Z_c k_{x\theta}] \tag{11.45}$$

The terms x_r and Z_c are defined in Figure 11.16. Similarly,

$$c_{\theta(g)} = \sum_{1}^{n} [c_\theta + c_z x_r^2 + c_x Z_c^2 - 2Z_c c_{x\theta}] \tag{11.46}$$

The stiffness $[k_{\theta(cap)}]$ and damping $[c_{\theta(cap)}]$ for the pile cap can be obtained from the following equations (Prakash and Puri, 1988):

$$k_{\theta(cap)} = G_s r_0^2 D_f \bar{S}_{\theta 1} + G_s r_0^2 D_f \left[\frac{\delta^2}{3} + \left(\frac{Z_c}{r_0} \right)^2 - \delta \left(\frac{Z_c}{r_0} \right) \right] \bar{S}_{x1} \tag{11.47}$$

and

$$c_{\theta(cap)} = \delta r_0^4 \sqrt{G_s \rho_s} \left\{ \bar{S}_{\theta 2} + \left[\frac{\delta^2}{3} + \left(\frac{Z_c}{r_0} \right)^2 - \delta \left(\frac{Z_c}{r_0} \right) \right] \bar{S}_{x2} \right\} \tag{11.48}$$

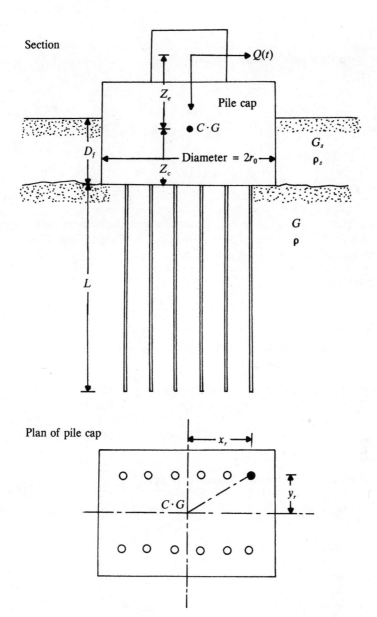

Figure 11.16   Definition of parameters in Eqs. (11.45) and (11.46)

where

r_0 = equivalent radius of the pile cap

$$\delta = \frac{D_f}{r_0} \tag{11.49}$$

Thus, the total stiffness $[k_{\theta(T)}]$ and damping $[c_{\theta(T)}]$ constants are

$$k_{\theta(T)} = k_{\theta(g)} + k_{\theta(cap)} \tag{11.50}$$

and

$$c_{\theta(T)} = c_{\theta(g)} + c_{\theta(cap)} \tag{11.51}$$

Once the magnitude of $k_{\theta(T)}$ and $c_{\theta(T)}$ are determined, the response of the system can be calculated in the same manner as outlined in Chapter 2 and Section 5.5. For convenience, this is outlined here as well.

a. *Damping ratio:*

$$D_\theta = \frac{c_{\theta(T)}}{2\sqrt{k_{\theta(T)}I_g}} \tag{11.52}$$

where I_g = mass moment of inertia for the pile cap and the machinery about the centroid of the block. Referring to Figure 11.17a,

I_g = mass moment of inertia about the y axis

$$= \frac{m}{12}(L^2 + h^2) \tag{11.53a}$$

and, similarly, referring to Figure 11.17b,

I_g = mass moment of inertia about the y axis

$$= \frac{m}{12}(3r_0^2 + h^2) \tag{11.53b}$$

b. *Undamped natural frequency:*

$$\omega_n = \sqrt{\frac{k_{\theta(T)}}{I_g}} \tag{11.54}$$

$$f_n = \frac{1}{2\pi}\sqrt{\frac{k_{\theta(T)}}{I_g}} \tag{11.55}$$

c. *Damped natural frequency:*

$$f_m = f_n\sqrt{1 - 2D_\theta^2} \qquad \text{(for constant force excitation)} \tag{11.56}$$

$$f_m = \frac{f_n}{\sqrt{1 - 2D_\theta^2}} \qquad \text{(for rotating mass–type excitation)} \tag{11.57}$$

The amplitude of vibration can be determined by using Eqs. (5.46), (5.47), (5.48), and (5.49).

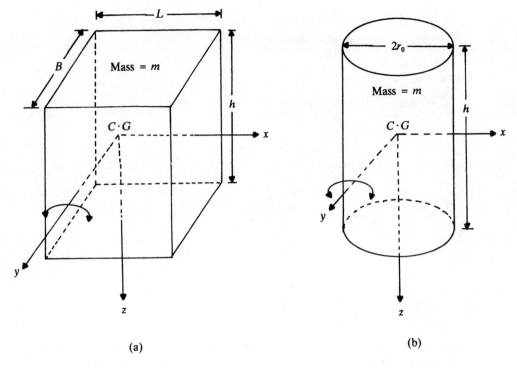

(a)   (b)

Figure 11.17   Mass moment of inertia I_g

Example 11.4

Refer to Example 11.2. Determine $k_{x(T)}$ and $c_{x(T)}$ for the sliding mode of vibration. Assume Poisson's ratio of soil, $\mu = 0.25$.

Solution

Stiffness and damping constants for single pile: From Eqs. (11.30) and (11.31),

$$k_x = \frac{E_P I_P}{R^3} f_{x1}$$

and

$$c_x = \frac{E_P I_P}{R^2 v_s} f_{x2}$$

Given $E_P = 3 \times 10^6$ lb/in.^{2}; $R = 6.77$ in.

$$I_P = \frac{\pi}{4} R^4 = \frac{\pi}{4}(6.77)^4 = 1649.8 \text{ in}^4$$

$$v_s = \sqrt{\frac{G}{\rho}} = \sqrt{\frac{4000}{(118/12^3)/(32.2 \times 12)}} = 4757.5 \text{ in./s}$$

$$\frac{E_P}{G} = \frac{3 \times 10^6}{4000} = 750$$

From Table 11.1, for $\mu = 0.25$ and $E_P/G = 750$, $f_{x1} = 0.027$ and $f_{x2} = 0.068$. So

$$k_x = \frac{E_P I_P}{R^3} f_{x1} = \frac{(3 \times 10^6)(1649.8)}{(6.77)^3}(0.027)$$

$$= 4.31 \times 10^5 \text{ lb/in.}$$

$$c_x = \frac{E_P I_P}{R^2 v_s} f_{x2} = \frac{(3 \times 10^6)(1649.8)}{(6.77)^2(4757.5)}(0.068)$$

$$= 1.543 \times 10^3 \text{ lb-s/in.}$$

Stiffness and damping constants for group pile: From Eqs. (11.32) and (11.33),

$$k_{x(g)} = \frac{\sum\limits_{1}^{n} k_x}{\sum\limits_{r=1}^{n} \alpha_{L(r)}}$$

and

$$c_{x(g)} = \frac{\sum\limits_{1}^{n} c_x}{\sum\limits_{r=1}^{n} \alpha_{L(r)}}$$

To find $\alpha_{L(r)}$ with A as the *reference pile*, the following table can be prepared using Figure 11.14. Assume piles to be *flexible*.

Reference pile→ Interacting pile ↓	β (deg)	A $\dfrac{S}{2R}$	$\alpha_{L(r)}$
A	0	0	1.0
B	0	4.43	0.32
C	45	6.27	0.24
D	90	4.43	0.18
			1.74

Similarly, for the other reference piles, $\sum \alpha_{L(r)}$ will be 1.74. So, the average value of $\sum\limits_{r=1}^{n} \alpha_{L(r)} = 1.74$. Thus,

$$k_{x(g)} = \frac{(4)(4.31 \times 10^5)}{1.74} = 9.91 \times 10^5 \text{ lb/in.}$$

$$c_{x(g)} = \frac{(4)(1.543 \times 10^3)}{1.74} = 3.547 \times 10^3 \text{ lb-s/in.}$$

Stiffness and damping for pile cap: From Eqs. (11.34) and (11.35),

$$k_{x(\text{cap})} = G_s D_f \bar{S}_{x1}$$

and

$$c_{x(\text{cap})} = D_f r_0 \bar{S}_{x2} \sqrt{G_s \rho_s}$$

From Chapter 5 with $\mu = 0.25$, $\bar{S}_{x1} = 4.0$ and $\bar{S}_{x2} = 9.10$. So

$$k_{x(cap)} = (4000)(5 \times 12)(4.0) = 9.6 \times 10^5 \text{ lb/in.}$$

$$c_{x(cap)} = (5 \times 12)(47.39)\left[\sqrt{\frac{(4000)(118/12^3)}{32.2 \times 12}}\right](9.1)$$

$$= 21.76 \times 10^3 \text{ lb-s/in.}$$

Total stiffness and damping:

$$k_{x(T)} = k_{x(g)} + k_{x(cap)} = 9.91 \times 10^5 + 9.6 \times 10^5$$
$$= 19.51 \times 10^5 \text{ lb/in.}$$

$$c_{x(T)} = c_{x(g)} + c_{x(cap)} = 3.547 \times 10^3 + 21.76 \times 10^3$$
$$= 25.307 \times 10^3 \text{ lb-s/in.}$$ ■

Example 11.5

In Example 11.4, if the weight of the machine being supported is 20,000 lb, determine the damping ratio.

Solution

Weight of the pile cap: 44,100 lb
Total weight of pile cap and machine:

$$44,100 + 20,000 = 64,100 \text{ lb}$$

From Eq. (11.38),

$$D_x = \frac{c_{x(T)}}{2\sqrt{k_{x(T)}m}} = \frac{25.307 \times 10^3}{2\sqrt{(19.51 \times 10^5)[64.100/(32.2 \times 12)]}} = 0.703$$ ■

Example 11.6

Refer to Example 11.2. Determine $k_{\theta(T)}$ and $c_{\theta(T)}$ for the rocking mode of vibration. Assume Poisson's ratio of soil to be 0.25.

Solution

Stiffness and damping constants for single pile: From Eqs. (11.41) and (11.42),

$$k_\theta = \frac{E_P I_P}{R} f_{\theta 1}$$

and

$$c_\theta = \frac{E_P I_P}{v_s} f_{\theta 2}$$

$$\frac{E_P}{G} = \frac{3 \times 10^6}{4000} = 750$$

From Table 11.2, for $\mu = 0.25$ and $E_P/G = 750$, the values of $f_{\theta 1}$ and $f_{\theta 2}$ are 0.39 and 0.275, respectively.

$$k_\theta = \frac{(3 \times 10^6)(1649.8)}{6.77}(0.39) = 285.12 \times 10^6 \text{ lb-in./rad}$$

$$c_\theta = \frac{(3 \times 10^6)(1649.8)}{4757.5}(0.275)$$

$$= 0.286 \times 10^6 \text{ lb-in.-s/rad}$$

Cross-stiffness and cross-damping constants: From Eqs. (11.43) and (11.44),

$$k_{x\theta} = \frac{E_P I_P}{R^2} f_{x\theta 1}$$

$$c_{x\theta} = \frac{E_P I_P}{R v_s} f_{x\theta 2}$$

From Table 11.3, $f_{x\theta 1} = -0.076$ and $f_{x\theta 2} = -0.115$. Thus,

$$k_{x\theta} = \frac{(3 \times 10^6)(1649.8)}{(6.77)^2}(-0.076) = -8.21 \times 10^6 \text{ lb/rad}$$

$$c_{x\theta} = \frac{(3 \times 10^6)(1649.8)}{(6.77)(4757.5)}(-0.115)$$

$$= -1.77 \times 10^4 \text{ lb-s/rad}$$

Stiffness and damping constants for pile group: From Eqs. (11.45) and (11.46),

$$k_{\theta(g)} = \sum_1^n [k_\theta + k_z x_r^2 + k_x Z_c^2 - 2Z_c k_{x\theta}]$$

and

$$c_{\theta(g)} = \sum_1^n [c_\theta + c_z x_r^2 + c_x Z_c^2 - 2Z_c c_{x\theta}]$$

For this problem,

$n = 4$

$k_\theta = 285.12 \times 10^6$ lb-in./rad

$k_z = 2.17 \times 10^6$ lb/in.   (from Problem 11.2)

$x_r = 2.5$ ft $= 30$ in.

$k_x = 4.31 \times 10^5$ lb/in.   (from Problem 11.4)

$k_{x\theta} = -8.21 \times 10^6$ lb/in.

$Z_c = 3$ ft $= 36$ in.

So

$$k_{\theta(g)} = 4[285.12 \times 10^6 + (2.17 \times 10^6)(30)^2$$

$$+ (4.31 \times 10^5)(36)^2 - (2)(36)(-8.21 \times 10^6)]$$

$$= 13.55 \times 10^9 \text{ lb-in./rad}$$

Similarly, with

$$n = 4$$

$$c_\theta = 0.286 \times 10^6 \text{ lb-in.-s/rad}$$

$$c_z = 0.545 \times 10^4 \text{ lb-s/in.} \qquad \text{(from Problem 11.2)}$$

$$x_r = 30 \text{ in.}$$

$$c_x = 1.543 \times 10^3 \text{ lb-s/in.} \qquad \text{(from Problem 11.4)}$$

$$c_{x\theta} = -1.77 \times 10^4 \text{ lb-s/rad}$$

$$Z_c = 36 \text{ in.}$$

the result is

$$c_{\theta(g)} = 4[(0.286 \times 10^6) + (0.545 \times 10^4)(30)^2$$
$$+ (1.543 \times 10^3)(36)^2 - (2)(36)(-1.77 \times 10^4)]$$
$$= 33.84 \times 10^6 \text{ lb-in.-s/rad}$$

Stiffness and damping of pile cap: From Eq. (11.47),

$$k_{\theta(\text{cap})} = G_s r_0^2 D_f \bar{S}_{\theta 1} + G_s r_0^2 D_f \left[\frac{\delta^2}{3} + \left(\frac{Z_c}{r_0} \right)^2 - \delta \left(\frac{Z_c}{r_0} \right) \right] \bar{S}_{x1}$$

where

$$\delta = \frac{D_f}{r_0} = \frac{5 \times 12}{47.39} = 1.266$$

So

$$k_{\theta(\text{cap})} = (4000)(47.39)^2 (5 \times 12)(2.5)$$
$$+ (4000)(47.39)^2 (5 \times 12) \left[\frac{(1.266)^2}{3} + \left(\frac{36}{47.39} \right)^2 - (1.266) \left(\frac{36}{47.39} \right) \right] (4)$$
$$= 1.347 \times 10^9 + 0.321 \times 10^9$$
$$= 1.668 \times 10^9 \text{ lb-in./rad}$$

Again, from Eq. (11.48)

$$c_{\theta(\text{cap})} = \delta r_0^4 \sqrt{G_s \rho_s} \left\{ \bar{S}_{\theta 2} + \left[\frac{\delta^2}{3} + \left(\frac{Z_c}{r_0} \right)^2 - \delta \left(\frac{Z_c}{r_0} \right) \right] \bar{S}_{x2} \right\}$$

or

$$c_{\theta(\text{cap})} = (1.266)(47.49)^4 \left[\sqrt{\frac{(4000)(118/12^3)}{32.2 \times 12}} \right] \left\{ 1.8 + \left[\frac{(1.266)^2}{3} + \left(\frac{36}{47.39} \right)^2 \right. \right.$$
$$\left. \left. - (1.266) \left(\frac{36}{47.39} \right) \right] (9.1) \right\}$$
$$= 16.94 \times 10^6 \text{ lb-in.-s/rad}$$

Total stiffness and damping:

$$k_{\theta(T)} = k_{\theta(g)} + k_{\theta(\text{cap})} = 13.55 \times 10^9 + 1.668 \times 10^9$$

$$= 15.218 \times 10^9 \text{ lb-in./rad}$$

$$c_{\theta(T)} = c_{\theta(g)} + c_{\theta(\text{cap})} = 33.84 \times 10^6 + 16.94 \times 10^6$$

$$= 50.78 \times 10^6 \text{ lb-in.-s/rad} \qquad \blacksquare$$

11.5   Torsional Vibration of Embedded Piles

Torsional vibration of an embedded pile was analyzed by Novak and Howell (1977) and Novak and El-Sharnouby (1983). According to these analyses, the pile (Figure 11.18) is assumed to be vertical, circular in cross section (radius = R), elastic, end-bearing, and perfectly connected to the soil. The soil is considered to be a linear, viscoelastic medium with frequency-independent material damping of the hysteretic type. Referring to Figure 11.18, the pile is undergoing a complex harmonic rotation around the vertical axis, which can be described as

$$\alpha(z, t) = \alpha(z)e^{i\omega t} \qquad (11.58)$$

where

$\alpha(z)$ = complex amplitude of pile rotation at a depth z

$i = \sqrt{-1}$

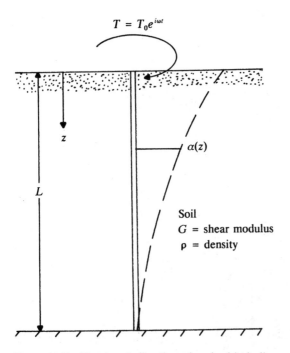

Figure 11.18 Torsional vibration of embedded pile

The motion of the pile is resisted by a torsional soil reaction. The elastic soil reaction setting on a pile element dz can then be given as

$$GR^2(S_{\alpha 1} + iS_{\alpha 2})[\alpha(z, t)]\, dz \tag{11.59}$$

where

$S_{\alpha 1}(a_0) =$ stiffness parameter

$$= 2\pi\left(2 - a_0 \frac{J_0 J_1 + Y_0 Y_1}{J_1^2 + Y_1^2}\right) \tag{11.60}$$

$S_{\alpha 2}(a_0) =$ damping parameter

$$= \frac{4}{J_1^2 + Y_1^2} \tag{11.61}$$

$a_0 =$ dimensionless frequency $= \omega R \sqrt{\dfrac{\rho}{G}}$

$R =$ pile radius

$G =$ shear modulus of soil

$\rho =$ density of soil

$J_0(a_0), J_1(a_0) =$ Bessel functions of the first kind and of order 0 and 1, respectively

$Y_0(a_0), Y_1(a_0) =$ Bessel functions of the second kind and of order 0 and 1, respectively

The parameters $S_{\alpha 1}$ and $S_{\alpha 2}$ also depend on the material damping of the soil. It was mentioned in Chapter 5 that the material damping is more important for torsional mode of vibration than any other. This damping can be included by addition of an out-of-phase complement to the soil shear modulus, or

$$G^* = G_1 + iG_2 = G_1(1 + i\tan\delta) \tag{11.62}$$

where

$$\tan\delta = \frac{G_2}{G_1}$$

$G_1, G_2 =$ real and imaginary parts, respectively, of the complex shear modulus

$\delta =$ loss angle

Thus, the term G in Eq. (11.59) can be replaced by G^*. Also G^* enters Eqs. (11.60) and (11.61) through the dimensionless frequency a_0. Using this method of analysis, Novak and Howell (1977) showed that the stiffness and damping constants of fixed-tip single piles can be given as

$$k_\alpha = \frac{G_P J}{R} f_{\alpha 1}$$

(11.63)

and

$$c_\alpha = \frac{G_P J}{\sqrt{\dfrac{G}{\rho}}} f_{\alpha 2}$$

(11.64)

where

G_P = shear modulus of the pile material

J = polar moment of inertia of the pile cross section

$f_{\alpha 1}, f_{\alpha 2}$ = nondimensional parameters

The variations of $f_{\alpha 1}$ and $f_{\alpha 2}$ for timber piles ($\rho/\rho_P = 2$) are shown in Figures 11.19 and 11.20. Figures 11.21 and 11.22 show similar variations for concrete piles ($\rho/\rho_P = 0.7$). It is important to note the following:

1. For a given type of pile, the nondimensional parameter $f_{\alpha 2}$ is relatively more frequency dependent than $f_{\alpha 1}$.

2. Novak and Howell (1977) showed that the displacement of slender piles rapidly diminishes with increasing depth and varies to a lesser degree with frequency. So the effect of the tip condition is less important for slender piles in which the tip is fixed by the soil.

3. The pronounced effect of material damping may be seen in Figures 11.20 and 11.22. The value of $\tan \delta = 0.1$ is of typical order in soil. At low frequencies the material damping significantly increases the torsional damping of the pile. (Compare $f_{\alpha 2}$ values for $\tan \delta = 0.1$ to those for $\tan \delta = 0$ for a given value of a_0.)

Group Piles Subjected to Torsional Vibration

If a group pile is subjected to torsional vibration as shown in Figure 11.23, the torsional stiffness $[k_{\alpha(g)}\}$ and damping $[c_{\alpha(g)}]$ constants can be expressed as

$$k_{\alpha(g)} = \sum_1^n [k_\alpha + k_x(x_r^2 + y_r^2)]$$

(11.65)

and

$$c_{\alpha(g)} = \sum_1^n [c_\alpha + c_x(x_r^2 + y_r^2)]$$

(11.66)

The expressions for k_α and c_α are given in Equations (11.63) and (11.64), and k_x and c_x are the stiffness and damping constants for sliding vibration [Eqs. (11.30)

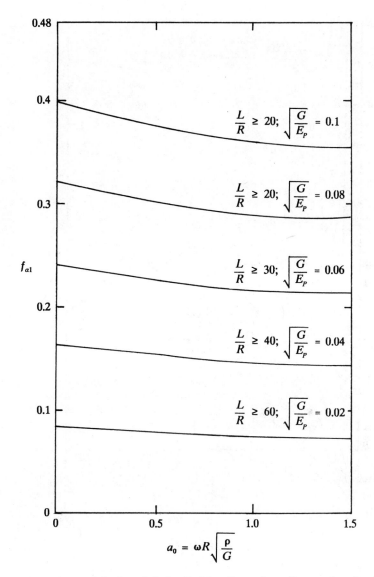

$$a_0 = \omega R \sqrt{\frac{\rho}{G}}$$

Figure 11.19 Variation of $f_{\alpha 1}$ for timber pile—$\rho/\rho_p = 2$; $\rho_p =$ density of pile material (after Novak and Howell, 1977)

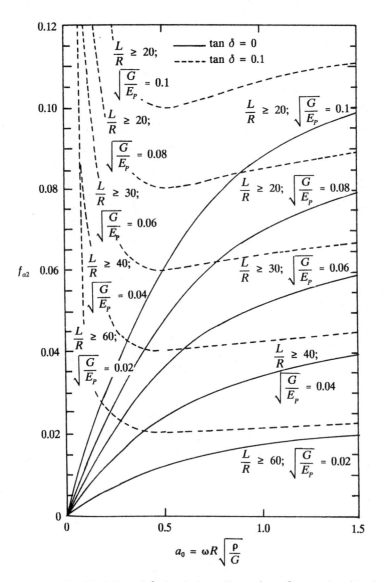

Figure 11.20 Variation of $f_{\alpha 2}$ for timber pile—$\rho/\rho_p = 2$; ρ_p = density of pile material (after Novak and Howell, 1977)

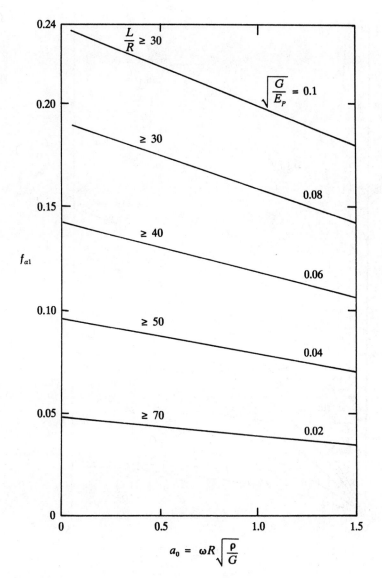

Figure 11.21 Variation of $f_{\alpha 1}$ for concrete pile—$\rho/\rho_p = 0.7$; ρ_p = density of pile material (after Novak and Howell, 1977)

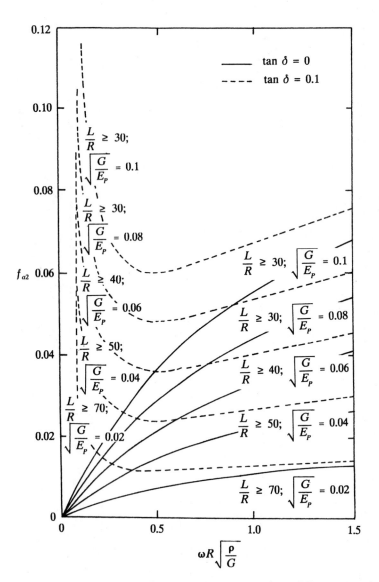

Figure 11.22   Variation of $f_{\alpha 2}$ for concrete pile—$\rho/\rho_p = 0.7$; ρ_p = density of pile material (after Novak and Howell, 1977)

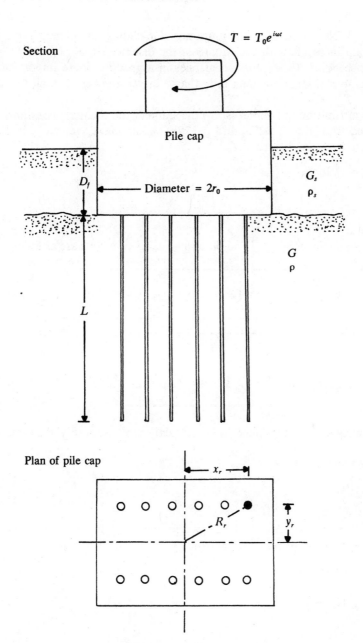

Figure 11.23   Group pile subjected to torsional vibration

and (11.31)]. Note that the contribution of the sliding component for a pile in the group increases with the square of the distance ($R_r = \sqrt{x_r^2 + y_r^2}$) from the reference point. So the torsion of piles in a group is more important for a small number of large-diameter piles than a larger number of small-diameter piles.

The contribution of the pile cap to the stiffness and damping constants can be obtained from Eqs. (5.127a) and (5.127b). Assuming $\overline{C}_{\alpha 1}$ and $\overline{C}_{\alpha 2}$ in those equations to be equal to zero

$$k_{\alpha(cap)} = D_f G_s r_0^2 \overline{S}_{\alpha 1} \tag{11.67}$$

and

$$c_{\alpha(cap)} = D_f r_0^3 \overline{S}_{\alpha 2} \sqrt{G_s \rho_s} \tag{11.68}$$

Thus, the total stiffness $[k_{\alpha(T)}]$ and damping $[c_{\alpha(T)}]$ constants are as follows.

$$k_{\alpha(T)} = k_{\alpha(g)} + k_{\alpha(cap)} = \sum_1^n [k_\alpha + k_x(x_r^2 + y_r^2)] + D_f G_s r_0^2 \overline{S}_{\alpha 1} \tag{11.69}$$

$$c_{\alpha(T)} = c_{\alpha(g)} + c_{\alpha(cap)}$$
$$= \sum_1^n [c_\alpha + c_x(x_r^2 + y_r^2)] + D_f r_0^3 \overline{S}_{\alpha 2} \sqrt{G_s \rho_s} \tag{11.70}$$

Following are the relationships for calculation of response of the system.

a. *Damping ratio:*

$$D_\alpha = \frac{c_{\alpha(T)}}{2\sqrt{k_{\theta(T)} J_{zz}}} \tag{11.71}$$

where J_{zz} = mass moment of inertia of the pile cap and machinery about a vertical axis passing through the centroid. Referring to Figure 11.24a,

J_{zz} = mass moment of inertia about the z axis

$$= \frac{m}{12}(L^2 + B^2) \tag{11.72}$$

Referring to Figure 11.24b,

J_{zz} = mass moment of inertia about the z axis $= \dfrac{m r_0^2}{2} \tag{11.73}$

b. *Undamped natural frequency:*

$$\omega_n = \sqrt{\frac{k_{\alpha(T)}}{J_{zz}}} \tag{11.74}$$

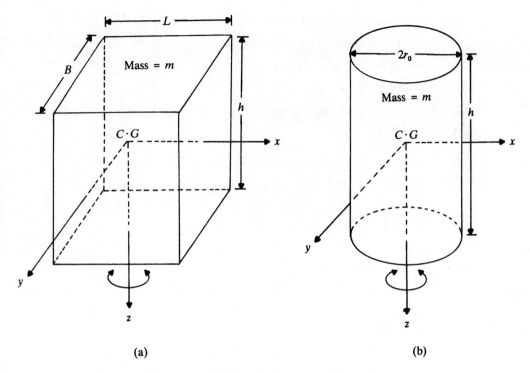

Figure 11.24 Mass moment of inertia J_{zz}

$$f_n = \frac{1}{2\pi}\sqrt{\frac{k_{\alpha(T)}}{J_{zz}}} \tag{11.75}$$

c. *Damped natural frequency*:

$$f_m = f_n\sqrt{1 - 2D_\alpha^2} \quad \text{(for constant force excitation)} \tag{11.76}$$

$$f_m = \frac{f_n}{\sqrt{1 - 2D_\alpha^2}} \quad \text{(for rotating mass–type excitation)} \tag{11.77}$$

d. *Amplitude of vibration at resonance*: Equations (5.68) and (5.69) can be used to calculate the amplitude of vibration. [Replace k_α in Eq. (5.68) by $k_{\alpha(T)}$.]

PROBLEMS

11.1 A machine foundation is supported by six piles, as shown in Figure P11.1. Given:
   Piles

   Type: concrete

   Size: 405 mm × 405 mm in cross section

   Length: 30 m

Unit weight of concrete $= 23 \text{ kN/m}^3$

Modulus of elasticity $= 21 \times 10^6 \text{ kN/m}^2$

Machine and foundation

Weight $= 2030 \text{ kN}$

Determine the natural frequency of the pile-foundation system for vertical vibration. Use the procedure outlined in Section 11.2.

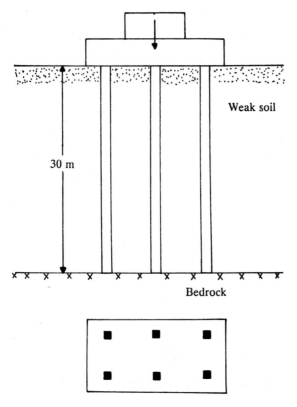

Weak soil

30 m

Bedrock

Figure P11.1

11.2   A wooden pile is shown in Figure P11.2. The pile has a diameter of 230 mm. Given: $E_P = 8.5 \times 10^6 \text{ kN/m}^2$. Determine its stiffness and damping constants for
a. vertical vibration,
b. sliding, and
c. rocking vibration.

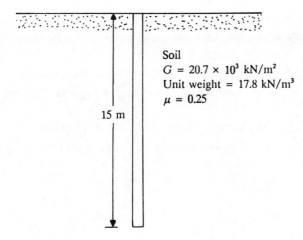

Figure P11.2

11.3 Refer to Problem 11.2. Assume that the Poisson's ratio for wooden piles is 0.35. Determine the approximate stiffness and damping constants for the pile for torsional vibration.

11.4 Solve Problem 11.2 assuming that the piles are made of concrete with $E_P = 21 \times 10^6$ kN/m².

11.5 Refer to Problem 11.4. Assume that the Poisson's ratio for concrete piles is 0.33. Determine the approximate stiffness and damping constants for the pile for torsional vibration.

For Problems 11.6–11.13, refer to the accompanying figure. Given:
Pile

$L = 25$ m

Size = 380 mm × 380 mm in cross section

$E_P = 21 \times 10^6$ kN/m²

Poisson's ration, $\mu_{\text{pile}} = 0.35$

Pile cap

$B = 3.4$ m

$x' = 0.5$ m

$D_f = 2$ m

$h = 3$ m

The pile cap is made of concrete. Unit weight of concrete is 23 kN/m³.
Soil

$$G = G_s = 24{,}500 \text{ kN/m}^2$$

Unit weight, $\gamma = \gamma_s = 18.5$ kN/m³

Poisson's ratio, $\mu = 0.25$

11.6 Determine $k_{z(g)}$ and $c_{z(g)}$ for the pile group for the vertical mode of vibration.

11.7 Determine the total stiffness and damping constants $k_{z(T)}$ and $c_{z(T)}$ for the vertical mode of vibration.

11.8 Determine $k_{x(g)}$ and $c_{x(g)}$ for the pile group for the horizontal mode of vibration.

11.9 Determine the total stiffness and damping constants $k_{x(T)}$ and $c_{x(T)}$ for the horizontal mode of vibration.

11.10 Determine $k_{\theta(g)}$ and $c_{\theta(g)}$ for the pile group for the rocking mode of vibration.

11.11 Determine the total stiffness and damping constants $k_{\theta(T)}$ and $c_{\theta(T)}$ for the rocking mode of vibration.

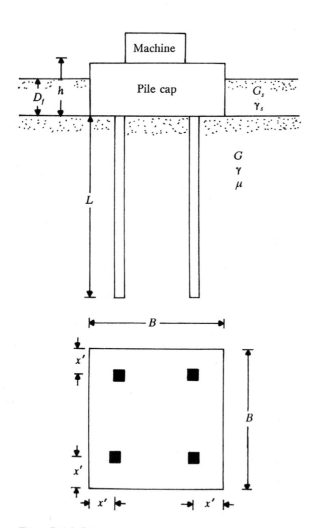

Figure P11.6–P11.11

REFERENCES

Novak, M. (1974). "Dynamic Stiffness and Damping of Piles," *Canadian Geotechnical Journal*, Vol. 11, No. 4, pp. 574–598.

Novak, M. (1977). "Vertical Vibration of Floating Piles," *Journal of the Engineering Mechanics Division*, ASCE, Vol. 103, No. EM1, pp. 153–168.

Novak, M., and El-Sharnouby, B. (1983). "Stiffness and Damping Constants of Single Piles," *Journal of the Geotechnical Engineering Division*, ASCE, Vol. 109, No. GT7, pp. 961–974.

Novak, M., and Howell, J. F. (1977). "Torsional Vibrations of Pile Foundations," *Journal of the Geotechnical Engineering Division*, ASCE, Vol. 103, No. GT4, pp. 271–285.

Poulos, H. G. (1968). "Analysis of the Settlement of Pile Groups," *Geotechnique*, Vol. 18, No. 4, pp. 449–471.

Poulos, H. G. (1971). "Behavior of Laterally Loaded Piles. II Pile Groups," *Journal of the Soil Mechanics and Foundations Division*, ASCE, Vol. 97, No. SM5, pp. 733–751.

Prakash, S. and Puri, V. K. (1988). *Foundation for Machines: Analysis and Design*, John Wiley and Sons, Inc., New York.

Richart, F. E., Jr. (1962). "Foundation Vibration," *Transactions*, ASCE, Vol. 127, Part I, pp. 863–898.

Chapter 12

SEISMIC STABILITY OF EARTH EMBANKMENTS

12.1   Introduction

Sudden ground displacement during earthquakes induces large inertia forces in embankments. As a result, the slope of an embankment is subjected to several cycles of alternating inertia force. There are several recorded cases in the past that show severe damage or collapse of earth embankment slopes due to earthquake-induced vibration (e.g., Ambraseys, 1960; Seed, Makdisi, and DeAlba, 1978). These damages include flow slides of saturated cohesionless soil slopes and slopes of cohesive soil with thin lenses of saturated sand inside them. Such flow slides are due to liquefaction of saturated sand deposits. Fundamental concepts of liquefaction were presented in Chapter 10. Other types of damages include collapse or deformation of dry or dense slopes in sand and also slopes in cohesive soils. In the following sections, the analysis for the stability of earth embankments for these types of slopes under earthquake loading conditions will be treated. It will be assumed that these soils experience very little reduction in strength due to cyclic loading.

In general, deformations suffered by an earth embankment during a strong earthquake may take several forms, such as those shown in Figure 12.1a, b, and c. Figure 12.1a shows a type of deformation pattern that may be concentrated in a narrow zone with a definite slip surface. However, substantial deformation may occur without the development of a slip surface, as shown in Figure 12.1b. In cohesionless slopes, the slip surface is usually a plane, as shown in Figure 12.1c (Seed and Goodman, 1964).

12.2   Free Vibration of Earth Embankments

For a proper evaluation of the seismic stability of earth embankments, it is necessary to have some knowledge of the vibration of embankments due to earthquakes, with some simplifying assumptions. This can be done by the use of one-dimensional *shear slice theory* (Mononobe, Takata, and Matumura, 1963; Seed and Martin, 1966). Figure 12.2 shows an earth embankment in the form of a triangular wedge. The height of the wedge is H. Now, the following assump-

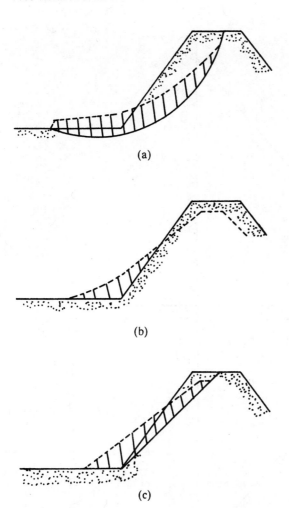

(a)

(b)

(c)

Figure 12.1 Deformation of earth embankments

tions will be made:

1. The earth embankment is infinitely long.

2. The foundation material is rigid.

3. The width-to-height ratio of the embankment is large. This means that the deformation of the embankment is due only to shear.

4. The shear stress on any horizontal strip is uniform.

Regarding the first assumption made, it can be shown that when the length-to-height ratio of an embankment is four or greater, then effect of end restraints on the natural frequencies of vibration is negligible. So, for all practical purposes most of the embankments can be assumed to be infinitely long.

Consider an elementary strip of thickness dz, as shown in Figure 12.2. The forces acting on this elementary strip (per unit length at right angles to the

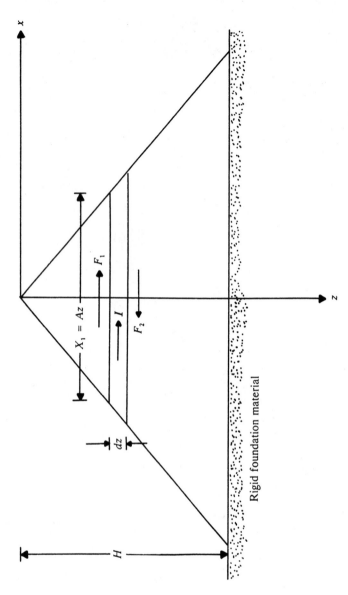

Figure 12.2   Free vibration of an earth embankment

section shown) are

a. Shear force:

$$F_1 = G\frac{\partial u}{\partial z}X_1$$

b. Shear force:

$$F_2 = F_1 + \frac{\partial F_1}{\partial z}\cdot dz$$

c. Inertia force:

$$I = (\text{mass})(\text{acceleration})$$

$$= \left(\frac{\gamma}{g}\cdot X_1\cdot dz\right)\frac{\partial^2 u}{\partial t^2} = \rho Az\frac{\partial^2 u}{\partial t^2}dz$$

where

G = shear modulus of the embankment material

u = displacement in the x direction

γ = unit weight of embankment material

g = acceleration due to gravity

A = a constant of proportionality

$\rho = \gamma/g$ = density of the embankment material

Note that

$$F_2 - F_1 = I$$

So

$$\left[AG\frac{\partial u}{\partial z}z + AG\frac{\partial}{\partial z}\left(z\frac{\partial u}{\partial z}\right)dz\right] - \left[AG\frac{\partial u}{\partial z}z\right] = \rho Az\frac{\partial^2 u}{\partial t^2}dz$$

or

$$\frac{\partial^2 u}{\partial t^2} = \frac{G}{\rho}\left(\frac{\partial^2 u}{\partial z^2} + \frac{1}{z}\frac{\partial u}{\partial z}\right) \tag{12.1}$$

In the preceding equation, the viscous damping force has been neglected, and the boundary conditions for solving it are as follows:

a. $\partial u/\partial z = 0$ at $z = 0$ for all values of t.

b. $u = 0$ at $z = H$ for all values of t.

The solution to Eq. 12.1 is

$$u(z,t) = \sum_{n=1}^{n=\infty} [A_n \sin \omega_n t + B_n \cos \omega_n t] J_0\left(\beta_n\frac{z}{H}\right) \tag{12.2}$$

where

A_n, B_n = constants

J_0 = Bessel function of first kind and order 0

β_n = the zero value of the frequency equation $J_0\left(\omega_n H \times \sqrt{\dfrac{\rho}{G}}\right) = 0$

(So $\beta_1 = 2.404$, $\beta_2 = 5.52$, $\beta_3 = 8.65$, ...)

ω_n = undamped natural circular frequency of dam in the

nth mode of vibration $= \dfrac{\beta_n}{H}\sqrt{\dfrac{G}{\rho}}$

Thus,

$$
\begin{aligned}
\omega_1 &= \frac{\beta_1}{H}\sqrt{\frac{G}{\rho}} = \frac{2.404}{H}\sqrt{\frac{G}{\rho}} \\[2mm]
\omega_2 &= \frac{5.52}{H}\sqrt{\frac{G}{\rho}} \\[2mm]
\omega_3 &= \frac{8.65}{H}\sqrt{\frac{G}{\rho}}
\end{aligned}
\tag{12.3}
$$

12.3   Forced Vibration of an Earth Embankment

Figure 12.3 shows a triangular earth embankment being subjected to a horizontal ground motion, $u_g(t)$. The eqution of motion for the analysis of the vibration of an embankment for such a case can be given as (Seed and Martin, 1966)

$$
\frac{\partial^2 u}{\partial t^2} - \frac{G}{\rho}\left[\frac{\partial^2 u}{\partial t^2} + \frac{1}{z}\frac{\partial u}{\partial t}\right] = -\frac{\partial^2 u_g}{\partial t^2}
\tag{12.4}
$$

The solution to Eq. (12.4) can be given as

$$
u(z,t) = \sum_{n=1}^{n=\infty} \frac{2J_0[\beta_n(z/H)]}{\omega_n \beta_n J_1(\beta_n)} \int_0^t \ddot{u}_g \sin[\omega_n(t-t')]\,dt'
\tag{12.5}
$$

where J_1 = Bessel function of the first order.

Like all materials, soil possesses the property of damping out vibration. The viscous damping factor of soil in the embankment has not been included in Eqs. (12.4) and (12.5). If viscous damping is included, then Eq. (12.5) will be modified to the form

$$
u(z,t) = \sum_{n=1}^{n=\infty} \frac{2J_0[\beta_n(z/H)]}{\omega_n \beta_n J_1(\beta_n)} \int_0^t \ddot{u}_g e^{-D_n \omega_n(t-t')} \sin[\omega_d(t-t')]\,dt'
\tag{12.6}
$$

where D_n = damping factor in the nth mode.

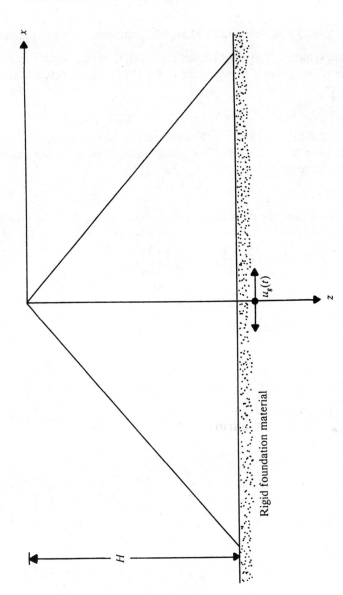

Figure 12.3   Forced vibration of an earth embankment

$\omega_d = \omega_n\sqrt{1 - D_n^2}$ = damped natural angular frequency in the nth mode

The relative velocity, $\dot{u}(z, t)$, and acceleration, $\ddot{u}(z, t)$, at any depth z and time t can be obtained by proper differentiation of Eq. (12.6). The absolute acceleration can be given by

$$\ddot{u}_a(z, t) = \ddot{u}(z, t) + \ddot{u}_g(t) \tag{12.7}$$

where $\ddot{u}_a(z, t)$ = absolute acceleration. For the case of zero damping ($D_n = 0$), it can be shown from Eq. (12.4) that the model contribution to the absolute acceleration can be given by

$$\ddot{u}_{an}(z, t) = \omega_n^2 u_n(z, t) \tag{12.8}$$

For small values of damping (that is, $D_n \approx 0$) $\omega_d \approx \omega_n$. Thus, from Eq. (12.6),

$$\ddot{u}_a(z, t) = \sum_{n=1}^{n=\infty} 2\omega_n \left[\frac{1}{\beta_n J_1(\beta_n)} \right] \left[J_0\left(\beta_n \frac{z}{H} \right) \right] \int_0^t \ddot{u}_g e^{-D_n\omega_n(t-t')}$$
$$\times \sin[\omega_n(t - t')]\, dt' \tag{12.9a}$$

The preceding equation can be rewritten as

$$\ddot{u}_a(z, t) = \sum_{n=1}^{n=\infty} \ddot{u}_{an}(z, t) \tag{12.9b}$$

where

$$\ddot{u}_{an}(z, t) = \omega_n \eta_n(z) V_n(t) \tag{12.10}$$

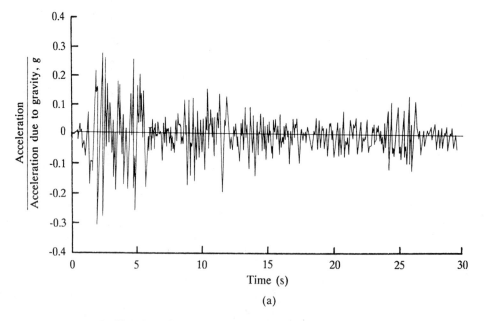

(a)

Figure 12.4 (Continued)

$$\omega_n = \frac{\beta_n}{H}\sqrt{\frac{G}{\rho}} \qquad (12.11)$$

$$\eta_n(z) = \frac{2J_0[\beta_n(z/H)]}{\beta_n J_1(\beta_n)} \qquad (12.12)$$

$$V_n(t) = \int_0^t \ddot{u}_g e^{-D_n\omega_n(t-t')}\sin[\omega_n(t-t')]\,dt' \qquad (12.13)$$

For a given ground acceleration record $[\ddot{u}_g(t)]$ and embankment, Eq. (12.9) can be programmed in a computer and the variation of the absolute acceleration with depth can be obtained. An example for such a case is shown in Figure 12.4. Figure 12.4b shows the variation of acceleration of a 100-ft-high embankment with time that has been subjected to a ground acceleration, as shown in Figure 12.4a.

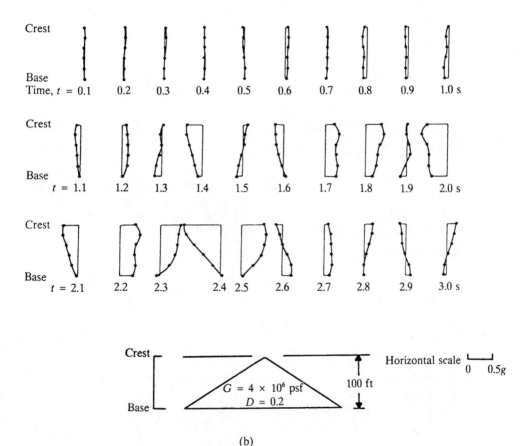

Crest

Base
Time, $t = 0.1$ 0.2   0.3   0.4   0.5   0.6   0.7   0.8   0.9   1.0 s

Crest

Base
$t = 1.1$ 1.2   1.3   1.4   1.5   1.6   1.7   1.8   1.9   2.0 s

Crest

Base
$t = 2.1$ 2.2   2.3   2.4 2.5   2.6   2.7   2.8   2.9   3.0 s

Crest

$G = 4 \times 10^6$ psf
$D = 0.2$ 100 ft

Horizontal scale
0   0.5g

Base

(b)

Figure 12.4   (a) Accelerogram of El Centro, California, earthquake, May 18, 1940—N-S component; (b) Acceleration distribution at 0.1-s intervals for 100-ft-high dam subjected to El Centro earthquake (after Seed and Martin, 1966)

12.4   Velocity and Acceleration Spectra

The term $V_n(t)$ given by Eq. (12.13) is a function of the ground acceleration ($\ddot{u}_g$), damping (D_n), natural frequency (ω_n) and the time (t). For a given earthquake record, the maximum value of $V_n(t)$ that will correspond to a given value of ω_n can easily be determined. This is referred to as the *spectral velocity*, S_{Vn}, where

$$S_{Vn} = \left\{ \int_0^t \ddot{u}_g e^{-D_n\omega_n(t-t')} \sin[\omega_n(t-t')]\, dt' \right\}_{max} \tag{12.14}$$

The spectral velocities, S_{Vn}, corresponding to various values of ω_n can be calculated and plotted in a graphical form (for a given value of D_n). This is called a *velocity spectrum*. Note that S_{Vn} has the units of velocity. Similar plots can be made for a number of values of the damping ratio. Figure 12.5 shows the nature of the plot of S_{Vn} with the natural period. Note that

$$T_n = \frac{2\pi}{\omega_n} \tag{12.15}$$

where T_n = natural period.

For a given mode of vibration of an embankment, the maximum value of the acceleration can be given as

$$[\ddot{u}_{an}(z)]_{max} = \omega_n \eta_n(z) S_{Vn} \tag{12.16}$$

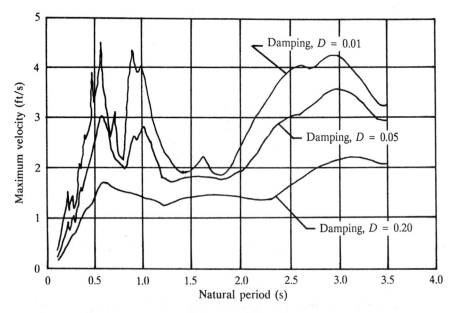

Figure 12.5 Velocity response spectra for El Centro (1940) earthquake (after Seed, 1975)

Again, keeping in mind that acceleration is equal to natural frequency times the velocity, Eq. (12.16) can be rewritten as

$$[\ddot{u}_{an}(z)]_{max} = \eta_n(z)S_{an} \tag{12.17}$$

where

$$S_{an} = \text{spectral acceleration} = \omega_n S_{Vn}$$

The expression for $\eta_n(z)$ is given by Eq. (12.12). Since the values of β_n (for $n = 1, 2, 3, \ldots$) are known, $\eta_n(z)$ can easily be calculated. The variation of $\eta_n(z)$ for $n = 1, 2, 3$ is shown in Figure 12.6. Thus, the spectral acceleration for a given

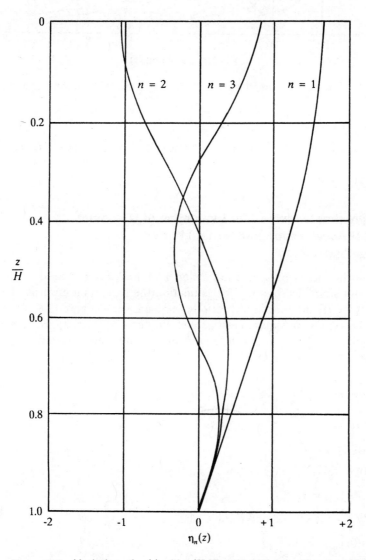

Figure 12.6   Variation of $\eta_n(z)$ with z/H [Eq. (12.12)] (after Seed and Martin, 1966)

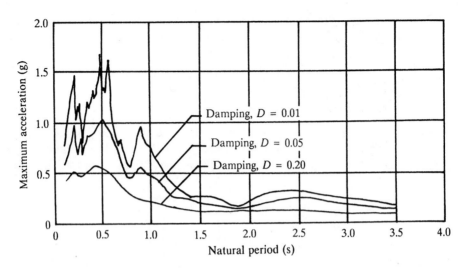

Figure 12.7   Acceleration response spectra for El Centro (1940) earthquake (after Seed, 1975)

value of D_n and ω_n (or T_n) can be calculated. A plot of S_{an} versus T_n is referred to as the *acceleration spectrum*. Figure 12.7 shows an example of an acceleration response spectra.

12.5   Approximate Method for Evaluation of Maximum Crest Acceleration and Natural Period of Embankments

Based on the theory presented in Section 12.3, Makdisi and Seed (1979) have presented a simplified method for estimating the maximum crest acceleration $[\ddot{u}_{a(max)}$ at $z = 0]$ and the natural period of embankment (Figure 12.8). According to this theory, the maximum crest acceleration can be given approximately by the square root of the sum of the square of maximum acceleration at the crest for the first three modes, or

$$\ddot{u}_{a(max)} \ (\text{at } z = 0) = \sqrt{\sum_{n=1}^{3} [\ddot{u}_{an}(0)]^2_{max}} \qquad (12.18)$$

From Eq. (12.17)

$$[\ddot{u}_{a1}(0)]^2_{max} = \eta_1(0)S_{a1}$$

But $\eta_1(0) = 1.6$ (Figure 12.6). So

$$[\ddot{u}_{a1}(0)]_{max} = 1.6S_{a1} \qquad (12.19)$$

Similarly, for the second and third modes,

$$[\ddot{u}_{a2}(0)]_{max} = 1.06S_{a2} \qquad (12.20)$$

$$[\ddot{u}_{a3}(0)]_{max} = 0.86S_{a3} \qquad (12.21)$$

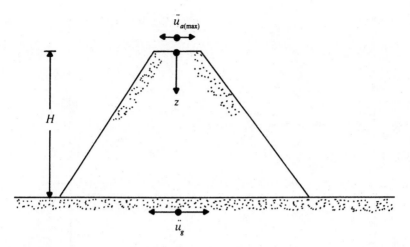

Figure 12.8   Approximate method for evaluation of maximum crest acceleration

Now, combining Eqs. (12.18) through (12.21),

$$[\ddot{u}_a(0)]_{max} = \sqrt{(1.6S_{a1})^2 + (1.06S_{a2})^2 + (0.86S_{a3})^2} \qquad (11.22)$$

The step-by-step procedure for obtaining the maximum crest acceleration is given next.

1. Plot graphs of the variations of G/G_{max} versus shear strain (γ') and D versus shear strain (γ') for the soil present in the embankment as shown in Figure 12.9. (*Note: G* = shear modulus, G_{max} = maximum shear modulus, D = damping ratio.) This can be done by using the principles outlined in Chapter 4.

2. Obtain an acceleration spectra for the design earthquake.

3. Assume a value of the shear modulus G and calculate G/G_{max}.

4. For the assumed value of G/G_{max} (Step 3), determine the shear strain γ' (Figure 12.9).

5. Corresponding to the shear strain obtained in Step 4, obtain the damping ratio D (Figure 12.9).

6. Calculate ω_n ($n = 1, 2,$ and 3) using Eq. (12.3). The value of the shear modulus to be used is from Step 3.

7. Using the damping ratio obtained in Step 5 and $\omega_1, \omega_2,$ and ω_3 obtained in Step 6, obtain spectral accelerations S_{a1}, S_{a2}, and S_{a3}. (This is from the acceleration spectra obtained in Step 2.)

8. Calculate the maximum crest acceleration using Eq. (12.22).

9. Calculate the average equivalent shear strain in the embankment as follows. The shear strain, $\gamma'(z, t)$, can be given by (Figure 12.8)

$$\gamma'(z, t) = \sum_{n=1}^{\infty} \frac{2J_1[\beta_n(z/H)]}{H\omega_n J_1(\beta_n)} V_n(t) \qquad (12.23)$$

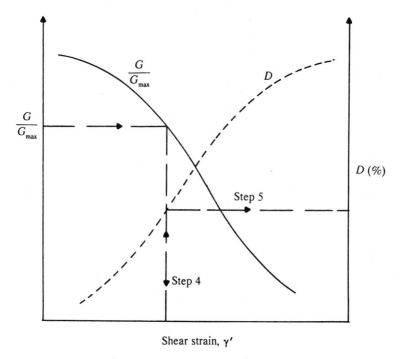

$$\text{Shear strain, } \gamma'$$

Figure 12.9 Nature of variation of G/G_{max} and D with shear strain

The terms in the right-hand side of Eq. (12.23) have all been defined in Sections 12.2 and 12.3. In Section 12.2, we have defined ω_n as

$$\omega_n = \frac{\beta_n}{H} \sqrt{\frac{G}{\rho}}$$

or

$$\omega_n^2 = \frac{\beta_n^2}{H^2} \times \frac{G}{\rho} \tag{12.24}$$

Substituting Eq. (12.24) into Eq. (12.23), we obtain

$$\gamma'(z,t) = H\frac{\rho}{G}\left[\sum_{n=1}^{\infty} \frac{2J_1(\beta_n(z/H))}{\beta_n^2 J_1(\beta_n)} \omega_n V_n(t) \right] \tag{12.25}$$

$$= H\frac{\rho}{G}\left[\sum_{n=1}^{\infty} \phi_n'(z)\omega_n V_n(t) \right] \tag{12.26}$$

where

$$\phi_n'(z) = \frac{2J_1[\beta_n(z/H)]}{\beta_n^2 J_1(\beta_n)} \tag{12.27}$$

The variations of ϕ_n' with depth (z) for $n = 1, 2,$ and 3 are given in Figure 12.10. The maximum shear strain at any depth z of the embankment can be approxi-

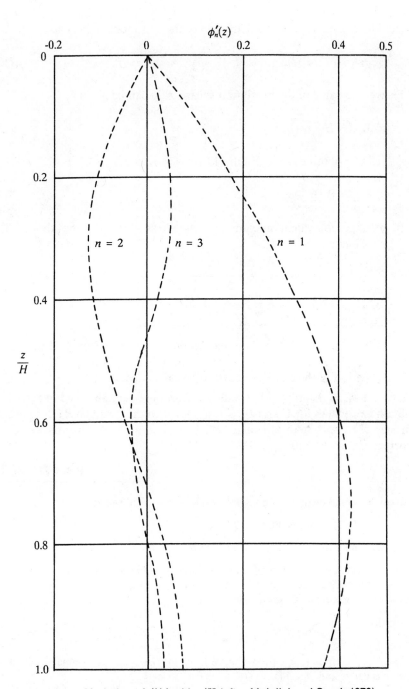

Figure 12.10   Variation of $\phi'_n(z)$ with z/H (after Makdisi and Seed, 1979)

mated by considering the contribution of the first mode only. Thus,

$$\gamma'_{max}(z) = H \frac{\rho}{G} \phi'_1(z) S_{a1} \tag{12.28}$$

The average value of the maximum shear strain can be given as

$$(\gamma'_{av})_{max} = H \frac{\rho}{G} (\phi'_1)_{av} S_{a1} \tag{12.29}$$

$(\phi'_1)_{av}$ can be obtained from Figure 12.10 as 0.3. So

$$(\gamma'_{av})_{max} = 0.3 H \frac{\rho}{G} S_{a1} \tag{12.30}$$

The average equivalent maximum cyclic shear strain can be about 65% of $(\gamma'_{av})_{max}$. Thus

$$(\gamma'_{av})_{eq} = (0.3)(0.65) H \frac{\rho}{G} S_{a1} = 0.195 H \frac{\rho}{G} S_{a1} \tag{12.31}$$

10. Compare $(\gamma'_{av})_{eq}$ to the shear strain obtained in Step 4. If they are the same, then the maximum crest acceleration obtained in Step 8 is correct. The natural period of the embankment can be calculated at $2\pi/\omega_1$.

11. If $(\gamma'_{av})_{eq}$ from Step 9 is different than the strain obtained in Step 4, then obtain new values for G and D corresponding to the strain level $(\gamma'_{av})_{eq}$ obtained in Step 9. Repeat Steps 6 through 10. A few iterations of this type will give the correct values of $[\ddot{u}_a(0)]_{max}$, G, D, and the natural period.

Example 12.1

An earth embankment is 30 m high. For the embankment soil, given:

$$\text{Unit weight, } \gamma = 19.65 \text{ kN/m}^3$$

$$\text{Maximum shear modulus} = 160{,}000 \text{ kN/m}^2$$

Figure 12.11 shows the nature of variation of G/G_{max} and D with shear strain. Figure 12.12 shows a normalized acceleration spectra (maximum ground acceleration is 0.25 g). Determine the maximum crest acceleration.

Solution

Iteration 1
Let G/G_{max} be equal to 0.4. From Figure 12.11, for $G/G_{max} = 0.4$, the magnitude of shear strain is 0.07% and $D \approx 14\%$. If $G/G_{max} = 0.4$,

$$G = (0.4)(160{,}000) = 64{,}000 \text{ kN/m}^2$$

From Eq. (12.3),

$$\omega_1 = \frac{2.404}{H} \sqrt{\frac{G}{\rho}} = \frac{2.404}{30} \sqrt{\frac{64{,}000}{(19.65/9.81)}} = 14.32 \text{ rad/s}$$

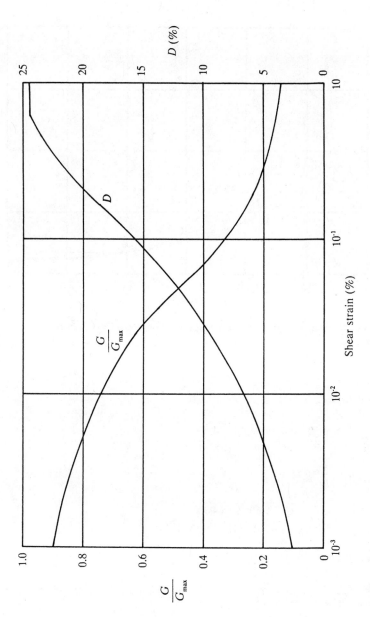

Figure 12.11

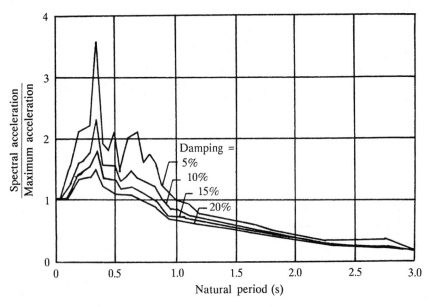

Figure 12.12   Normalized acceleration response spectra—Taft Record, N-S component (after Makdisi and Seed, 1979)

So

$$\text{Period } T_1 = \frac{2\pi}{\omega_1} = 0.435 \text{ s}$$

$$\omega_2 = \frac{5.52}{H}\sqrt{\frac{G}{\rho}} = \frac{5.52}{30}\sqrt{\frac{64,000}{(19.65/9.81)}} = 32.89 \text{ rad/s}$$

$$\text{Period } T_2 = \frac{2\pi}{\omega_2} = 0.191 \text{ s}$$

$$\omega_3 = \frac{8.65}{H}\sqrt{\frac{G}{\rho}} = \frac{8.65}{30}\sqrt{\frac{64,000}{(19.65/9.81)}} = 51.54 \text{ rad/s}$$

$$\text{Period } T_3 = \frac{2\pi}{\omega_3} = 0.122 \text{ s}$$

From Figure 12.12, for these values of T_1, T_2, and T_3 and $D \approx 14\%$, the spectral accelerations are as follows:

$$S_{a1} = (1.35)(0.25 \text{ g}) = 0.3375 \text{ g}$$

$$S_{a2} = (1.41)(0.25 \text{ g}) = 0.3525 \text{ g}$$

$$S_{a3} = (1.18)(0.25 \text{ g}) = 0.295 \text{ g}$$

From Eq. (12.22)

$$[\ddot{u}_a(0)]_{\max} = \sqrt{(1.6S_{a1})^2 + (1.06S_{a2})^2 + (0.86S_{a3})^2}$$

$$= \sqrt{(1.6 \times 0.3375 \text{ g})^2 + (1.06 \times 0.3525 \text{ g})^2 + (0.86 \times 0.295 \text{ g})^2}$$

$$= 0.704 \text{ g}$$

Using Eq. (12.31),

$$(\gamma'_{av})_{eq} = 0.195H\frac{\rho}{G}S_{a1}$$

$$= (0.195)(30)\left(\frac{19.65/9.81}{64,000}\right)(0.3375 \times 9.81) = 0.061\%$$

The above value of $(\gamma'_{av})_{eq}$ is approximately the same as the assumed value. So

$$[\ddot{u}_a(0)]_{max} \approx \underline{0.70\ g.} \qquad \blacksquare$$

12.6 Fundamental Concepts of Stability Analysis

Until the mid-1960s, most of the earth embankment slopes were analyzed by the so-called pseudostatic method. According to this method, a trial failure surface *ABC*, as shown in Figure 12.13 is chosen. *ABC* is an arc of a circle with its center at *O*. Considering the unit length of the embankment at a right angle to the cross section shown, the forces acting on the trial failure surface are as follows:

a. Weight of the wedge, *W*.

b. Inertia force on the wedge, k_hW, which accounts for the effect of an earthquake on the trial wedge. The factor k_h is the average coefficient of horizontal acceleration.

c. Resisting force per unit area, *s*, which is the shear strength of the soil acting along the trial failure surface, *ABC*.

The factor of safety with respect to strength, F_s, is calculated as

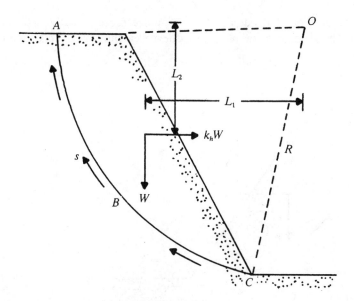

Figure 12.13 Stability analysis for slope

$$F_s = \frac{\text{resisting moment about } O}{\text{overturning moment about } O} = \frac{s(\overset{\frown}{ABC})R}{WL_1 + k_h WL_2}$$

This procedure is repeated with several trial failure surfaces to determine the minimum values of F_s. It is assumed that if the minimum value of F_s is equal to or greater than 1, the slope is stable.

The magnitude of k_h used for the design of many dams in the past ranged from 0.05 to 0.15 in the United States. In Japan, this value has been less than 0.2. Following are some examples of this type of assumption in the design of earth dams (Seed, 1981).

Dam	Country	Horizontal seismic coefficient, k_h	Minimum factor of safety, F_s
Aviemore	New Zealand	0.1	1.5
Bersemisnoi	Canada	0.1	1.25
Digma	Chile	0.1	1.15
Globocica	Yugoslavia	0.01	1.0
Karamauri	Turkey	0.1	1.2
Kisenyama	Japan	0.12	1.15
Mica	Canada	0.1	1.25
Misakubo	Japan	0.12	—
Netzahualcoyote	Mexico	0.15	1.35
Oroville	United States	0.1	1.2
Paloma	Chile	0.12 to 0.2	1.25 to 1.2
Ramganga	India	0.12	1.2
Tercan	Turkey	0.15	1.2
Yeso	Chile	0.12	1.5

A second method that has gained acceptance more recently is the determination of the displacement of slopes due to earthquakes. This method is

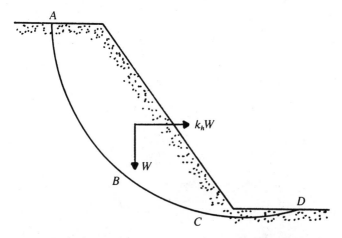

Figure 12.14 Soil slope

primarily based on the original concept proposed by Newmark (1965) and can be explained in the following manner.

Consider a slope as shown in Figure 12.14. When this slope is subjected to an earthquake, the stability of the slope will depend on the shear strength of the soil and the average coefficient of horizontal acceleration. The factor of safety of the soil mass located above the most critical surface ABC will become equal to 1 when k_h becomes equal to k_y. This value of $k_h = k_y$ may be defined as the *coefficient of yield acceleration*. Now refer to Figure 12.15a, which shows a plot of the horizontal acceleration with time to which the soil wedge $ABCD$ is being subjected (Figure 12.14). At time $t = t_1$, the horizontal acceleration is $k_y g$ ($g =$ acceleration due to gravity). Between time $t = t_1 = t_2$, the velocity of the sliding

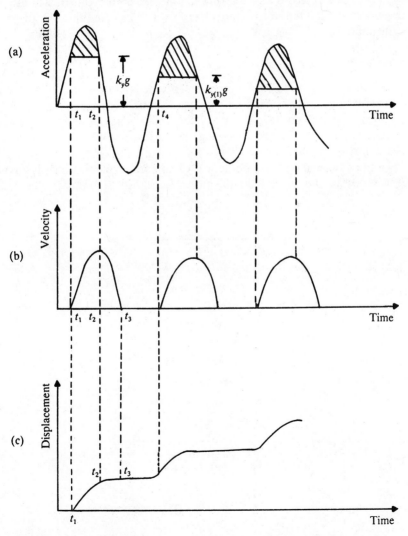

Figure 12.15 Integration method to determine downslope displacement

wedge will increase. This velocity can be determined by integration of the shaded area. The velocity will gradually decrease and become equal to zero at $t = t_3$ (Figure 12.15b). The displacement of the soil wedge can now be determined by integration of the area under the velocity versus time plot between $t = t_1$ and $t = t_3$ (Figure 12.15c). It is important to note that the peak shear strength of the soil along the critical surface $ABCD$ has now been mobilized. Hence, when the horizontal acceleration reaches $k_{y(1)}g$ (which is less than $k_y g$) at time $t = t_4$, the velocity of the sliding wedge will again increase, since the post-peak strength will be mobilized. As before, we can determine the velocity and the displacement of the sliding wedge by using the integration method. Hence, with time, the displacement of the wedge gradually increases. In most cases of embankment stability considerations, it can be shown (Seed, 1981) that where the crest acceleration does not exceed $0.75g$, deformation of such embankments will usually be acceptably small if the embankment has $F_s = 1.15$ as determined by the pseudostatic analysis.

Average Value of k_h

In reference to Figure 12.13, it has been mentioned in this section that an average value of k_h is usually assumed for the pseudostatic method of analysis of slopes. It is now essential to have a general theoretical background as to what this average value of k_h is. The following theoretical derivation has been recommended by Seed and Martin (1966).

Figure 12.16 shows a hypothetical earth embankment, which is triangular in cross section. Let us consider the inertia force on an arbitrary soil wedge Oac.

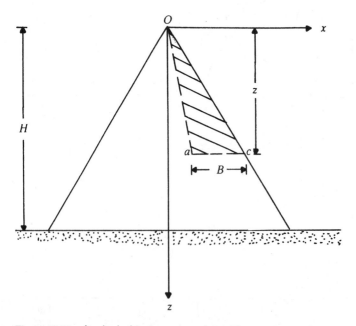

Figure 12.16   Analysis for average value of k_h

The displacement of the embankment at a depth z can be given as [Eqs. (12.6) and (12.13)]

$$u(z,t) = \sum_{n=1}^{n=\infty} \frac{2J_0[\beta_n(z/H)]}{\omega_n \beta_n J_0(\beta_n)} \cdot V_n(t) \qquad (12.32)$$

So, the distribution of shear strain can be obtained as

$$\frac{\partial u}{\partial z}(z,t) = \sum_{n=1}^{n=\infty} \frac{2J_1[\beta_n(z/H)]}{H\omega_n J_1(\beta_n)} V_n(t) \qquad (12.33)$$

Hence, the distribution shear stress, $\tau(z,t)$ is

$$\tau(z,t) = G\frac{\partial u}{\partial z}(z,t) = \sum_{n=1}^{n=\infty} \frac{2GJ_1[\beta_n(z/H)]}{H\omega_n J_1(\beta_n)} V_n(t) \qquad (12.34)$$

The shear force, $F(z,t)$, acting on the base of the wedge *Oac* is

$$F(z,t) = \tau(z,t)B \qquad (12.35)$$

However,

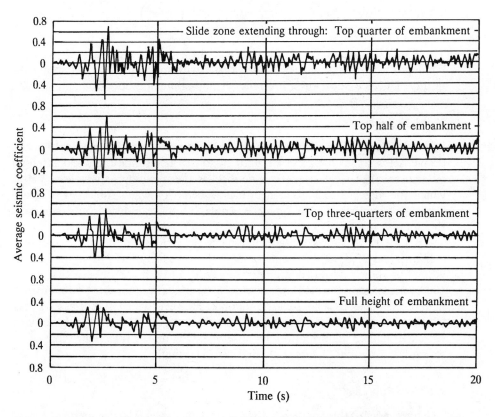

Figure 12.17 Values of average seismic coefficient for 100-ft-high embankment subjected to El Centro earthquake (shear wave velocity, $v_s = 1000$ ft/s, 20% critical damping) (after Seed and Martin, 1966)

$$F(z, t) = (\text{mass of the wedge } Oac)\ddot{u}_a(t)_{av}$$

$$= (\tfrac{1}{2}\rho Bz)\ddot{u}_a(t)_{av} \tag{12.36}$$

where

$$\ddot{u}_a(t)_{av} = \text{average lateral acceleration}$$

$$\rho = \text{density of the soil in the wedge}$$

So

$$\ddot{u}_a(t)_{av} = \frac{2F(z, t)}{\rho Bz} = \frac{2\tau(z, t)B}{\rho Bz} = \frac{2\tau(z, t)}{\rho z} \tag{12.37}$$

Combining Eqs. (12.34) and (12.37),

$$\ddot{u}_a(t)_{av} = \sum_{n=1}^{n=\infty} \frac{4GJ_1[\beta_n(z/H)]}{\rho H\omega_n z J_1(\beta_n)} V_n(t) \tag{12.38}$$

So

$$\boxed{k_h = \frac{1}{g}\ddot{u}_a(t)_{av} = \sum_{n=1}^{n=\infty} \frac{4GJ_1[\beta_n(z/H)]}{g\rho H\omega_n z J_1(\beta_n)} \cdot V_n(t)} \tag{12.39}$$

The value of k_h is a function of time. Since the average acceleration varies with the depth z, the magnitude of k_h also varies with time. Figure 12.17 shows the results of a calculation for k_h for a model embankment at four different levels.

Yield Strength

In the analysis of stability for earth embankments, it is important to make proper selection of the yield strength of soil to determine the shear strength parameters. The yield strength is defined as the maximum stress level below which the material exhibits a near-elastic behavior when subjected to cyclic stresses of numbers and frequencies similar to those induced by earthquake shaking.

Figure 12.18 shows the concept of cyclic yield strength of a clayey soil (Makdisi and Seed, 1978). The material in this case has a yield strength of about 90% of its static undrained strength. In Figure 12.18, it can be seen that under 100 cycles of stress, which amounts to 80% of static undrained strength, the material behaves in a near-elastic manner. However when 10 cycles of stress, which amounts to 95% of static undrained strength, is applied, substantially large permanent deformation is observed (Figure 12.18). Hence the yield strength is about 90% of its static undrained strength.

Stability Analysis

The remainder of this chapter is divided into two parts. The first part is devoted to the *pseudostatic methods of stability analysis* and the second part, to the *determination of the deformation of slopes*.

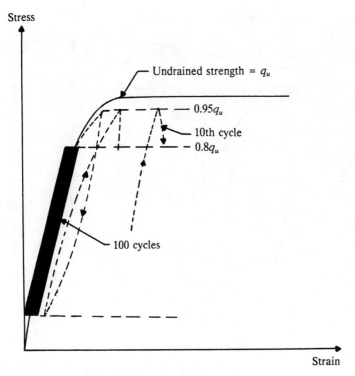

Figure 12.18   Concept of cyclic yield strength

PSEUDOSTATIC ANALYSIS

12.7   Clay Slopes ($\phi = 0$ Condition)— Koppula's Analysis

A clay ($\phi = 0$ condition) slope of height H is shown in Figure 12.19a. In order to determine the minimum factor of safety of the slope with respect to strength, we consider a trial failure surface ABC, which is an arc of a circle with its center located at O. Let the saturated unit weight and the undrained cohesion of the clay soil be equal to γ and c_u, respectively. The undrained cohesion c_u may increase with depth z measured from the top of the slope and can be expressed as

$$c_u = c_0 + a_0 z \tag{12.40}$$

where $a_0 = a$ constant.

Per unit length of the slope, for consideration of the stability of the soil mass located above the trial failure surface, the following forces need to be considered:

a. Weight of the soil mass, W

b. Undrained cohesion, c_u, per unit area along the trial failure surface ABC

c. Inertia force on the soil mass, $k_h W$ (where $k_h = $ *average* horizontal acceleration of the mass)

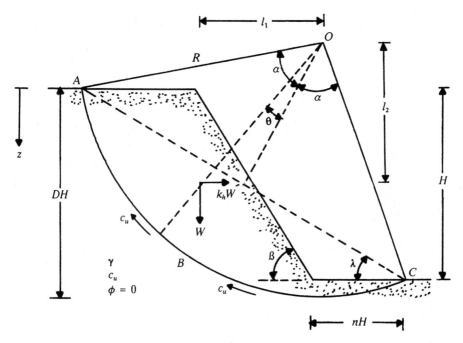

Figure 12.19 Koppula's analysis for clay slopes ($\phi = 0$ condition)

The overturning moment, M_D, about point O can be given as

$$M_D = \underbrace{Wl_1}_{M_W} + \underbrace{k_h Wl_2}_{M_E} \tag{12.41}$$

Koppula (1984) has expressed M_W and M_E in the following forms:

$$M_W = \frac{\gamma H^3}{12}(1 - 2\cot^2\beta - 3\cot\alpha\cot\beta + 3\cot\beta\cot\lambda$$
$$+ 3\cot\lambda\cot\alpha - 6n\cot\beta - 6n^2 - 6n\cot\alpha + 6n\cot\lambda) \tag{12.42}$$

$$M_E = \frac{k_h \gamma H^3}{12}(\cot\beta + \cot^3\lambda + 3\cot\alpha\cot^2\lambda$$
$$- 3\cot\alpha\cot\beta\cot\lambda - 6n\cot\alpha\cot\lambda) \tag{12.43}$$

The restoring moment M_R about O is

$$M_R = R\int_{-\alpha}^{+\alpha} c_u R\, d\theta \tag{12.44}$$

where R = radius of the circular arc

$$c_u = c_0 + a_0[R\cos(\lambda + \theta) - R\cos(\alpha - \lambda) + H] \tag{12.45}$$

Combining Eqs. (12.44) and (12.45)

$$M_R = \frac{a_0 H^3}{4 \sin^2 \alpha \sin^2 \lambda} [\alpha(1 - \cot \alpha \cos \lambda) + \cot \lambda] + \frac{c_0 H^2 \alpha}{2 \sin^2 \alpha \sin^2 \lambda} \tag{12.46}$$

The factor of safety F_s against sliding can be given as

$$F_s = \frac{M_R}{M_W + M_E} \tag{12.47}$$

The minimum value of F_s has to be determined by considering several trial failure surfaces. Koppula (1984) has expressed a minimum factor of safety in the form

$$F_s = \frac{a_0}{\gamma} N_1 + \frac{c_0}{\gamma H} N_2 \tag{12.48}$$

where N_1 and N_2 = stability numbers.

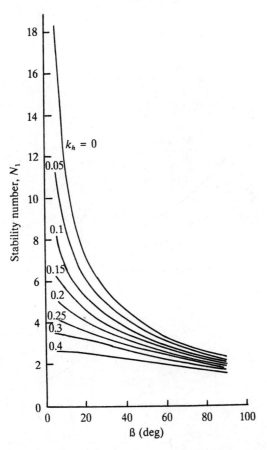

Figure 12.20 Variation of N_1 with slope angle ß (after Koppula, 1984)

The stability numbers are functions of k_h, the slope angle β, and also the depth factor, D (for the definition of depth factor, see Figure 12.19). Figure 12.20 shows the variation of N_1 and k_h varying from 0 to 0.4 and β varying from 0° to 90°. In a similar manner, the variation of N_2 with k_h and D for $\beta \leq 50°$ is shown in Figure 12.21. For $\beta \geq 55°$, the variation of N_2 with k_h is given in Figure 12.22. In order to use Eq. (12.48) and Figures 12.20, 12.21, and 12.22, the following points need to be kept in mind.

1. If the undrained shear strength of the soil increases linearly from zero at the top, then

$$c_u = a_0 z \tag{12.49}$$

For this case, the critical slip surface associated with minimum F_s passes through the toe of the slope and lies within the slope. So

$$\left. \begin{array}{l} n = 0 \\ D = 0 \end{array} \right\} \tag{12.50}$$

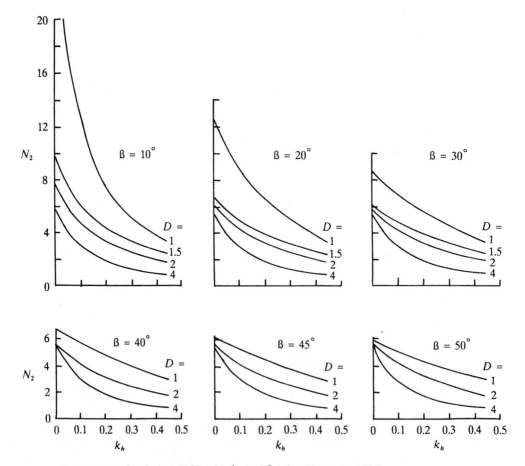

Figure 12.21   Variation of N_2 with k_h and ß (after Koppula, 1984)

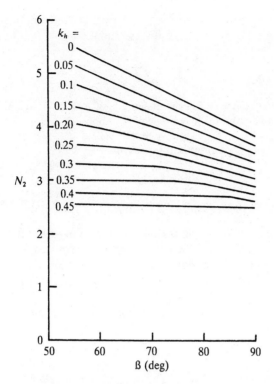

Figure 12.22   Variation of N_2 with $\beta \geq 55°$ and k_h (after Koppula, 1984)

Also

$$F_s = \frac{a_0}{\gamma} N_1 \tag{12.51}$$

2. If the magnitude of c_u is constant with depth, then

$$\left.\begin{array}{l} a_0 = 0 \\ c_u = c_0 \end{array}\right\} \tag{12.52}$$

For this case

$$F_s = \frac{c_0}{\gamma H} N_2 \tag{12.53}$$

The stability number N_2 is a function of β, D, and k_h if $\beta \leq 53°$. However if $\beta \geq 53°$, then N_2 is a function of β and k_h only (that is, $n = 0$ and $D = 0$).

Example 12.2

Refer to the slope shown in Figure 12.19. Given:

$$H = 15\,\text{m} \qquad \gamma = 18\,\text{kN/m}^3$$

$$\beta = 60° \qquad c_u = 48 + 3z \,(\text{kN/m}^2)$$

Determine the factor of safety F_s for $k_h = 0.3$.

Solution

From Eq. (12.48)

$$F_s = \frac{a_0}{\gamma} N_1 + \frac{c_0}{\gamma H} N_2$$

From Figure 12.20, for $\beta = 60°$ and $k_h = 0.3$, the magnitude of $N_1 \approx 2.38$. Again, from Figure 12.22, for $\beta = 60°$ and $k_h = 0.3$, the magnitude of N_2 is about 3.28. So

$$F_s = \left(\frac{3}{18}\right)(2.38) + \frac{48}{(18)(15)}(3.28) = 0.397 + 0.583 = \underline{0.98} \qquad \blacksquare$$

12.8   Slopes with $c - \phi$ Soil—Majumdar's Analysis

Taylor (1937) proposed the friction circle method of analyzing the stability of slopes with $c - \phi$ soils. In this analysis, the effect of earthquakes was not taken into consideration. The details of this slope stability analysis can be found in most soil mechanics textbooks (for example, Das, 1990). However, it can be summarized as follows.

Figure 12.23 shows a slope made of a soil having a shear strength that can be given as

$$\tau_f = c + \sigma' \tan \phi \qquad (12.54)$$

where

τ_f = shear strength

c = cohesion

σ' = effective normal stress

ϕ = drained friction angle

For ϕ greater than about 3°, the *critical circle* for stability analysis always passes through the toe, as shown in Figure 12.23. For stability analysis, one can define three different factors of safety for the soil at any point along the critical surface:

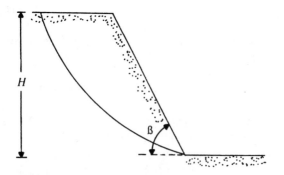

Figure 12.23   Slopes with $c - \phi$ soil

1. Factor of safety with respect to friction:

$$F_\phi = \frac{\tan \phi}{\tan \phi_d}$$

where ϕ_d = developed friction angle ($\leq \phi$).

2. Factor of safety with respect to cohesion:

$$F_c = \frac{c}{c_d}$$

where c_d = developed cohesion ($\leq c$).

3. Factor of safety with respect to strength:

$$F_s = \frac{c + \sigma' \tan \phi}{c_d + \sigma' \tan \phi_d} \tag{12.55}$$

It is obvious from the preceding definitions that if

$$\frac{c}{c_d} = \frac{\tan \phi}{\tan \phi_d}$$

then

$$F_c = F_\phi = F_s \tag{12.56}$$

It is important to note that the relationship for the factor of safety developed in Section 12.7 is the factor of safety with respect to strength.

Using the preceding concepts for factor of safety, Taylor's analysis (1937) for the stability of slopes by the friction circle method can be given in a graphical form, the nature of which is shown in Figure 12.24. Note that in Figure 12.24 the term m is defined as

$$m = \frac{c_d}{\gamma H}$$

Majumdar (1971) expanded Taylor's analysis of slope by taking into consideration the horizontal earthquake forces as shown in Figure 12.25. By simple mathematical manipulations, Majumdar showed that if the actual effective friction angle ϕ of the soil can be modified to ϕ_m, it can then be used in Taylor's analysis to determine the factor of safety with respect to strength (F_s) for the critical surface of a slope. The relationship between ϕ and ϕ_m can be expressed as

$$\phi_m = \tan^{-1}(M \tan \phi) \tag{12.57}$$

The term M in Eq. (12.57) is a function of the slope angle (β) and the horizontal coefficient of acceleration (k_h). Figure 12.26 shows the variation of M with k_h for $\beta = 15°, 30°, 45°, 60°$, and $75°$.

Figure 12.27 shows the modified plot of Taylor's chart (that is, m versus β) for use of stability analysis. It is important to note that ϕ_d for a given soil is always less than or equal to ϕ_m. In order to determine the factor of safety for a given slope, the following step-by-step procedure can be applied.

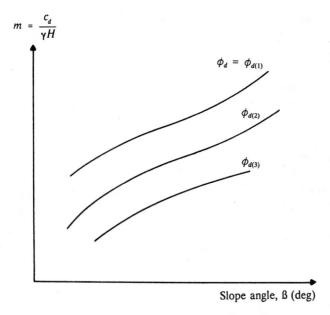

Figure 12.24   Nature of variation of m with β and ϕ_d

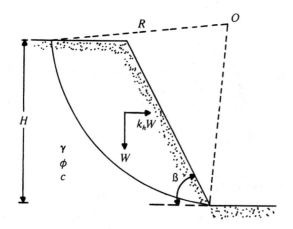

Figure 12.25   Analysis of slopes with $c - \phi$ soil

1. Determine the soil parameters ϕ and c and the unit weight γ.

2. Determine the parameters for the slope, that is, β and H.

3. For given values of ϕ, β, and k_h, determine the factor M from Figure 12.26.

4. Assume several values for the developed friction angle ϕ_d (such as $\phi_{d(1)}$, $\phi_{d(2)}$, $\phi_{d(3)}$, ...). Note that $\phi_d \leq \phi_m$.

5. For each assumed value of ϕ_d, determine the factor of safety with respect to friction, or

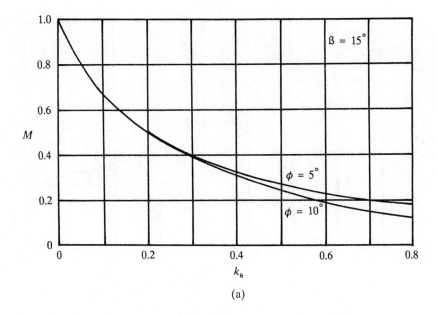

(a)

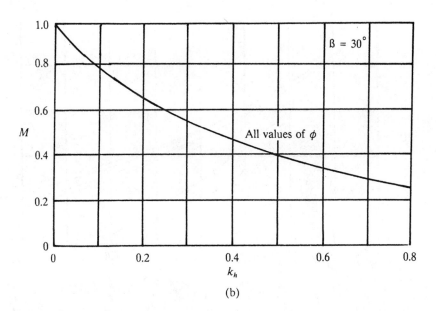

(b)

Figure 12.26   (Continued)

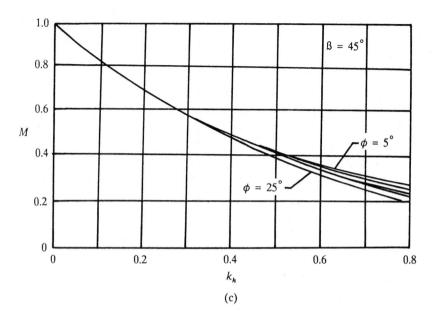

(c)

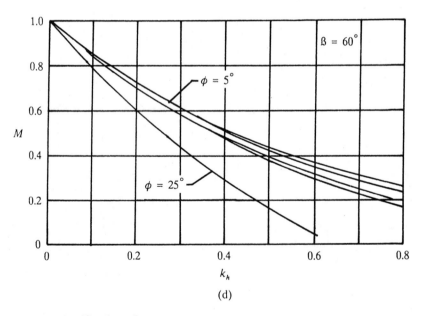

(d)

Figure 12.26 (Continued)

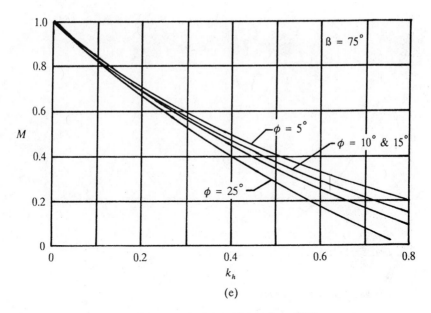

(e)

Figure 12.26   Variation of M with k_h (after Majumdar, 1971)

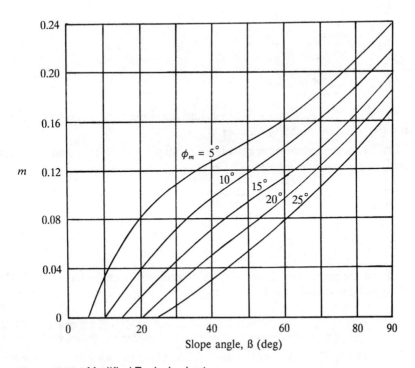

Figure 12.27   Modified Taylor's chart

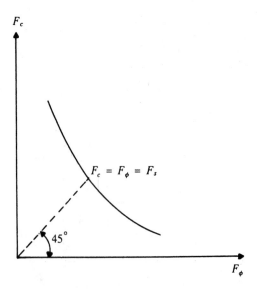

Figure 12.28   Calculation of F_s

$$F_{\phi(1)} = \frac{\tan \phi_m}{\tan \phi_{d(1)}}$$

$$F_{\phi(2)} = \frac{\tan \phi_m}{\tan \phi_{d(2)}}$$

$$F_{\phi(3)} = \frac{\tan \phi_m}{\tan \phi_{d(3)}}$$

6. With each assumed value of ϕ_d and the slope angle β, go to Figure 12.27 and determine the stability number m.

7. From the values of m calculated in Step 6, calculate c_d and the factor of safety with respect to cohesion (F_c) as

$$c_{d(1)} = m_1 \gamma H; \qquad F_{c(1)} = \frac{c}{c_{d(1)}}$$

$$c_{d(2)} = m_2 \gamma H; \qquad F_{c(2)} = \frac{c}{c_{d(2)}}$$

8. Plot a graph of F_ϕ versus F_c as determined from Steps 5 and 7 (Figure 12.28) and determine $F_s = F_c = F_\phi$.

Example 12.3

A homogeneous slope is shown in Figure 12.29a. Using the procedure described in this section, determine the factor of safety with respect to strength. Use $k_h = 0.3$.

Solution

Given $H = 12$ m, $\beta = 30°$, $\gamma = 16$ kN/m³, $c = 20$ kN/m², and $\phi = 34°$. Now, referring to Figure 12.26b, for $k_h = 0.3$, $M \approx 0.54$. So

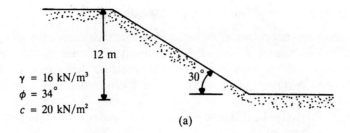

(a)

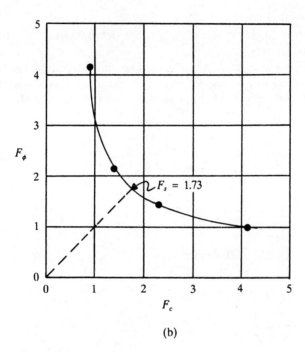

(b)

Figure 12.29

$$\phi_m = \tan^{-1}(M \tan \phi) = \tan^{-1}[(0.54)(\tan 34)] = 20°$$

The following table can be prepared.

Assumed developed friction angle, ϕ_d (deg)	$\tan \phi_d$	$F_\phi = \dfrac{\tan \phi_m}{\tan \phi_d}$	m (Figure 12.27)	$F_c = \dfrac{c}{c_d} = \dfrac{c}{m\gamma H}$
5	0.0875	4.16	0.11	0.95
10	0.176	2.07	0.075	1.39
15	0.268	1.36	0.046	2.26
20	0.364	1	0.025	4.17

A plot of F_ϕ versus F_c is shown in Figure 12.29b, from which $F_s = F_c = F_\phi = \underline{1.73}$.   ∎

12.9   Slopes with $c - \phi$ Soil—Prater's Analysis

It was mentioned in Section 12.6 that, due to earthquakes, slopes may undergo permanent deformation. Prater (1979) has analyzed slopes with $c - \phi$ soils to determine the *yield horizontal acceleration*, which is defined as the *threshold horizontal acceleration*, $k_h = k_y$, acting upon a sliding mass, above which permanent deformation occurs. It corresponds to a factor of safety with respect to strength (F_s) of unity. In this analysis the failure surface was assumed to be an arc of a logarithmic spiral defined by the equation (Figure 12.30)

$$r = r_0 e^{\theta \tan \phi} \tag{12.58}$$

where ϕ = soil friction angle. Prater's analysis for determination of the yield acceleration is summarized next.

Figure 12.31 shows a homogeneous slope. The unit weight, cohesion, and angle of friction of the soil are, respectively, γ, c, and ϕ. ABC is a trial failure surface that is the arc of a logarithmic spiral. Referring to Figure 12.31,

$$m = e^{p \tan \phi} \tag{12.59}$$

$$d = \frac{r_0}{H} = [\sin t \sqrt{1 + m^2 - 2m \cos p}]^{-1} \tag{12.60}$$

$$j = t + \sin^{-1}\left(\frac{\sin p}{\sqrt{1 + m^2 - 2m \cos p}}\right) \tag{12.61}$$

$$q = \pi - p - j \tag{12.62}$$

Considering a unit length of the embankment at right angles to the cross section shown, the overturning moment about O can be given as

$$M_D = M_W + M_E = (1 \mp k_v)M_g + M_E \tag{12.63}$$

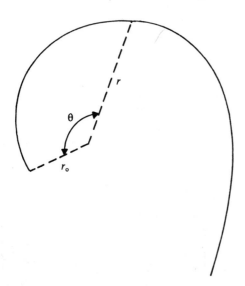

Figure 12.30   Log spiral

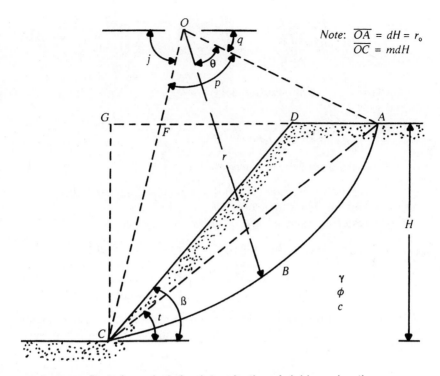

Figure 12.31   Prater's analysis for determination of yield acceleration

where

M_g = moment due to gravity force

$$= (M_1 - M_2 - M_3) \tag{12.64}$$

M_1 = moment of the soil weight in the area $OABC$ about O

$$= \frac{\gamma d^3 H^3}{3(9 \tan^2 \phi + 1)} [(m^3 \sin j - \sin q) - 3 \tan \phi (m^3 \cos j + \cos q)] \tag{12.65}$$

M_2 = moment of the soil weight in the area OAF about O

$$= \frac{\gamma d^3 H^3}{6} [\sin^3 q(\cot^2 q - \cot^2 j)] \tag{12.66}$$

M_3 = moment of the soil weight in the area CDF about O

$$= \frac{\gamma H^3}{6} [\cot^2 \beta - \cot^2 j - 3md \cos j(\cot \beta - \cot j)] \tag{12.67}$$

k_v = average vertical acceleration

M_E = moment due to horizontal inertia force

$$= M_e k_h = (M_4 - M_5 - M_6)k_h \tag{12.68}$$

$$M_4 = \frac{\gamma d^3 H^3}{3(9 \tan^2 \phi + 1)} [(m^3 \cos j + \cos q) + 3 \tan \phi (m^3 \sin j - \sin q)]$$

(12.69)

$$M_5 = \frac{\gamma d^3 H^3}{3} [\sin^3 q (\cot q + \cot j)]$$

(12.70)

$$M_6 = \frac{\gamma H^3}{6} (3d \sin q + 1)(\cot \beta - \cot j)$$

(12.71)

k_h = average horizontal acceleration

Hence, combining Eqs. (12.63) through (12.71), an expression for the overturning moment, M_D, can be obtained.

The restoring moment, M_R, can now be expressed as

$$M_R = \begin{bmatrix} \text{moment of the cohesive} \\ \text{force developed along} \\ \text{the trial failure} \\ \text{surface } ABC, M_c \end{bmatrix} + \begin{bmatrix} \text{moment of the frictional} \\ \text{force developed along} \\ \text{the trial failure} \\ \text{surface } ABC, M_f \end{bmatrix}$$

However, based on the property of logarithmic spiral, the line of action of the resultant frictional force at any given point along the trial failure surface will pass through the origin O. Hence

$$M_f = 0$$

and

$$M_c = \frac{cd^2 H^2 (m^2 - 1)}{2 \tan \phi}$$

(12.72)

So, for equilibrium of the soil mass located above the trial failure surface

$$M_D - M_c = 0$$

$$M_g (1 \mp k_v) + M_e k_h - M_c = 0$$

$$M_g \mp M_g k_v + M_e k_h - M_c = 0$$

or

$$\boxed{k_h = \frac{M_c - M_g}{M_e \mp b M_g}}$$

(12.73)

where

$$b = \frac{k_v}{k_h}$$

(12.74)

Prater (1979) has suggested that a realistic value of b would be 0.3. The yield acceleration k_h for the most critical surface can be determined by trial and error. Table 12.1 shows the magnitudes of the yield acceleration determined in this manner with $b = 0$.

Table 12.1   Yield acceleration, $k_h = k_y$

β (deg)	$\tan \phi$	$c/\gamma H$			
		0.05	0.10	0.15	0.20
15	0.1	0	0.08	0.15	0.20
	0.2	0.1	0.20	0.27	0.33
	0.3	0.2	0.31	0.39	0.44
	0.4	0.3	0.41	0.50	0.55
	0.5	0.4	0.51	0.60	0.66
	0.6	0.49	0.61	0.70	0.76
	0.7	0.58	0.70	0.80	0.87
	0.8	0.66	0.79	0.89	0.97
	0.9	0.74	0.87	0.98	1.07
30	0.1	—	0	0.13	0.20
	0.2	0.00	0.11	0.25	0.35
	0.3	0.05	0.22	0.37	0.46
	0.4	0.14	0.32	0.46	0.56
	0.5	0.24	0.41	0.55	0.66
	0.6	0.32	0.50	0.63	0.75
	0.7	0.40	0.57	0.72	0.83
	0.8	0.47	0.65	0.79	0.91
	0.9	0.53	0.71	0.86	0.98
45	0.1	—	—	0.07	0.22
	0.2	—	0.00	0.18	0.33
	0.3	—	0.11	0.28	0.42
	0.4	0.00	0.20	0.37	0.51
	0.5	0.06	0.29	0.46	0.59
	0.6	0.14	0.36	0.53	0.67
	0.7	0.21	0.43	0.59	0.74
	0.8	0.27	0.49	0.66	0.80
	0.9	0.33	0.54	0.71	0.84
60	0.1	—	—	0.00	0.16
	0.2	—	—	0.08	0.26
	0.3	—	0.00	0.18	0.34
	0.4	—	0.05	0.26	0.42
	0.5	—	0.13	0.33	0.49
	0.6	0.00	0.20	0.39	0.55
	0.7	0.01	0.26	0.45	0.60
	0.8	0.07	0.32	0.50	0.65
	0.9	0.13	0.36	0.54	0.69
75	0.1	—	—	—	0.04
	0.2	—	—	0.00	0.14
	0.3	—	—	0.02	0.22
	0.4	—	—	0.10	0.29
	0.5	—	0.00	0.17	0.35
	0.6	—	0.01	0.23	0.40
	0.7	—	0.07	0.28	0.44
	0.8	—	0.12	0.32	0.38
	0.9	—	0.16	0.35	0.51

Note: b = 0.

12.10   Slopes with $c - \phi$ Soil—Conventional Method of Slices

In the analysis for the stability of slopes provided in Sections 12.7, 12.8, and 12.9, it is assumed that the soil is homogeneous. However, in a given slope, layered soil can be encountered. The method of slices is a general method that can easily account for the change of γ, c, and ϕ in the soil layers.

In order to explain this method, let us consider a slope as shown in Figure 12.32. Let ABC be a trial failure surface. Note that ABC is an arc of a circle with its center at O. The soil above the trial failure surface is divided into several slices. The length of each slice need not be the same. For the nth slice, consider a unit thickness at right angles to the cross section shown. The weight and the inertia forces are, respectively, W_n and $k_h W_n$. The forces P_n and P_{n+1} are the normal forces acting on the sides of the slice. Similarly, the shearing forces acting on the sides of the slice are T_n and T_{n+1}. The forces P_n, P_{n+1}, T_n, and T_{n+1} are difficult to determine. However we can make an approximate assumption that the resultants of P_n and T_n are equal in magnitude to the resultants of P_{n+1} and T_{n+1} and also their lines of action coincide. The normal reaction at the base of the slice is $N_r = W_n \cos \alpha_n$. It is assumed that the inertia force $k_h W_n$ has no effect on the magnitude of N_r. So the resisting tangential force T_r can be given as

$$T_r = \frac{1}{F_s}(cB_n \sec \alpha_n + N_r \tan \phi)$$

$$= \frac{1}{F_s}(cB_n \sec \alpha_n + W_n \cos \alpha_n \tan \phi) \tag{12.75}$$

Now, taking the moment about O for all the slices,

$$\sum_{n=1}^{p}(W_n R \sin \alpha_n + k_h W_n L_n) = \sum_{n=1}^{p} \frac{R}{F_s}(cB_n \sec \alpha_n + W_n \cos \alpha_n \tan \phi) \tag{12.76}$$

or

$$F_s = \frac{\displaystyle\sum_{n=1}^{p}(cB_n \sec \alpha_n + W_n \cos \alpha_n \tan \phi)}{\displaystyle\sum_{n=1}^{p}[W_n \sin \alpha_n + k_h W_n(L_n/R)]} \tag{12.77}$$

Note that the value of α_n may be either positive or negative. The value of α_n is positive when the slope of the arc is in the same quadrant as the ground slope. To find the minimum factor of safety—that is, the factor of safety with reference to the critical circle—several trials have to be made, each time changing the center of the trial circle.

For convenience, a slope in homogeneous soil is shown in Figure 12.32. However, the method of slices can be extended to slopes of layered soil, as shown

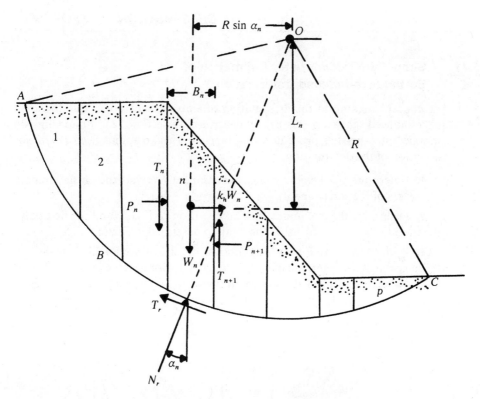

Figure 12.32 Conventional method of slices

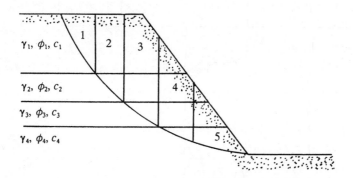

Figure 12.33 Method of slices for slopes in layered soil

in Figure 12.33. The general procedure of stability analysis is the same; however, some minor points need to be kept in mind. While using Equation (12.77), the values of ϕ and c will not be the same for all slices. For example, for slice 2, one has to use $\phi = \phi_2$ and $c = c_2$; similarly, for slice 3, $\phi = \phi_3$ and $c = c_3$ will need to be used.

DEFORMATION OF SLOPES

12.11   Simplified Procedure for Estimation of Earthquake-Induced Deformation

The concept relating to the deformation of embankment slopes due to earth-quake-induced vibration was briefly described in Section 12.6. Following is a simplified step-by-step procedure developed by Makdisi and Seed (1978) for estimation of the deformation.

1. Determine the height of the embankment (H) and the shear strength parameters of the soil (c and ϕ).

2. Determine the maximum crest acceleration $[\ddot{u}_a(0)]_{max}$ and the first natural period ($T_1 = 2\pi/\omega_1$) by using the method described in Section 12.5.

3. With reference to Figure 12.34, choose the critical section likely to deform and determine the magnitude of $k_{h(max)}g/[\ddot{u}_a(0)]_{max}$ from Figure 12.34. Note that $k_{h(max)}$ is the coefficient of the maximum average horizontal acceleration for a

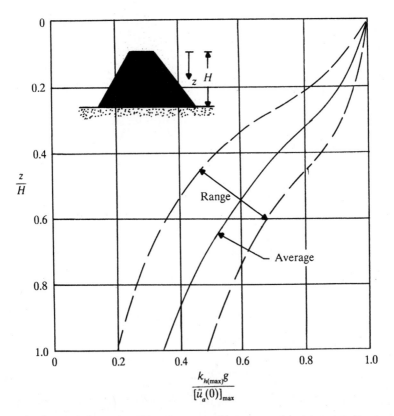

Figure 12.34 Variation of maximum acceleration ratio with depth of sliding mass (after Makdisi and Seed, 1978)

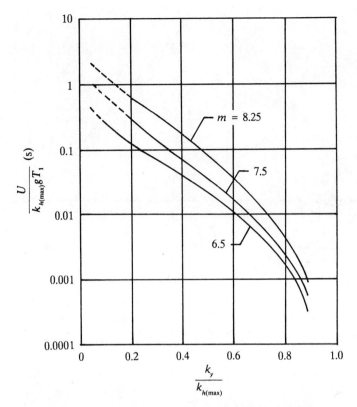

Figure 12.35   Variation of $U/[k_{h(\max)}gT_1]$ with $k_y/k_{h(\max)}$ (after Makdisi and Seed, 1978)

given value of z/H. The concept of the coefficient of average acceleration was explained in Section 12.6. Now determine the magnitude of $k_{h(\max)}g$.

4. Determine the yield acceleration—that is, the acceleration k_yg (see Section 12.9) for which the sliding mass has $F_s = 1$.

5. Determine $[k_y/k_{h(\max)}]$ and the magnitude of the earthquake (M). With these values go to Figure 12.35 to obtain $[U/k_{h(\max)}gT_1]$ (in seconds). With the known values of $k_{h(\max)}g$ (Step 3) and T_1 (Step 2), the magnitude U can be determined. Note that U is the deformation of the slope in the horizontal direction.

Example   12.4

Refer to the soil embankment in Example 12.1 (Figure 12.36).
a. Calculate the yield acceleration k_yg by using the concept described in Section 12.9. Use Table 12.1 with $b = 0$.
b. For the critical failure surface passing through the toe of the embankment, calculate the slope deformation using the procedure described in Section 12.11. Use the magnitude of earthquake $M = 7.0$. Also use the results of Example 12.1 for maximum crest acceleration.

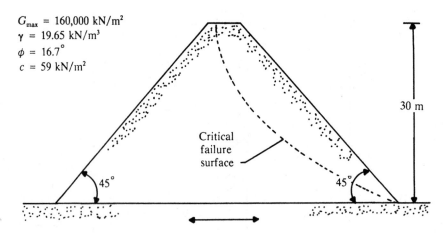

$G_{max} = 160,000$ kN/m^2
$\gamma = 19.65$ kN/m^3
$\phi = 16.7°$
$c = 59$ kN/m^2

30 m

Critical
failure
surface

45°   45°

Figure 12.36

Solution

a. Referring to Table 12.1, for $\beta = 45°$, $\tan\phi = \tan 16.7 = 0.3$, and $c/\gamma H = 59/[(19.65)(30)] = 0.1$, the magnitude of k_y is 0.11. So, the yield acceleration is 0.11 g.

b. Referring to Figure 12.34, for $z/H = 1$, the average value of $[k_{h(max)}g]/[\ddot{u}_a(0)]_{max}$ is about 0.34. From Example 12.1, $[\ddot{u}_a(0)]_{max} = 0.70$ g. So

$$\frac{k_{h(max)}g}{[\ddot{u}_a(0)]_{max}} = 0.34$$

$$k_{h(max)} = \frac{(0.34)(0.70\ \text{g})}{\text{g}} = 0.238$$

Thus,

$$\frac{k_y}{k_{h(max)}} = \frac{0.11}{0.238} = 0.462$$

Also, from Example 12.1, $T_1 = 0.435$ sec. For earthquake magnitude $M = 7.0$, referring to Figure 12.35,

$$\frac{U}{k_{h(max)}g T_1} \approx 0.036$$

So

$$U = (0.036)(0.238)(9.81)(0.435) = 0.0366\ \text{m} = \underline{36.6\ \text{mm}}$$ ∎

PROBLEMS

12.1   An earth embankment is 25 m high. For the embankment soil,

Unit weight $= 18$ kN/m^3

At a certain shear strain level

$G = 50,000$ kN/m^2 and $D = 15\%$

Using the acceleration spectra given in Figure 12.12 (maximum ground acceleration is 0.23 g), estimate the maximum crest acceleration of the embankment.

12.2 An earth embankment is 60 ft high. For the embankment soil,

$$\text{Unit weight} = 115 \text{ lb/ft}^3$$

$$\text{Maximum shear modulus} = 3.4 \times 10^6 \text{ lb/ft}^2$$

Using the variation of G/G_{max} and D with shear strain as given in Figure 12.11 and the acceleration spectra given in Figure 12.12 (maximum ground acceleration = 0.2g), determine the maximum crest acceleration.

12.3 A clay slope ($\phi = 0°$) is built over a layer of rock. For the slope,

$$\text{Height} = 20 \text{ m}$$

$$\text{Slope angle, } \beta = 30°$$

$$\text{Saturated unit weight of soil} = 17.8 \text{ kN/m}^3$$

$$\text{Undrained shear strength, } c_u = 5z \text{ kN/m}^2 \qquad (z = \text{depth measured from the top of the slope})$$

Determine the factor of safety F_s if $k_h = 0.4$. Use the procedure outlined in Section 12.7.

12.4 Redo Problem 12.3 assuming $c_u = 40 + 5z$ kN/m^2 and other parameters remain the same.

12.5 Refer to Problem 12.3. Other parameters remaining the same, let the slope angle β be changed from 30° to 75°. Calculate and plot the variation of the factor of safety (F_s) with β. Use the procedure outlined in Section 12.7.

12.6 For a homogenous soil,

$$\text{Slope angle, } \beta = 30°$$

$$\text{Height, } h = 50 \text{ ft}$$

$$\text{Soil cohesion} = 1220 \text{ lb/ft}^2$$

$$\text{Soil friction angle, } \phi = 25°$$

$$\text{Unit weight of soil} = 122 \text{ lb/ft}^3$$

$$k_h = 0.25$$

Determine the factor of safety with respect to strength. Use the procedure described in Section 12.8.

12.7 Repeat Problem 12.6 with the following:

$$\beta = 45°, \qquad H = 25 \text{ m}, \qquad c = 60 \text{ kN/m}^2$$

$$\text{Soil friction angle} = 20°$$

$$\text{Unit weight of soil} = 19 \text{ kN/m}^3$$

$$k_h = 0.3$$

12.8 For the slope described in Problem 12.6, what would be the yield acceleration (that is, $k_h g$)? Use the procedure described in Section 12.9. Use $b = 0$.

12.9 The properties of the soil of a given slope 15.3 m high are as follows:

Unit weight, $\gamma = 18.5$ kN/m^3

Cohesion, $c = 42.5$ kN/m^2

Angle of friction, $\phi = 20°$

The yield acceleration for the slope was estimated to be 0.36. This was done using the procedure described in Section 12.9 with $b = 0$. Estimate the slope angle β.

12.10 A homogeneous slope is shown in Figure P12.10. For the trial failure surface shown, determine the factor of safety with respect to strength. Use the method of slices. (*Note:* The slope angle is $\beta = 30°$.)

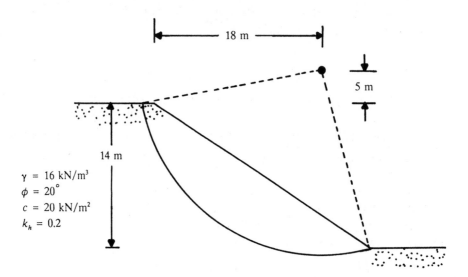

$\gamma = 16$ kN/m^3
$\phi = 20°$
$c = 20$ kN/m^2
$k_h = 0.2$

Figure P12.10

12.11 A 25-mm-high embankment (c-ϕ soil) is constructed over a hard stratum. The critical failure circle passes through the toe of the slope and the average yield acceleration of the slope is 0.15 g. The maximum crest acceleration due to an earthquake of magnitude $M = 7.5$ has been estimated to be $0.6g$. The first natural period is 0.8 s. Estimate the slope deformation. Use the procedure described in Section 12.11.

12.12 Refer to Example 12.4. Other quantities remaining the same, if the soil friction angle ϕ is changed to 21.8°, estimate the slope deformation.

REFERENCES

Ambraseys, N. N. (1960). "On the Seismic Behavior of Earth Dams," *Proceedings*, 2nd World Conference on Earthquake Engineering, Tokyo, Vol. I, pp. 331–345.

Das, B. M. (1990). *Principles of Geotechnical Engineering*, 2nd ed., PWS-KENT, Boston.

Koppula, S. D. (1984). "Pseudo-Static Analysis of Clay Slopes Subjected to Earthquakes," *Geotechnique*, Institute of Civil Engineers, London, Vol. 34, No. 1, pp. 71–79.

Majumdar, D. K. (1971). "Stability of Soil Slopes under Horizontal Earthquake Force," *Geotechnique*, Institute of Civil Engineers, London, Vol. 21, No. 1, pp. 372–378.

Makdisi, F. I., and Seed, H. B. (1978). "Simplified Procedure for Estimating Dam and Embankment Earthquake-Induced Deformations," *Journal of the Geotechnical Engineering Division*, ASCE, Vol. 104, No. GT7, pp. 849–859.

Makdisi, F. I., and Seed, H. B. (1979). "Simplified Procedure for Evaluating Embankment Response," *Journal of the Geotechnical Engineering Division*, ASCE, Vol. 105, No. GT12, pp. 1427–1434.

Mononobe, N., Takata, A., and Matumura, M. (1936). "Seismic Stability of Earth Dams," *Proceedings*, 2nd Congress of Large Dams, Vol. 4, Washington, D.C.

Newmark, N. M. (1965). "Effect of Earthquakes on Dams and Embankments," *Geotechnique*, Institute of Civil Engineers, London, Vol. 15, No. 2, p. 139.

Prater, E. G. (1979). "Yield Acceleration for Seismic Stability of Slopes," *Journal of the Geotechnical Engineering Division*, ASCE, Vol. 105, No. GT5, pp. 682–687.

Seed, H. B. (1975). "Earthquake Effects on Soil-Foundation System" in *Foundation Engineering Handbook* (H. F. Winterkorn and H. Y. Faud, eds.), Van Nostrand Reinhold, New York.

Seed, H. B. (1981). "Earthquake-Resistant Design of Earth Dams," *Proceedings*, International Conference on Recent Advances in Geotechnical Earthquake Engineering and Soil Dynamics, University of Missouri-Rolla, Vol. III, pp. 1157–1177.

Seed, H. B., and Goodman, R. E. (1964). "Earthquake Stability of Slopes of Cohesionless Soils," *Journal of the Soil Mechanics and Foundations Division*, ASCE, Vol. 90, No. SM6, pp. 43–56.

Seed, H. B., and Martin, G. R. (1966). "The Seismic Coefficient of Earth Dam Design," *Journal of the Soil Mechanics and Foundations Division*, ASCE, Vol. 92, No. SM3, pp. 25–58.

Seed, H. B., Makdisi, F. I., and DeAlba, P. (1978). "Performance of Earth Dams During Earthquakes," *Journal of the Geotechnical Engineering Division*, ASCE, Vol. 104, No. GT7, pp. 967–994.

Taylor, D. W. (1937). "Stability of Earth Slopes," *Journal of the Boston Society of Civil Engineers*, Vol. 24, No. 3, pp. 197–246.

PRIMARY AND SECONDARY FORCES OF SINGLE-CYLINDER ENGINES

Machineries involving a crank mechanism produce a reciprocating force. This mechanism is shown in Figure A.1a, in which

OA = crank length = r_1

AB = length of the connecting rod = r_2

Let the crank rotate at a constant angular velocity ω. At time $t = 0$, the vertical distance between O and B (Figure A.1b) is equal to $r_1 + r_2$. At time t, the vertical distance between O and B is equal to $r_1 + r_2 - z$, or

$$z = (r_1 + r_2) - (r_2 \cos \alpha + r_1 \cos \omega t) \tag{A.1}$$

But,

$$r_2 \sin \alpha = r_1 \sin \omega t \tag{A.2}$$

Now,

$$\cos \alpha = \sqrt{1 - \sin^2 \alpha} = \sqrt{1 - \left(\frac{r_1}{r_2}\right)^2 \sin^2 \omega t}$$

$$\approx 1 - \frac{1}{2}\left(\frac{r_1}{r_2}\right)^2 \sin^2 \omega t \tag{A.3}$$

Substituting Eq. (A.3) into Eq. (A.1),

$$z = (r_1 + r_2) - (r_2 \cos \alpha + r_1 \cos \omega t)$$

$$= r_2(1 - \cos \alpha) + r_1(1 - \cos \omega t)$$

$$= r_2\left[1 - 1 + \frac{1}{2}\left(\frac{r_1}{r_2}\right)^2 \sin^2 \omega t\right] + r_1(1 - \cos \omega t)$$

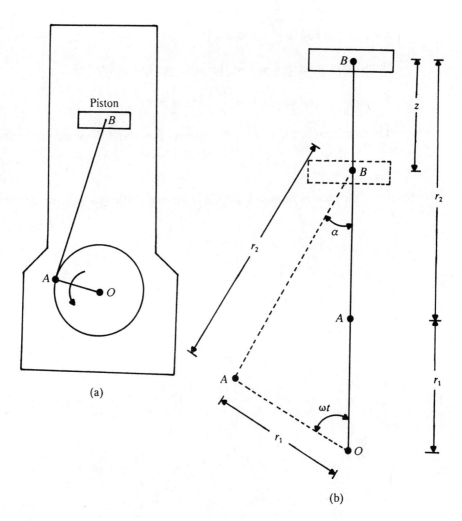

Figure A.1

$$= \left(\frac{\frac{1}{2}r_1^2}{r_2}\right)\sin^2 \omega t + r_1(1 - \cos \omega t) \tag{A.4}$$

However,

$$\sin^2 \omega t = \tfrac{1}{2}(1 - \cos 2\omega t) \tag{A.5}$$

Substituting Eq. (A.5) into Eq. (A.4), one obtains

$$z = \left(\frac{\frac{1}{4}r_1^2}{r_2}\right)(1 - \cos 2\omega t) + r_1(1 - \cos \omega t)$$

$$= \left(r_1 + \frac{\frac{1}{4}r_1^2}{r_2}\right) - \left[r_1 \cos \omega t + \left(\frac{\frac{1}{4}r_1^2}{r_2}\right)\cos 2\omega t\right] \tag{A.6}$$

The acceleration of the piston can be given by

$$\ddot{z} = r_1\omega^2\left[\cos\omega t + \left(\frac{r_1}{r_2}\right)\cos 2\omega t\right] \tag{A.7}$$

If the mass of the piston is m, the force can be obtained as

$$F = m\ddot{z} = mr_1\omega^2\cos\omega t + m\left(\frac{r_1^2}{r_2}\right)\omega^2\cos 2\omega t \tag{A.8}$$

The first term of Eq. (A.8) is the primary force, and the maximum primary force

$$F_{max(prim)} = mr_1\omega^2 \tag{A.9}$$

Similarly, the second term of Eq. (A.8) is generally referred to as the secondary force, and

$$F_{max(sec)} = m\left(\frac{r_1^2}{r_2}\right)\omega^2 \tag{A.10}$$

CONVERSION FACTORS

B.1 Conversion Factors from English to SI Units

Length:	1 ft $= 0.3048$ m
	1 ft $= 30.48$ cm
	1 ft $= 304.8$ mm
	1 in. $= 0.0254$ m
	1 in. $= 2.54$ cm
	1 in. $= 25.4$ mm
Area:	1 ft^2 $= 929.03 \times 10^{-4}$ m^2
	1 ft^2 $= 929.03$ cm^2
	1 ft^2 $= 929.03 \times 10^2$ mm^2
	1 in.^{2} $= 6.452 \times 10^{-4}$ m^2
	1 in.^{2} $= 6.452$ cm^2
	1 in.^{2} $= 645.16$ mm^2
Volume:	1 ft^3 $= 28.317 \times 10^{-3}$ m^3
	1 ft^3 $= 28.317$ cm^3
	1 in.^{3} $= 16.387 \times 10^{-6}$ m^3
	1 in.^{3} $= 16.387$ cm^3
Coefficient of permeability:	1 ft/min $= 0.3048$ m/min
	1 ft/min $= 30.48$ cm/min
	1 ft/min $= 304.8$ mm/min
	1 ft/s $= 0.3048$ m/s
	1 ft/s $= 304.8$ mm/s
	1 in./min $= 0.0254$ m/min
	1 in./s $= 2.54$ cm/s
	1 in./s $= 25.4$ mm/s
Coefficient of consolidation:	1 in.^{2}/s $= 6.452$ cm^2/s
	1 in.^{2}/s $= 20.346 \times 10^3$ m^2/y
	1 ft^2/s $= 929.03$ cm^2/s

Force: 1 lb = 4.448 N
                    1 lb = 4.448 × 10^{-3} kN
                    1 lb = 0.4536 kgf
                    1 kip = 4.448 kN
                    1 U.S. ton = 8.896 kN
                    1 lb = 0.4536 × 10^{-3} metric ton
                    1 lb/ft = 14.593 N/m

Stress: 1 lb/ft^2 = 47.88 N/m^2
                    1 lb/ft^2 = 0.04788 kN/m^2
                    1 U.S. ton/ft^2 = 95.76 kN/m^2
                    1 kip/ft^2 = 47.88 kN/m^2
                    1 lb/in.^{2} = 6.895 kN/m^2

Unit weight: 1 lb/ft^3 = 0.1572 kN/m^3
                    1 lb/in.^{3} = 271.43 kN/m^3

Moment: 1 lb-ft = 1.3558 N·m
                    1 lb-in. = 0.11298 N·m

Energy: 1 ft-lb = 1.3558 J

B.2   Conversion Factors from SI to English Units

Length: 1 m   = 3.281 ft
                    1 cm = 3.281 × 10^{-2} ft
                    1 mm = 3.281 × 10^{-3} ft
                    1 m   = 39.37 in.
                    1 cm = 0.3937 in.
                    1 mm = 0.03937 in.

Area: 1 m^2 = 10.764 ft^2
                    1 cm^2 = 10.764 × 10^{-4} ft^2
                    1 mm^2 = 10.764 × 10^{-6} ft^2
                    1 m^2 = 1550 in.^{2}
                    1 cm^2 = 0.155 in.^{2}
                    1 mm^2 = 0.155 × 10^{-2} in.^{2}

Volume: 1 m^3 = 35.32 ft^3
                    1 cm^3 = 35.32 × 10^{-4} ft^3
                    1 m^3 = 61,023.4 in.^{3}
                    1 cm^3 = 0.061023 in.^{3}

Coefficient of   1 m/min   = 3.281 ft/min
permeability: 1 cm/min = 0.03281 ft/min
                    1 mm/min = 0.003281 ft/min
                    1 m/s   = 3.281 ft/s
                    1 mm/s   = 0.03281 ft/s
                    1 m/min   = 39.37 in./min
                    1 cm/s   = 0.3937 in./s
                    1 mm/s   = 0.03937 in./s

Coefficient of $1\ cm^2/s = 0.155\ in.^2/s$
Consolidation: $1\ m^2/y\ = 4.915 \times 10^{-5}\ in.^2/s$
$1\ cm^2/s = 1.0764 \times 10^{-3}\ ft^2/s$

Force: 1 N $= 0.2248$ lb
1 kN $= 224.8$ lb
1 kgf $= 2.2046$ lb
1 kN $= 0.2248$ kip
1 kN $= 0.1124$ U.S. ton
1 metric ton $= 2204.6$ lb
1 N/m $= 0.0685$ lb/ft

Stress: $1\ N/m^2\ = 20.885 \times 10^{-3}\ lb/ft^2$
$1\ kN/m^2 = 20.885\ lb/ft^2$
$1\ kN/m^2 = 0.01044$ U.S. ton/ft^2
$1\ kN/m^2 = 20.885 \times 10^{-3}\ kip/ft^2$
$1\ kN/m^2 = 0.145\ lb/in.^2$

Unit weight: $1\ kN/m^3 = 6.361\ lb/ft^3$
$1\ kN/m^3 = 0.003682\ lb/in.^3$

Moment: $1\ N \cdot m = 0.7375$ lb-ft
$1\ N \cdot m = 8.851$ lb-in.

Energy: $1\ J = 0.7375$ ft-lb

ANSWERS TO SELECTED PROBLEMS

Chapter 2

2.2 7.88 cps; 0.127 s

2.3 0.004 m

2.4 **a.** 0.101 s
b. $-9799 \sin[(83.77 \text{ rad/s})(t \text{ s})]$ lb
c. $13,207 \sin[(62.161 \text{ rad/s})(t \text{ s})]$ lb

2.5 $\dfrac{1}{2\pi}\sqrt{\dfrac{k_1 k_2}{(k_1 + k_2)m}}$

2.6 $\dfrac{1}{2\pi}\sqrt{\dfrac{k_1 + k_2}{m}}$

2.7 $f_n = 8.37$ cps, $T = 0.119$ s

2.8 0.897 mm

2.9 **a.** 2.211 cps   **b.** 0.028   **c.** 1.19   **d.** 0.415 in.

2.10 Maximum force = 86,419 lb; minimum force = 43,581 lb

2.11 Maximum force = 637 lb; minimum force = 168 lb

2.12 **a.** 8077 kN-s/m   **b.** 0.29   **c.** 1.9   **d.** 7.54 cps

2.13 **a.** 3.795×10^{-2} mm   **b.** 11.68 kN

2.14 $\omega_{n_1}, \omega_{n_2} = \dfrac{1}{\sqrt{2}}\left\{\left(\dfrac{k_1 + k_2}{m_1} + \dfrac{k_2 + k_3}{m_2}\right) \mp \left[\left(\dfrac{k_1 + k_2}{m_1} - \dfrac{k_2 + k_3}{m_2}\right)^2 + 4\dfrac{k_2^2}{m_1 m_2}\right]^{0.5}\right\}^{0.5}$

2.15 57.7 N/mm

Chapter 4

4.1   **a.** 0.4   **b.** $E = 38.37 \times 10^3$ lb/in.^{2}; $G = 13.72 \times 10^3$ lb/in.^{2}

4.2 3.84×10^4 kN/m^2

4.3   1374 cps

4.4 $v_{p_1} = 678$ ft/s; $v_{p_2} = 1748$ ft/s; $z_1 = 35.3$ ft

4.5 $v_{p_1} = 520$ m/s; $v_{p_2} = 3819$ m/s; $z_1 = 26.2$ m

4.6 $v_{p_1} = 240$ m/s; $v_{p_2} = 1102$ m/s; $z_1 = 7.75$ m; $z_2 = 24.9$ m

4.7   **a.** $v_{p_1} = 1005$ ft/s; $v_{p_2} = 4582$ ft/s   **b.** $z' = 25.2$ ft; $z'' = 82.4$ ft
  c. $9°$

4.8 $v_{p_1} = 800$ ft/s; $z = 12$ m

4.9 $\beta = 0.426°$; $z' = 7.64$ m

4.10

Depth (m)	v_r (m/s)
0.122	219.6
0.278	222.4
0.555	222
1.099	219.8
1.887	339.7
2.174	326.1
2.825	339
3.401	340

4.11

R_D(%)	G (kN/m^2)				
	$\sigma_0 = 50$ kN/m^2	100 kN/m^2	150 kN/m^2	200 kN/m^2	300 kN/m^2
0	38,032	53,705	65,873	76,064	93,159
20	44,995	63,632	77,933	89,990	110,215
40	53,066	75,047	91,914	106,133	129,868
60	62,453	88,323	108,173	124,907	152,980
80	73,418	103,828	127,163	146,836	179,836
100	86,295	122,039	149,467	172,590	211,379

4.12   95,940 kN/m^2

4.13 $G \approx 40,080$ kN/m^2; $D \approx 12\%$

4.14   14,700 lb/in.^{2}

4.15

Depth from ground surface (m)	G_{max} (kN/m^2)
0	0
2	19,470
	17,430
10	31,770
	11,980
13	13,100

4.16

Depth from ground surface (ft)	G_{max} (lb/in^2)
0	0
20	9,450
	11,047
40	14,268
	5,154
60	5,890

4.17 $G = 2760$ lb/in.^{2}; $D = 18.8\%$

4.18 $G = 12,400$ lb/in.^{2}; $D = 5.6\%$

4.19

Shear strain (%)	G (kN/m^2)	D (%)
10^{-3}	91,800	2.5
10^{-2}	43,500	5
10^{-1}	15,500	9.5
1	4,350	20.5

4.20 3.18%

Chapter 5

5.1 **a.** 877 cpm **b.** 4.97×10^{-3} in. **c.** 3.368×10^{-3} in.

5.2 **a.** 878 cpm **b.** 5.27×10^{-3} in. **c.** 3.45×10^{-3} in.

5.3 **a.** 572 cpm **b.** 3.68×10^{-2} mm

5.4 **a.** 1075 cpm **b.** 0.039 mm

5.5 0.035 mm

5.6 $f_m = 968$ cpm
$A_z = 0.0473$ mm

5.7 **a.** 827 cpm **b.** 0.186×10^{-3} rad

5.8 **a.** 827 cpm **b.** 0.42×10^{-3} rad

5.9 **a.** 605 cpm **b.** 7.52×10^{-2} mm

5.10 **a.** 747 cpm **b.** 0.168 mm

5.11 **a.** 748 cpm **b.** 0.59×10^{-5} rad

5.12 Anvil: 8.98×10^{-2} in.; foundation: 8.48×10^{-2} in.

5.13 **a.** 8.63 cps **b.** 0.00228 in. **c.** 0.00158 in.

5.14 **a.** 33.05 cps **b.** 0.00057 in. **c.** 0.000193 in.

5.15 **a.** 5.88 cps **b.** 0.0369 mm **c.** 0.0339 mm

5.16 **a.** 6.33 cps **b.** 1.9×10^{-5} rad

Chapter 6

6.1 35,340 lb/ft^2

6.2 40,670 lb/ft^2

6.3 926 kN/m²

6.4 1 m

6.5 961 kN/m²

6.6 1001 kN/m²

6.7 $T = 2.53$ s

6.8 1.63°

6.9 1.19°

6.10 $Q = 47,210$ lb/ft

6.11 1 ft

6.12 1.2 m

6.13 1979 kN

Chapter 8

8.1 9738 lb/ft

8.2 8.5 ft

8.3 $P_{AE} = 41.39$ kN/m; $\overline{H} = 1.38$ m

8.4 1.65 m

8.5 5400 lb/ft

8.6 9700 lb/ft

8.7 41 kN/m

8.8 52,726 lb/ft

8.9 456.3 kN/m

8.10 **a.** 3226 lb/ft   **b.** 10,928 lb/ft   **c.** 8422 lb/ft

Chapter 9

9.1 $\gamma'_{xz}(\%) = 3.674[\varepsilon_z\,(\%)]^{1.586}$

9.2 0.61 in.

9.3 11.51 mm

Chapter 10

10.2 R_D (%)	$\dfrac{a_{max}}{g}$	**10.3** R_D (%)	$\dfrac{a_{max}}{g}$
20	0.056	20	0.054
30	0.083	30	0.074
40	0.111	40	0.099
50	0.145	50	0.129
60	0.188	60	0.168
70	0.237	70	0.212
80	0.290	80	0.259

10.4

R_D (%)	$\dfrac{a_{max}}{g}$
20	0.045
30	0.068
40	0.091
50	0.119
60	0.154
70	0.194
80	0.237

10.5

R_D (%)	$\dfrac{a_{max}}{g}$
20	0.054
30	0.081
40	0.108
50	0.142
60	0.183
70	0.231
80	0.282

10.6   **a.**

Depth (ft)	τ_{av} (lb/ft^2)
10	133.8
20	280.8
30	421.6
40	544
50	646
60	641.5
70	643.5

b.

Depth (ft)	τ_h (lb/ft^2)
10	176.8
20	273.8
30	370.8
40	467.8
50	564.8
60	661.8
70	758.8

c. 19 ft to 60 ft

10.7   **a.**

Depth (m)	τ_{av} (kN/m^2)
5	7.3
10	14.9
15	21.2
20	19.4
21.3	19.3

b.

Depth (m)	τ_h (kN/m^2)
5	9.53
10	14.92
15	20.32
20	25.72
21.3	27.12

c. 13 m to 16.5 m

10.8   No

10.9   No

10.10   1.52 m

10.11   1.48 m

Chapter 11

11.1   525 cpm

11.2   **a.** $k_z = 1.41 \times 10^{-5}$ kN/m; $c_z = 264.5$ kN-s/m
　　　b. $k_x = 3.8 \times 10^{-4}$ kN/m; $c_x = 96.98$ kN-s/m
　　　c. $k_\theta = 4796$ kN-m/rad; $c_\theta = 3.39$ kN-m-s/rad

11.3 $k_\alpha = 752$ kN-m/rad; $c_\alpha = 0.16$ kN-m-s/rad

11.4   **a.** $k_z = 113.8 \times 10^3$ kN/m; $c_z = 204$ kN-s/m
　　　b. $k_x = 44.76 \times 10^3$ kN/m; $c_x = 118.25$ kN-s/m
　　　c. $k_\theta = 8.38 \times 10^{-3}$ kN-m/rad; $c_\theta = 7.02$ kN-m-s/rad

11.5 $k_\alpha = 754$ kN-m/rad; $c_\alpha = 0.3$ kN-m-s/rad

11.6 $k_{z(g)} = 810 \times 10^3$ kN/m; $c_{z(g)} = 2533$ kN-s/m

11.7 $k_{z(T)} = 942 \times 10^3$ kN/m; $c_{z(T)} = 6063$ kN-s/m

11.8 $k_{x(g)} = 287 \times 10^3$ kN/m; $c_{x(g)} = 1134$ kN-s/m

11.9 $k_{x(T)} = 483 \times 10^3$ kN/m; $c_{x(T)} = 8845$ kN-s/m

11.10 $k_{\theta(g)} = 4434 \times 10^3$ kN-m/rad; $c_{\theta(g)} = 11,885$ kN-m-s/rad

11.11 $k_{\theta(T)} = 500 \times 10^4$ kN-m/rad; $c_{\theta(T)} = 24,720$ kN-m-s/rad

Chapter 12

12.1 0.637 g

12.2 0.6 g

12.3 0.677

12.4 1.065

12.5

β (deg)	F_s
30	0.677
40	0.653
60	0.555
70	0.489

12.6 2.3

12.7 1.15

12.8 0.2 g

12.9 ≈ 40

12.10 1.18

12.11 7.5 mm

12.12 1.16 mm

Index